VOLUME EIGHTY TWO

Current Topics in DEVELOPMENTAL BIOLOGY

Red Cell Development

Series Editors

Paul M. Wassarman
Department of Molecular, Cell and Developmental Biology
Mount Sinai School of Medicine
New York, NY, USA

Olivier Pourquié
Investigator Howard Hughes Medical Institute
Stowers Institute for Medical Research
Kansas City, MO, USA

VOLUME EIGHTY TWO

CURRENT TOPICS IN DEVELOPMENTAL BIOLOGY

Red Cell Development

Edited by

JAMES J. BIEKER

Mount Sinai School of Medicine
Department of Molecular, Cell and Developmental Biology
New York, NY, USA

AMSTERDAM • BOSTON • HEIDELBERG • LONDON
NEW YORK • OXFORD • PARIS • SAN DIEGO
SAN FRANCISCO • SINGAPORE • SYDNEY • TOKYO
Academic Press is an imprint of Elsevier

Academic Press is an imprint of Elsevier
84 Theobald's Road, London WC1X 8RR, UK
Radarweg 29, PO Box 211, 1000 AE Amsterdam, The Netherlands
Linacre House, Jordan Hill, Oxford OX2 8DP, UK
30 Corporate Drive, Suite 400, Burlington, MA 01803, USA
525 B Street, Suite 1900, San Diego, CA 92101-4495, USA

First edition 2008

ISBN: 978-0-12-374366-4
ISSN: 0070-2153

For information on all Academic Press publications
visit our website at books.elsevier.com

Printed and bound in USA

08 09 10 11 12 10 9 8 7 6 5 4 3 2 1

Contents

Contributors

Nancy C. Andrews
Children's Hospital Boston and Harvard Medical School, Boston, Massachusetts 02115

James J. Bieker
The Brookdale Department of Molecular, Cell and Developmental Biology, Mount Sinai School of Medicine, New York 10029

Emery H. Bresnick
Department of Pharmacology, University of Wisconsin School of Medicine and Public Health, 385 Medical Sciences Center, Madison, Wisconsin 53706

Vladan P. Čokić
Laboratory of Experimental Hematology, Institute for Medical Research, Belgrade, Serbia

Steven R. Ellis
Department of Biochemistry and Molecular Biology, University of Louisville, Louisville, Kentucky

Gordon D. Ginder
Department of Internal Medicine, Massey Cancer Center, Virginia Commonwealth University, Richmond, Virginia 23298

Merlin N. Gnanapragasam
Department of Internal Medicine, Massey Cancer Center, Virginia Commonwealth University, Richmond, Virginia 23298

Frank Grosveld
Department of Cell Biology and Genetics, Erasmus MC, PO Box 2040, 3000 CA Rotterdam, The Netherlands

Petra Klous
Department of Cell Biology and Genetics, Erasmus MC, PO Box 2040, 3000 CA Rotterdam, The Netherlands

Jurgen Kooren
Department of Cell Biology and Genetics, Erasmus MC, PO Box 2040, 3000 CA Rotterdam, The Netherlands

Wouter de Laat
Department of Cell Biology and Genetics, Erasmus MC, PO Box 2040, 3000 CA Rotterdam, The Netherlands

Jeffrey M. Lipton
Division of Hematology/Oncology and Stem Cell Transplantation, Schneider Children's Hospital, Albert Einstein College of Medicine, Long Island Jewish Medical Center, New Hyde Park, New York, and The Feinstein Institute for Medical Research, Manhasset, New York

Deepa Manwani
Schneider Children's Hospital, New York 11040, and The Brookdale Department of Molecular, Cell and Developmental Biology, Mount Sinai School of Medicine, New York 10029

Kathleen McGrath
Department of Pediatrics, Center for Pediatric Biomedical Research, University of Rochester, Rochester, New York 14642

Omar Y. Mian
Department of Internal Medicine, Massey Cancer Center, Virginia Commonwealth University, Richmond, Virginia 23298

Daan Noordermeer
Department of Cell Biology and Genetics, Erasmus MC, PO Box 2040, 3000 CA Rotterdam, The Netherlands

James Palis
Department of Pediatrics, Center for Pediatric Biomedical Research, University of Rochester, Rochester, New York 14642

Robert-Jan Palstra
Department of Cell Biology and Genetics, Erasmus MC, PO Box 2040, 3000 CA Rotterdam, The Netherlands

Alan N. Schechter
Molecular Medicine Branch, NIDDK, National Institutes of Health, Bethesda, Maryland 20892

Marieke Simonis
Department of Cell Biology and Genetics, Erasmus MC, PO Box 2040, 3000 CA Rotterdam, The Netherlands

Erik Splinter
Department of Cell Biology and Genetics, Erasmus MC, PO Box 2040, 3000 CA Rotterdam, The Netherlands

Ryan J. Wozniak
Department of Pharmacology, University of Wisconsin School of Medicine and Public Health, 385 Medical Sciences Center, Madison, Wisconsin 53706

Diedra M. Wrighting
Children's Hospital Boston and Harvard Medical School, Boston, Massachusetts 02115

Preface

Analyses of erythroid gene expression and regulation have historically been performed at the cutting edge of basic and clinical research, whether they be the first description of developmentally regulated changes in sites of hemoglobinogenesis (Jordan and Spiedel, 1923; McCutcheon, 1936), determination of sickle-cell anemia as an aberration of a single genetic locus (Pauling *et al.*, 1949), analysis of the quaternary, three-dimensional structure of hemoglobin (Perutz, 1987), the early successes of molecular cloning and gene structure determination (Lawn *et al.*, 1978), or the introduction of foreign genes into mice (Costantini and Lacy, 1981).

These analyses have also led to an extraordinary compendium of genetic aberrations that alter the proper regulation of red blood cell expression patterns (Stamatoyannopoulos *et al.*, 2001). Those related to globin regulation run the gamut of causes and provide clinical examples of molecular mutations that cover all aspects of transcriptional regulatory mechanisms, whether they be located within the enhancers and promoters, or whether they affect posttranscriptional controls such as splicing and processing/polyadenylation (Bunn, 2001; Weatherall, 1994). It is likely true that for every molecular mechanism that has been discovered since the advent of molecular biology, there is a clinically relevant example to be found within the globin regulatory cascade.

Such an expansion of the biological horizon has also been realized by studying the cellular aspects of red blood cell onset as it follows from hematopoietic differentiative mechanisms (Metcalf and Moore, 1971; Morrison *et al.*, 1995; Till and McCulloch, 1980). The recognition of a stem cell, and the conceptual details that define such a cell that are so pervasive today in many systems, has its origins within studies originally performed in circulating blood cells that aimed to identify how these varied cells were continuously established and propagated during the life of an individual (McCulloch and Till, 2005). Indeed, well into the fourth decade during which the self-renewal and multipotential concepts of hematopoiesis have been experimentally described and refined, the complexities of the biology have forced some of the initially posited questions to remain outstanding, for example whether multiple embryonic origins arise independently or not (Samokhvalov *et al.*, 2007).

Not surprisingly, unanticipated directions in cellular and molecular mechanisms demanded novel approaches to address these compelling issues of red gene regulation. For example, insights on the phenomena of distal locus control elements and their role in erythroid-specific and developmentally appropriate expression (Behringer *et al.*, 1990; Enver *et al.*, 1990; Grosveld *et al.*, 1987) have been abetted by the ability to analyze large constructs that contain the complete genomic globin cluster in transgenic mice (Gaensler *et al.*, 1993; Peterson *et al.*, 1993; Strouboulis *et al.*, 1992).

One might have expected that, after so many decades of serious study, there would be few surprises in store for those who study red blood cell biology. As will be seen by perusal of this volume, this is clearly not the case, and it is in the spirit of these pioneering studies that this volume has been assembled. We begin with the embryological origins of the red blood cell, and McGrath and Palis describe the important role of primitive erythropoiesis in establishing the first population of red blood cells in the embryo. Intensive study of this has led to surprising observations that challenge the dogma of primitive red blood cell properties and their life in circulation. In addition, the close relationship between erythroid and megakaryocyte development has unexpectedly been found within these primitive cells.

Manwani and Bieker then focus in Chapter 2 on the immediate environment within which these red blood cells are first formed, known as the erythroblastic island. Even though these were first described almost 50 years ago, it is only recently that cellular and molecular details have begun to illuminate their importance for the normal process of erythroid development, and reinforces the important issue that studying cells in culture may not yield the complete picture on what actually occurs within the developing embryo. Such a concept has become more appreciated very recently, especially with the idea of the "niche" (itself first described almost 30 years ago within the context of hematopoietic development (Schofield, 1978)), a configuration of heterologous cells localized in the fetal liver or the bone marrow within which cell–cell interactions between its residents play a critical role in proper maturation of maturing hematopoietic cells that are eventually released into circulation (Scadden, 2006). In this context, the erythroblast island can be considered a specialized micro-niche.

We then zero in on the genetic control mechanisms for erythropoiesis. However, rather than describing the transcription factors and *cis*-acting control elements important for red blood cell gene expression, we describe novel observations related to epigenetic mechanisms of gene control. Wozniak and Bresnick begin in Chapter 3 by covering the important regulatory role of histone and their modifications, particularly as they relate to the regulation of the β-like globin locus and the GATA2 gene. Of particular interest has been the role of the histone code to this process, and how small differences in histone acetylation and methylation lead to significant changes in developmental regulation of these genes. The details of how tissue-restricted transcription factors play critical roles in helping to properly coordinate these histone marks in the proper way, and the issues that remain unresolved, are discussed.

The concept of epigenesis is further amplified by Ginder, Gnanapragasam, and Mian, who describe DNA methylation and its effects on red blood cell gene control in a range of species. Although the inverse correlation of DNA methylation and gene expression of the β-like globin regulatory domain was first noted almost 30 years ago, it is only recently that the molecular details have become apparent. Globin regulation continues to provide a fertile ground in which to address the mechanistic effects of DNA methylation, and Chapter 4 highlights the recent observations that link histone modifications with DNA methylation, and the potential role of specific proteins that recognize DNA modifications in developmental regulation of the globin and other red blood cell genes.

Attempts to obtain a three-dimensional idea of these molecular observations required a novel technology, and de Laat, Klous, Kooren, Noordermeer, Palstra, Simonis, Splinter, and Grosveld describe the detailed conformations and molecular dynamics that can be deduced by 3C and 4C approaches. These have helped to distinguish between models of long-range erythroid gene expression control mechanisms, and have moved the focus away from the one-dimensional regulatory observations of the past. The 3C technology can address local mechanisms one at a time, and indeed has also been used to demonstrate the importance of the EKLF and GATA1 transcription factors for establishing the proper structure at the β-like globin locus. But the expansion of this technology to the 4C approach described in Chapter 5 can lead to a more global understanding of the spatial relationships between loci at a given time within a particular cell type. In combination with high-resolution fluorescent microscopy, the nuclear architecture that follows from such long-range interactions can be ascertained to a resolution never previously seen.

One of the major end results of these genetic control mechanisms is to produce a red blood cell that expresses the proper level of hemoglobin to ensure adequate oxygen transport throughout the body. Iron is a critical component for this process, and Wrighting and Andrews describe the cellular and organismal control mechanisms that come into play to ensure that adequate levels of iron are properly incorporated and metabolized by the red blood cell. Of particular importance are the roles of hepcidin and ferroportin in this process, and Chapter 6 integrates the biochemical properties of these molecules and their interacting partners with disease states that result from their mutation or altered regulation.

Although oxygen transport is the most well-described property of red blood cells, nitric oxide is another colorless gas that interacts with hemoglobin and becomes chemically altered as a result. Čokić and Schechter provide a comprehensive review of the synthesis and regulation of nitric oxide and place this in the context of red blood cell differentiation. The role of accessory cells, such as endothelial and stromal cells, in this process is part of the wider framework in which nitric oxide exerts its effects of erythroid cells at a variety of points. As detailed in Chapter 7, these impinge on both extracellular (e.g., vasculogenesis) and intracellular (e.g., hemoglobin function and expression) control mechanisms, leading to an unanticipated intersection with therapies used to increase fetal hemoglobin.

We then end with a detailed description by Ellis and Lipton of Diamond-Blackfan anemia, a red blood cell disorder that results from a specific type of genetic aberration, as a particularly cogent example of how the combination of basic and clinical observations has helped to unravel its molecular basis. Of particular surprise has been the determination that Diamond-Blackfan anemia is a disorder of ribosome synthesis or function, and the potential mechanisms by which this might lead to compromised red blood cell function are evaluated in Chapter 8. Interestingly, this situation is no longer unique, as there are now other red blood cell (and non-red blood cell) disorders that are also linked to ribosomal dysfunction. Given the pediatric nature of this disease, the authors discuss how established treatment options are being considered and reevaluated in the light of these novel genetic causes and the molecular mechanisms that are altered as a result.

It is hoped that the reader will appreciate why the authors of these chapters are still excited about the study of red blood cell development, even after all of the years of intensive efforts. The synergistic merge between basic research and clinical observations that characterize this field has no peer in science, and the "bench to bedside" road has been well traversed in both directions for decades. Its continuous history of novel observations illuminated by innovative approaches that exert a wide range of influence on other fields of study progresses unabated, and indeed will continue to lead to a deeper understanding and appreciation of the multilayered complexities of eukaryotic biology.

James J Bieker, PhD
New York, NY

REFERENCES

Behringer, R. R., Ryan, T. M., Palmiter, R. D., Brinster, R. L., and Townes, T. M. (1990). Human γ- to β-globin gene switching in transgenic mice. *Genes Dev.* **4,** 380–389.

Bunn, H. F. (2001). Human hemoglobins: Sickle hemoglobin and other mutants. *In* "The Molecular Bases of Blood Diseases," (G. Stamatoyannopoulos, P. W. Majerus, R. M. Perlmutter, and H. Varmus, eds.), pp. 227–273. W. B. Saunders Co., Philadelphia.

Costantini, F., and Lacy, E. (1981). Introduction of a rabbit beta-globin gene into the mouse germ line. *Nature* **294,** 92–94.

Enver, T., Raich, N., Ebens, A. J., Papayannopoulou, T., Costantini, F., and Stamatoyannopoulos, G. (1990). Developmental regulation of human fetal-to-adult globin gene switching in transgenic mice. *Nature* **344,** 309–313.

Gaensler, K. M., Kitamura, M., and Kan, Y. W. (1993). Germ-line transmission and developmental regulation of a 150-kb yeast artificial chromosome containing the human beta-globin locus in transgenic mice. *Proc. Natl. Acad. Sci. USA* **90,** 11381–11385.

Grosveld, F., Assendelft, G. B. V., Greaves, D. R., and Kollias, B. (1987). Position-independent, high-level expression of the human β-globin gene in transgenic mice. *Cell* **51,** 975–985.

Jordan, H. E., and Spiedel, C. C. (1923). Blood cell formation and distribution in relation to the mechanism of thyroid accelerated metamorphosis in the larval frog. *J. Exp. Med.* **38,** 529–539.

Lawn, R. M., Fritsch, E. F., Parker, R. C., Blake, G., and Maniatis, T. (1978). The isolation and characterization of linked delta- and beta-globin genes from a cloned library of human DNA. *Cell* **15,** 1157–1174.

McCulloch, E. A., and Till, J. E. (2005). Perspectives on the properties of stem cells. *Nat. Med.* **11,** 1026–1028.

McCutcheon, F. H. (1936). Hemoglobin function during the life history of the bullfrog. *J. Cell. Comp. Physiol.* **8,** 63–81.

Metcalf, D., and Moore, M. A. S. (1971). Haemopoietic Cells. North-Holland, Amsterdam.

Morrison, S. J., Uchida, N., and Weissman, I. L. (1995). The biology of hematopoietic stem cells. *Annu. Rev. Cell Dev. Biol.* **11,** 35–71.

Pauling, L., Itano, H., Singer, S. J., and Wells, I. C. (1949). Sickle cell anemia: A molecular disease. *Science* **110,** 543.

Perutz, M. F. (1987). Molecular anatomy, physiology, and pathology of hemoglobin. *In* "The Molecular Bases of Blood Diseases," (G. Stamatoyannopoulos, A. W. Nienhuis, P. Leder, and P. W. Majerus, eds.). W. B. Saunders Co., Philadelphia.

Peterson, K. R., Clegg, C. H., Huxley, C., Josephson, B. M., Haugen, H. S., Furukawa, T., and Stamatoyannopoulos, G. (1993). Transgenic mice containing a 248-kb yeast artificial chromosome carrying the human beta-globin locus display proper developmental control of human globin genes. *Proc. Natl. Acad. Sci. USA* **90,** 7593–7597.

Samokhvalov, I. M., Samokhvalova, N. I., and Nishikawa, S. (2007). Cell tracing shows the contribution of the yolk sac to adult haematopoiesis. *Nature* **446,** 1056–1061.

Scadden, D. T. (2006). The stem-cell niche as an entity of action. *Nature* **441,** 1075–1079.

Schofield, R. (1978). The relationship between the spleen colony-forming cell and the haemopoietic stem cell. *Blood Cells* **4,** 7–25.

Stamatoyannopoulos, G., Majerus, P. W., Perlmutter, R. M., and Varmus, H. (2001). The Molecular Bases of Blood Diseases. W. B. Saunders Co., Philadelphia.

Strouboulis, J., Dillon, N., and Grosveld, F. (1992). Developmental regulation of a complete 70-kb human beta-globin locus in transgenic mice. *Genes Dev.* **6,** 1857–1864.

Till, J. E., and McCulloch, E. A. (1980). Hemopoietic stem cell differentiation. *Biochim. Biophys. Acta* **605,** 431–459.

Weatherall, D. J. (1994). The Thalassemias. *In* "The Molecular Bases of Blood Diseases," (G. Stamatoyannopoulos, A. W. Nienhuis, P. W. Majerus, and H. Varmus, eds.), pp. 157–205. W. B. Saunders Co., Philadelphia.

CHAPTER ONE

Ontogeny of Erythropoiesis in the Mammalian Embryo

Kathleen McGrath* *and* James Palis*

Contents

Abstract

Red cells are required not only for adult well-being but also for survival and growth of the mammalian embryo beyond early postimplantation stages of development. The embryo's first "primitive" erythroid cells, derived from a transient wave of committed progenitors, emerge from the yolk sac as immature precursors and differentiate as a semisynchronous cohort in the bloodstream. Surprisingly, this maturational process in the mammalian embryo is characterized by globin gene switching and ultimately by enucleation. The yolk sac also synthesizes a second transient wave of "definitive" erythroid progenitors that enter the bloodstream and seed the liver of the fetus. At the same time, hematopoietic stem cells within the embryo also seed the liver and are the

* Department of Pediatrics, Center for Pediatric Biomedical Research, University of Rochester, Rochester, New York 14642

Current Topics in Developmental Biology, Volume 82
ISSN 0070-2153, DOI: 10.1016/S0070-2153(07)00001-4

presumed source of long-term erythroid potential. Fetal definitive erythroid precursors mature in macrophage islands within the liver, enucleate, and enter the bloodstream as erythrocytes. Toward the end of gestation, definitive erythropoiesis shifts to its final location, the bone marrow. It has recently been recognized that the yolk sac-derived primitive and fetal definitive erythroid lineages, like their adult definitive erythroid counterpart, are each hierarchically associated with the megakaryocyte lineage. Continued comparative studies of primitive and definitive erythropoiesis in mammalian and nonmammalian embryos will lead to an improved understanding of terminal erythroid maturation and globin gene regulation.

1. Introduction

The red cells of mammals are unique in the animal kingdom because they circulate as enucleated cells, while the red cells of fish, amphibians, and birds remain nucleated (Gulliver, 1875). Nearly 100 years ago, it was recognized that two distinct populations of red cells circulate in the bloodstream of early mammalian embryos (Maximow, 1909). The first population consisted of extremely large, nucleated red cells that originated in the yolk sac (Fig. 1.1). These "primitive" red cells were subsequently superseded by a second "definitive" population of smaller, enucleated red cells

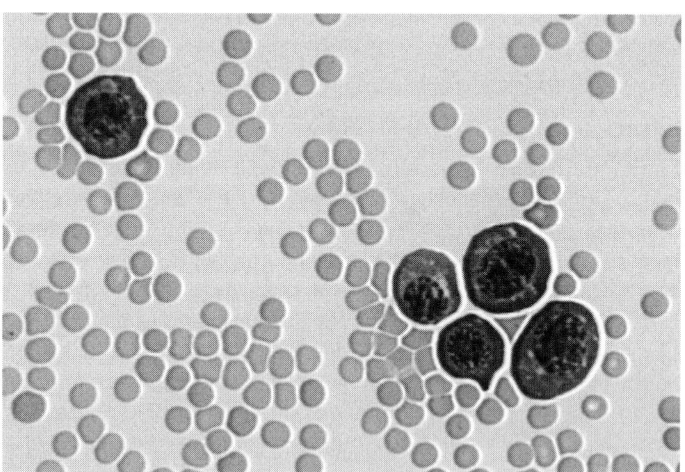

Figure 1.1 Visual comparison of primitive and definitive erythroid cells in the mouse. Circulating blood cells were isolated both from E9.5 embryos and from adult mice, mixed together, cytospun, and stained with Wright-Giemsa. The nucleated primitive erythroid cells are at the basophilic stage of maturation and markedly larger than the definitive erythrocytes. (See Color Insert.)

that continue to circulate during fetal and postnatal life. Because primitive red cells are nucleated and confined to the embryo, they were thought to share many characteristics with the nucleated red cells of nonmammalian vertebrates (Tavassoli, 1991). However, recent findings have challenged this long-held paradigm and have demonstrated that primitive erythroblasts fully mature, like their definitive counterparts, into enucleated erythrocytes. Furthermore, studies in the murine embryo suggest that definitive erythroid cells are derived from more than one source during embryogenesis. This chapter will discuss the current understanding of the ontogeny of the primitive and definitive erythroid lineages in the mammalian embryo.

Embryonic erythropoiesis is best understood in the context of adult erythropoiesis. In the adult human, maintenance of a steady-state normal red cell mass requires the synthesis of two million erythrocytes every second. This red cell production is sustained by the continued generation from hematopoietic stem cells (HSC) of committed progenitors that are assayed by their ability to form colonies in semisolid media supplemented with cytokines. Clonal colonies containing both erythroid cells and platelet-forming megakaryocytes support the concept that the erythroid and megakaryocyte lineages share a common bipotential progenitor (Debili *et al.*, 1996; McCleod *et al.*, 1976; Suda *et al.*, 1983). Such bipotential progenitors have been prospectively isolated from the bone marrow by flow cytometry (Akashi *et al.*, 2000). Downstream of these bipotential progenitors lay unipotential erythroid and megakaryocyte progenitors. The most immature erythroid-restricted progenitor is the burst-forming unit erythroid (BFU-E) that generates the more mature colony-forming unit erythroid (CFU-E) (Heath *et al.*, 1976; Stephenson *et al.*, 1971). These progenitors, in turn, give rise to nucleated erythroblasts that undergo a limited number of cell divisions as they decrease their cell size, accumulate hemoglobin, and undergo progressive nuclear condensation. Erythroid precursors mature in association with macrophage cells in "erythroblast islands" which serve as a stromal microenvironment within the bone marrow cavity (Bessis *et al.*, 1978). Following enucleation, young erythrocytes remove ribosomes and mitochondria, assume a biconcave shape, and enter the bloodstream where they function to provide oxygen to all tissues of the body. Erythropoiesis in the embryo differs from erythropoiesis in the adult because of two significant dilemmas faced by the embryo. First, functional red cells are required before long-term HSC and their microenvironmental niches are established. Second, the mammalian embryo grows extremely rapidly and embryonic erythropoiesis must generate ever-increasing numbers of red cells to accommodate this growth. In fact, a 70-fold increase in the red cell mass has been estimated to occur in fetal mice between embryonic day 12.5 (E12.5) and E16.5 of gestation (Russell *et al.*, 1968). This contrasts with erythropoiesis in the adult, which is in steady state unless perturbed by some stress such as bleeding or hemolysis.

2. Primitive Erythropoiesis

2.1. Emergence of blood islands in the yolk sac

Studies in multiple organisms indicate that the initial generation of blood cells in the embryo depends on the formation of mesoderm cells during gastrulation. In the mouse embryo, mesoderm cells begin to traverse the primitive streak and occupy an intermediate position between primitive ectoderm and visceral endoderm germ layers at E6.5–7.0. Cell tracking studies indicate that mesoderm cells migrate through the posterior streak and contribute to the formation of all the extraembryonic structures, including the yolk sac, the chorion, and the amnion (Kinder *et al.*, 1999). The yolk sac is a bilayer structure, composed of mesoderm-derived and visceral endoderm-derived cell layers. Within the mesoderm layer, pools of primitive erythroid cells, the so-called "blood islands," rapidly emerge between E7.5 and E8.0 in the mouse conceptus (Ferkowicz *et al.*, 2003; Haar and Ackerman, 1971; Silver and Palis, 1997). These blood islands become enveloped by endothelial cells which form the initial vascular plexus of the yolk sac (reviewed by Ferkowicz and Yoder, 2005). The emergence of primitive erythroid cells and endothelial cells at the same time (early gastrulation) and place (yolk sac mesoderm) within the early conceptus has long suggested that these lineages share a common developmental origin.

Pioneering experiments in the chick embryo led to the concept that signals from the visceral endoderm layer of the yolk sac induce the formation of blood and endothelium in adjacent yolk sac mesoderm (Wilt, 1965). Visceral endoderm was also found to be important for blood vessel formation in yolk sac mesoderm in mouse embryos (Palis *et al.*, 1995). Support for the role of visceral endoderm in endothelial cell and blood cell formation comes from GATA-4-null embryoid bodies that lack visceral endoderm and display markedly reduced blood island formation (Bielinska *et al.*, 1996). Multiple signaling cascades have been implicated in the initiation of both blood cell and blood vessel development. Alterations in vascular endothelial growth factor (VEGF) signaling cause defects in numbers and migration of hematopoietic and vascular precursors (Hamada *et al.*, 2000; Shalaby *et al.*, 1997). Furthermore, Indian hedgehog signaling has been found to be an important component of the induction both for blood cells (Belaousoff *et al.*, 1999; Dyer *et al.*, 2001) and for endothelium (Byrd *et al.*, 2002). The bone morphogenetic protein and Wnt signaling cascades may also combine to specify hematopoietic regions in the vertebrate embryo (Marvin *et al.*, 2001; Wang *et al.*, 2007). Finally, observations in frog and chick embryos suggest that spatial restriction of blood islands within the yolk sac occurs through inhibitory fibroblast growth factor signals (Kumano and Smith, 2000; Nakazawa *et al.*, 2006).

2.2. Primitive erythroid cell maturation

Primitive red cells arise in yolk sac blood islands shortly after gastrulation, beginning at E7.5 in the mouse (Haar and Ackerman, 1971; Silver and Palis, 1997). Primitive erythroblasts are produced by unique progenitors, termed primitive erythroid colony-forming cells (EryP-CFC), that generate colonies distinguishable from definitive erythroid colonies by their intermediate maturation time, colony morphology, and the unique pattern of globin gene expression (Palis *et al.*, 1999; Wong *et al.*, 1986; Table 1.1). EryP-CFC are first found in the murine yolk sac at E7.25, soon after the start of gastrulation but before the formation of blood islands. EryP-CFC then rapidly expand in numbers peaking at E8.0, after which they decrease until no longer found by E9.0 (Palis *et al.*, 1999). The transient nature of primitive erythropoiesis is because EryP-CFC emerge in such a narrow temporal wave. The progenies of these progenitors are the exclusive red cells in the embryo until the newly formed fetal liver releases the first definitive red cells into the circulation at E12 (Brotherton *et al.*, 1979; Kingsley *et al.*, 2004; Steiner and Vogel, 1973). Therefore, anemia observed in the fetus before E13 must be due to loss or decreased synthesis of primitive erythroid cells. Disruption of genes necessary for the emergence (SCL, LMO2) or maturation (GATA-1) of the primitive erythroid lineage indicates that primitive erythroblasts are necessary for survival of the embryo beyond E9.5–10.5 (Fujiwara *et al.*, 1996; Shivdasani *et al.*, 1995; Warren *et al.*, 1994). In contrast, mouse embryos specifically lacking definitive erythrocytes, as occurs following the targeted disruption of the c-myb transcription factor, survive until E15.5 (Mucenski *et al.*, 1991). These results indicate that the primitive erythroid lineage provides a sufficient source of red cells to ensure embryonic survival until relatively late stages of development.

Morphologic analysis indicates that primitive erythroblasts mature in a semisynchronous cohort as they circulate in the bloodstream (Fraser *et al.*, 2007; Kingsley *et al.*, 2004; Morioka and Minamikawa-Tachino, 1993). Within 24 h of the appearance of EryP-CFC, primitive proerythroblasts can be observed ensheathed by a primary vascular plexus. These immature primitive erythroblasts begin to circulate at E8.25 coincident with, or soon after, the onset of cardiac contractions (Ji *et al.*, 2003; Lucitti *et al.*, 2007; McGrath *et al.*, 2003). They continue to divide in the bloodstream until E13, as evidenced by the presence of circulating mitotic figures (Bethlenfalvay and Block, 1970), thymidine incorporation (de la Chapelle *et al.*, 1969), and cell cycle studies (Sangiorgi *et al.*, 1990). Primitive erythroblasts accumulate increasing amounts of hemoglobin and become progressively less basophilic (Sasaki and Matsumura, 1986; Steiner and Vogel, 1973). Hemoglobin synthesis continues until replication ceases (Fantoni *et al.*, 1968), and primitive red cells reach their steady-state hemoglobin content of 80–100 pg per cell, approximately six times the amount of hemoglobin found in adult murine

Table 1.1 Comparison of primitive and definitive erythropoiesis in the mouse

		Definitive	
	Primitive	Fetal	Adult
Progenitors[a] (colony formation n days)	EryP-CFC (5 days)	CFU-E (2 days) BFU-E (7–10 days)	CFU-E (2 days) BFU-E (7–10 days)
Sites of maturation	Yolk sac, bloodstream	Fetal liver	Bone marrow, spleen
Erythroblast islands	No	Yes	Yes
Cell size[b] (MCV)	400 fl	150 fl	70 fl
Hemoglobin (Hb) accumulation[c]	80–100 pg	25 pg	12 pg
β-Globin transcription[d]	βH1, ϵy, β1, β2	β1, β2	β1, β2
α-Globin transcription[d]	ζ, α1, α2	α1, α2	α1, α2
Cytokines	EPO (relative)	EPO (absolute), SCF	EPO (absolute), SCF
Transcription factors[e]	SCL, LMO2, GATA-2, GATA-1, EKLF, KLF-2	SCL, LMO2, GATA-2, GATA-1, EKLF, c-myb, Gfi-1b	SCL, LMO2, GATA-2, GATA-1, EKLF, c-myb

[a] CFU-E, BFU-E (Heath *et al.*, 1976; Stephenson *et al.*, 1971), EryP-CFC (Palis *et al.*, 1995).
[b] Kingsley *et al.* (2004).
[c] Steiner and Vogel (1973).
[d] Trimborn *et al.* (1999); Kingsley *et al.* (2004).
[e] GATA-1 (Fujiwara *et al.*, 1996; Pevny *et al.*, 1991), EKLF (Nuez *et al.*, 1995; Perkins *et al.*, 1995), KLF2 (Basu *et al.*, 2005), c-myb (Mucenski, 1991), Gfi-1b (Saleque *et al.*, 2002), SCL (Porcher *et al.*, 1996; Robb *et al.*, 1995; Shivdasani *et al.*, 1995), LMO2 (Warren *et al.*, 1994; Yamada *et al.*, 1998), GATA-2 (Tsai *et al.*, 1994).

erythrocytes (Table 1.1; Steiner and Vogel, 1973). This correlates with the finding that primitive erythroblasts are approximately six times larger than adult erythrocytes (Fig. 1.1; Kingsley *et al.*, 2004).

2.3. Globin gene expression

Hemoglobin molecules contain globin chains derived from both the α- and β-globin gene loci. While definitive erythroid cells in the mouse express α1-, α2- β1-, and β2-globins, primitive erythroid cells in addition express ζ-, βH1-, and ϵy-globins (Trimborn *et al.*, 1999). These latter embryonic globin genes are differentially expressed in primitive erythroid cells (Farace *et al.*, 1984; Kingsley *et al.*, 2006; Whitelaw *et al.*, 1990). The initially expressed ζ- and βH1-globin genes are superseded by the α1-, α2-, and ϵy-globin genes, a process termed "maturational" globin switching since this globin switching occurs as primitive erythroid precursors terminally differentiate (Kingsley *et al.*, 2006). These changes in globin transcript levels are associated with changes in RNA polymerase II density at their promoters. Furthermore, the βH1- and ϵy-globin genes in primitive erythroid cells reside in a single large hyperacetylated domain, suggesting that the maturational globin switching is regulated by altered transcription factor presence instead of chromatin accessibility as postulated in the adult (Kingsley *et al.*, 2006). In contrast, the regions containing these genes are not associated with histone hyperacetylation (Bulger *et al.*, 2003) and they are not expressed in definitive erythroid cells (Kingsley *et al.*, 2006; Trimborn *et al.*, 1999). Primitive erythroid cells in human embryos also appear to undergo maturational globin switching. Both ζ- to α-globin and ϵ- to γ-globin gene switches have been described between 5 and 7 weeks gestation (Peschle *et al.*, 1985). Differentiating human embryonic stem cells have recently been used to model embryonic hematopoiesis and their study has led to a renewed interest in globin gene expression and regulation in primitive and fetal definitive erythroid cells (Chang *et al.*, 2006; Olivier *et al.*, 2006; Zambidis *et al.*, 2005). A better understanding of the mechanisms regulating embryonic versus fetal/adult globin gene expression may ultimately lead to novel approaches for the treatment of the thalassemia syndromes and sickle cell anemia. Intriguingly, reactivation of the embryonic ζ-globin gene has been shown to ameliorate an adult mouse model of sickle cell disease (He and Russell, 2004).

2.4. Differences and commonalities between primitive and definitive erythropoiesis

As seen with globins, there are additional gene usage differences identified between primitive and definitive erythropoiesis (Table 1.1). In particular, mice lacking c-myb fail to generate definitive erythrocytes but appear to have a normal primitive red cell mass (Mucenski *et al.*, 1991). Targeted disruption of the transcriptional repressor Gfi-1b causes a block in the synthesis of definitive erythroid cells, while primitive erythroid cells are present but have a delay in maturation (Saleque *et al.*, 2002). While Runx1 is expressed both by primitive and by definitive erythroid cells, targeted disruption of Runx1 and its partner core binding factor β each leads to defects only in the latter (Okuda *et al.*, 1996; Sasaki *et al.*, 1996; Wang *et al.*, 1996). Targeted disruption of erythroid Kruppel-like factor (EKLF) leads to a complete block in definitive erythroid cell maturation in the fetal liver and fetal death at E15.5 (Nuez *et al.*, 1995; Perkins *et al.*, 1995). However, it has recently been recognized that EKLF regulates many erythroid-specific genes and EKLF-null fetuses display significant abnormalities of primitive erythroblasts (Hodge *et al.*, 2006).

Erythropoiesis in the adult is critically dependent on erythropoietin, a cytokine that promotes late-stage erythroid progenitor and immature precursor survival (reviewed by Koury, 2005). Addition of erythropoietin to yolk sac tissues explanted *in vitro* leads to an expansion of primitive erythroid cells containing hemoglobin and an increase in embryonic globin transcripts (Kimura *et al.*, 2000; McGann *et al.*, 1997). Furthermore, immature primitive erythroblasts express erythropoietin receptor transcripts (McGann *et al.*, 1997) and protein on their cell surface (Boussios *et al.*, 1989). Exogenous erythropoietin abrogates apoptosis of immature primitive erythroid cells cultured *in vitro* (Kimura *et al.*, 2000). Targeted disruption of erythropoietin or the erythropoietin receptor in mice leads to a 5- to 20-fold reduction in primitive erythroid cells by E11.5 and fetal demise from severe anemia by E13.5 (Kieran *et al.*, 1996; Lin *et al.*, 1996; Wu *et al.*, 1995). These results, taken together, indicate that erythropoietin signaling is critical for the survival and maturation of primitive erythroid precursors.

While definitive erythropoiesis in the fetus liver is completely blocked by the lack of erythropoietin signaling, some primitive erythroid cells continue to mature, suggesting that other cytokine signaling cascades may be differentially active in primitive versus definitive erythropoiesis. The cytokine stem cell factor (SCF), by signaling through the c-kit receptor, potentiates erythropoietin signaling and plays an important role in the proliferation of definitive erythroid precursors (reviewed by Munugalavadla and Kapur, 2005). Mice lacking c-kit signaling die of severe anemia between E14.5 and E16.5. These mice have defects in definitive hematopoiesis; however, the role of c-kit signaling in primitive erythroid cell maturation is unclear (Goldman *et al.*, 2006; Russell *et al.*, 1968). It remains to be determined which other cytokines regulate primitive erythroid cell maturation.

Despite these biological differences in transcriptional regulation and cytokine dependence, it is important to note that primitive and definitive erythropoiesis share many fundamental characteristics of mammalian erythroid differentiation. Both originate from unipotential progenitors and depend on the action of multiple transcription factors for maturation, including SCL, LMO2, and GATA-1 (Table 1.1 and references therein). The maturation process in both primitive and definitive erythropoiesis is characterized by downregulation of vimentin intermediate filaments (Sangiorgi *et al.*, 1990), the accumulation of hemoglobin at similar rates (Steiner and Vogel, 1973), and the upregulation of bcl-x to prevent apoptosis (Motoyama *et al.*, 1999). Finally, it has recently been shown that primitive erythroblasts in the mouse ultimately enucleate and, like definitive cells, circulate as erythrocytes (Kingsley *et al.*, 2004). Studies using embryonic-specific globin antibodies that distinguish primitive from definitive erythroid cells revealed that primitive erythroid cells enucleate between E12.5 and E17.5. These findings have recently been corroborated in mice with GFP expressed in primitive erythroid cells under control of an embryonic globin promoter (Fraser *et al.*, 2007). It is not known where and by what mechanism late-stage primitive erythroblasts enucleate since they are actively circulating, unlike their adult counterparts that mature and enucleate extravascularly attached to macrophage cells in erythroblast islands of the fetal liver and postnatal bone marrow (reviewed by Chasis, 2006, and by Manwani and Bieker, Chapter 2, in this volume).

3. "Definitive" Erythropoiesis in the Fetus

3.1. Characteristics of definitive erythropoiesis in the fetus

Primitive erythropoiesis fulfills the erythroid functions critical for early postimplantation embryonic survival and growth; however, the fetus requires increasing numbers of red cells throughout gestation. Prior to the formation of the bone marrow cavity, the liver serves as the site of maturation of definitive erythroid cells in the fetus. Soon after the liver begins to form as an organ at E9.5, it is colonized by external hematopoietic elements. Experiments with carefully staged embryos indicate that hematopoietic progenitors enter the liver at 28–30 sp (Houssaint, 1981; Johnson and Moore, 1975). BFU-E and CFU-E are found in the early fetal liver and their numbers expand exponentially for several days and peak at E14.5–15.5 (Kurata *et al.*, 1998; Palis *et al.*, 1999; Rich and Kubanek, 1979). Subsequently, there is gradual transition of hematopoietic activity to the bone marrow cavity and the liver ceases to be a hematopoietic organ in both the mouse and the human soon after birth. While fundamentally similar, there are some differences between fetal progenitors and their adult bone marrow counterparts. CFU-E in the murine fetus are more sensitive to erythropoietin (Rich and Kubanek, 1976). Fetal BFU-E have a greater and more rapid proliferative capacity. Unlike adult marrow-derived BFU-E, fetal liver-derived BFU-E are capable of proliferating in response to erythropoietin in the absence of added colony-stimulating factors (Emerson *et al.*, 1989; Migliaccio and Migliaccio, 1988; Valtieri *et al.*, 1989).

Morphologic examination of the fetal liver in the mouse reveals the presence of immature erythroid precursors at E11.5–12.5 (Marks and Rifkind, 1972). As development proceeds, these precursors associate with macrophage cells to form erythroblast islands similar to those in the bone marrow (Sasaki and Sonoda, 2000). Furthermore, PDGFR-alpha-expressing stromal cells appear to play a role in the fetal liver but not in the embryonic yolk sac microenvironment (Li *et al.*, 2006). There is a gradual transition to more mature erythroid precursor populations as development proceeds (Marks and Rifkind, 1972). Enucleated definitive red cells begin to emerge from the liver at E12 of mouse gestation (Kingsley *et al.*, 2004; Rifkind *et al.*, 1969). Over the next several days, the number of definitive erythroid cells expands exponentially in the circulation concomitant with the continued rapid growth of the fetus (Kingsley *et al.*, 2004; Russell *et al.*, 1968). Fetal erythrocytes in the mouse are approximately twice as large and contain twice the hemoglobin compared with their adult counterparts (Kingsley *et al.*, 2004; Steiner and Vogel, 1973). In the human, fetal erythrocytes can also be distinguished from adult erythrocytes by the accumulation of fetal hemoglobin (HbF, $\alpha2\gamma2$) rather than adult hemoglobin (HbA, $\alpha2\beta2$). A "switch" from fetal to adult hemoglobin synthesis begins at 32 weeks gestation and is completed after birth. Unlike primates, rodent red cells do not synthesize a distinct fetal form of hemoglobin (Fantoni *et al.*, 1967; Wong *et al.*, 1983). Even though the mouse does not have a unique fetal globin, human fetal globin genes are accurately expressed when introduced into the mouse fetus (Enver *et al.*, 1990; Stamatoyannopoulos, 2005). These results suggest that a conserved transcriptional difference exists during fetal erythropoiesis that regulates other fetal-specific characteristics and has been co-opted by primates to specify globin expression.

3.2. Fetal erythropoiesis and "stress" erythropoiesis

The mechanisms responsible for these differences observed in fetal and adult erythropoiesis remain unclear. Possibilities include hematopoietic cell intrinsic differences or microenvironmental differences between the fetal liver and the postnatal marrow (Muench and Namikawa, 2001). Critical environmental differences may be the relative hypoxia of the fetus, coupled with the need to increase red cell mass due to the expanding blood volume from growth. These factors may create signals and responses similar to those found in the adult where acute hypoxia elicits a "stress" response characterized by the rapid synthesis of large erythrocytes expressing increased amounts of fetal hemoglobin (Alter, 1979). The link between stress and fetal erythropoiesis is further supported by the phenotype of Stat5-null and flexed tail mice that each have normal steady-state adult erythropoiesis but display a transient fetal anemia and a blunted response as adults to acute erythroid stress induced by acute anemia (Lenox *et al.*, 2005; Socolovsky *et al.*, 1999). Therefore, adult stress erythropoiesis in adults may represent a reactivation of a fetal erythroid program that is distinct from adult steady-state erythropoiesis and is first used to rapidly expand the number of definitive erythrocytes during embryogenesis. Analysis of the flexed tail mutant implicates BMP4 signaling in stress erythropoiesis (Lenox *et al.*, 2005). It is not known if BMP4 signaling also plays a role in fetal erythropoiesis.

4. Developmental Origins of Erythropoiesis

4.1. Hemangioblast

The concept that the hematopoietic and vascular lineages emerge from common "hemangioblast" precursors has existed for over 100 years and is based, in part, on the close spatial and temporal emergence of primitive erythroid and endothelial cells in the yolk sac. These lineages also share the expression of many genes, including transcription factors and cell surface proteins (reviewed by Park *et al.*, 2005). Recent evidence suggests that hematopoietic potential arises from mesoderm cells expressing many markers associated with endothelium (Ema *et al.*, 2006). A unique blast colony-forming cell (blast-CFC) containing both hematopoietic and endothelial cell potential has been identified both in cultured embryonic stem cells and in mouse embryos (Choi *et al.*, 1998; Huber *et al.*, 2004). Consistent with this unique potential, blast-CFC express Flk-1 and are regulated by several transcription factors, including endoglin and GATA-2, expressed by hematopoietic and endothelial lineages (Perlingeiro, 2007; Lugus *et al.*, 2007). These hemangioblast precursors are primarily confined to the region of the primitive streak in gastrulating mouse embryos. Blood islands, composed of primitive erythroid precursors, arise in a ring along the mesometrial edge of the mouse conceptus (Drake *et al.*, 2000; Ferkowicz *et al.*, 2003; McGrath *et al.*, 2003). Since hemangioblast precursors are found primarily in the primitive streak and not in the yolk sac, it is thought that they rapidly commit to hematopoietic and vascular fates soon after their emergence during early gastrulation. There is increasing evidence to suggest that many, if not most, yolk sac vascular cells arise from unilineage angioblast precursors and not from hemangioblasts (Furuta *et al.*, 2006; Ueno and Weissman, 2006). In contrast, hemangioblast precursors contain primitive erythroid, definitive erythroid, and multilineage myeloid potential (Choi *et al.*, 1998; Huber *et al.*, 2004). These findings support the concept that all primitive erythroid and the first definitive erythroid cells in the embryo are ultimately derived from hemangioblast precursors.

4.2. Hematopoietic stem cell

A hallmark of adult hematopoiesis is the continuous generation of mature blood cells from HSC. The developmental origin of long-term HSC during murine embryogenesis capable of engrafting adult recipients is associated with the appearance of cell clusters arising from the dorsal aorta in the aorta-gonad-mesonephros (AGM) region at E10.5 (de Bruijn *et al.*, 2000; Muller *et al.*, 1994). The placenta serves as a site of HSC expansion (reviewed by Mikkola *et al.*, 2005) and may also be a site of HSC origin given that the allantois and chorion contain hematopoietic potential when cultured *in vitro* (Zeigler *et al.*, 2006). HSC are first found within the fetal liver at E11, consistent with their migration from these vascular sites of "hemogenic" endothelium (Ema and Nakauchi, 2000; Kumaravelu *et al.*, 2002; Muller *et al.*, 1994). However, definitive erythroid potential is found in the conceptus before long-term HSC formation. Specifically, BFU-E emerge in the yolk sac at E8.25 before the onset of circulation (Palis *et al.*, 1999; Wong *et al.*, 1986). Once circulation begins, BFU-E are found in increasing numbers in the bloodstream and then concentrated in the fetal liver by E10 (Palis *et al.*, 1999). These spatiotemporal kinetics suggest that yolk sac-derived BFU-E colonize the fetal liver. Interestingly, similar kinetics have been described for BFU-E in human embryos that emerge from the yolk sac at 4.5 weeks gestation, enter the bloodstream, and are found in increasing numbers in the liver by 6 weeks gestation (Migliaccio *et al.*, 1986). It is hypothesized that once these reach the liver's hematopoietic environment, these yolk sac-derived BFU-E complete their maturation to produce the first definitive red cells of the embryo. Co-organ culture of yolk sac and fetal liver primordial taken from <28 sp mouse embryos indicates that the liver contains soluble factors that promote the differentiation of definitive erythroid potential present in the yolk sac (Cudennec *et al.*, 1981). As there is no current method to distinguish the progeny of yolk sac definitive progenitors from those arising from later HSC sources, this hypothesis is not yet proven. However, Ncx1-null mouse embryos, lacking a heartbeat and systemic circulation, synthesize normal numbers of primitive and definitive erythroid progenitors in the yolk sac but fail to redistribute primitive erythroblasts and definitive erythroid progenitors to the embryo proper (Lux *et al.*, 2007). These recent results support the notion that the definitive erythroid progenitors that initially seed the fetal liver are entirely derived from the yolk sac. Furthermore, HSC do not mature at their site of synthesis (Godin *et al.*, 1999) and do not colonize the liver until E11. Thus, they have insufficient time to generate the mature erythrocytes that emerge from the liver beginning at E12.

Taken together, these data support a model of erythroid ontogeny in the embryo whereby three distinct waves of erythroid progenitors generate maturing precursors in three different microenvironments (Fig. 1.2). The first wave consists of EryP-CFC that generate primitive erythroid cells that mature in the bloodstream. The second wave consists of BFU-E that emerge from the yolk sac, colonize the fetal liver, and generate the first fetal definitive erythrocytes that enter the circulation. The third wave consists of long-term HSC-derived BFU-E that are responsible for continued synthesis of fetal erythrocytes within the liver, and ultimately adult erythrocytes within the bone marrow (Fig. 1.2). These three waves of erythropoiesis also are associated with distinct hematopoietic potentials. While the yolk sac-derived primitive erythroid wave was initially thought to be only erythroid, its onset is coincident with that of the megakaryocyte lineage (Palis *et al.*, 1999; Tober *et al.*, 2007; Xu *et al.*, 2001). A hierarchical association of these lineages is supported by the recent discovery of a unique bipotential primitive erythroid/megakaryocyte progenitor (Tober *et al.*, 2007). Like primitive erythroid progenitors, these bipotential primitive erythroid/megakaryocyte progenitors originate from hemangioblast precursors and expand transiently only within the yolk sac (Tober *et al.*, 2007).

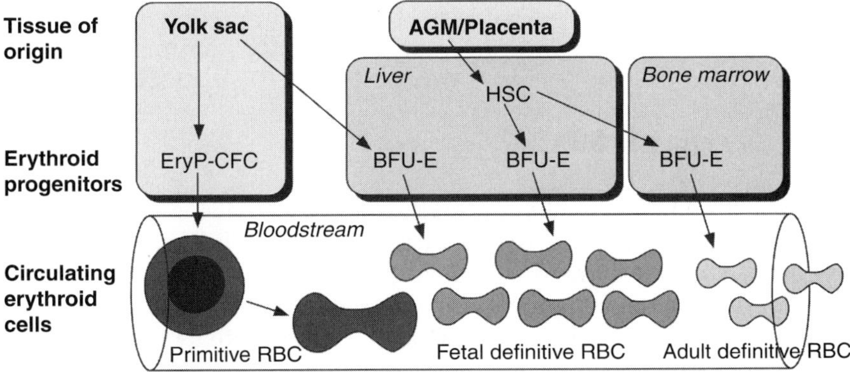

Figure 1.2 Simplified model of erythroid ontogeny in the mammalian embryo. Current data support a model whereby three waves of erythroid progenitors emerge in the mammalian embryo. The first wave consists of primitive erythroid progenitors (EryP-CFC) that originate in the yolk sac during early gastrulation and generate primitive erythroid precursors that mature to become enucleated erythrocytes in the bloodstream. The second wave consists of definitive erythroid progenitors (BFU-E) that emerge from the yolk sac and seed the emerging fetal liver. There they generate maturing definitive erythroid precursors that enucleate to become the first circulating definitive erythrocytes of the fetus. The third wave consists of definitive erythroid progenitors that originate from long-term hematopoietic stem cells (HSC) and mature initially in the fetal liver and subsequently in the postnatal bone marrow to later generate fetal and adult red blood cells (RBC). AGM, aorta-gonad-mesonephros region.

Macrophage progenitors first emerge within the yolk sac at the same developmental time as the primitive erythroid and megakaryocyte lineages (Palis *et al.*, 1999), suggesting that "primitive" hematopoiesis in mammals is in fact trilineage in nature. The second wave of yolk sac-derived erythropoiesis consists of the definitive erythroid lineage, which arises temporally and spatially in conjunction with the macrophage, mast cell, granulocyte and megakaryocyte lineages, as well as multipotential high-proliferative potential colony-forming cells (HPP-CFC; Palis *et al.*, 1999, 2001; Xie *et al.*, 2003). Furthermore, recent evidence indicates that the definitive erythroid lineage emerging from the yolk sac shares a common bipotential progenitor with the megakaryocyte lineage (Tober *et al.*, 2007). Thus, both primitive and definitive erythropoiesis arising in the early mammalian embryo during gastrulation, like later definitive erythropoiesis in the marrow, are each hierarchically associated with the megakaryocyte lineage. It remains controversial whether this second wave of erythropoiesis is associated with B lymphoid potential (Cumano *et al.*, 1993, 1996; Sugiyama *et al.*, 2007; Yokota *et al.*, 2006) or HSC capable of engrafting newborn but not adult mice (Yoder *et al.*, 1997). Interestingly, the contribution of yolk sac-derived hematopoietic potential to adult hematopoiesis has recently received some experimental support (Samokhvalov *et al.*, 2007). Finally, the third wave of perinatal and postnatal erythropoiesis is associated with the complete myeloid and lymphoid potential of the HSC.

5. Conclusions

The paradigm of embryonic erythropoiesis has been extensively modified from the simple two-tiered system of an evolutionarily primitive erythroid cell replaced by the adult type of definitive erythroid cell. First, there is a complexity of the definitive forms between the fetal and the adult states that may reflect the erythropoietic stress of the fetus. Second, the yolk sac-derived primitive erythroid lineage has all the hallmarks of mammalian erythropoiesis, including enucleation. However, primitive erythroid precursors begin to circulate and function as immature forms and mature in the bloodstream. Third, the yolk sac also provides a wave of definitive erythroid progenitors that are proposed to colonize the fetal liver and mature there. Thus, the yolk sac appears to provide needed erythropoietic functions to the embryo before HSC-derived hematopoiesis is fully functional. Fourth, the primitive and the definitive erythroid waves that emerge in the yolk sac are each hierarchically associated with megakaryocyte potential. It is not known if embryonic erythropoiesis in nonmammalian organisms is closely associated with thrombopoiesis. It is also not known whether there are two distinct waves of definitive erythropoiesis in nonmammalian organisms, such as the much-studied zebrafish, frog, and chick embryos, because of the lack of readily available CFC assays in these systems (Samarut *et al.*, 1979). Caution must be exercised when interpreting definitive erythroid potential as inherently downstream of an HSC in these organisms. Similarly, definitive erythropoiesis observed in murine and human embryonic stem cell maturation systems likely reflects the second wave of yolk sac-derived definitive erythropoiesis. Ultimately, a better understanding of the ontogeny of erythropoiesis in mammalian and nonmammalian species will continue to lead to novel insights regarding globin regulation and erythroid maturation.

ACKNOWLEDGMENTS

The authors thank coworkers and collaborators for many enjoyable discussions and their insights. We apologize to those whose work, through oversight and space limitations, has not been cited. The authors are funded by the National Institutes of Health, United States.

REFERENCES

Akashi, K., Traver, D., Miyamoto, T., and Weissman, I. L. (2000). A clonogenic common myeloid progenitor that gives rise to all myeloid lineages. *Nature* **404,** 193–197.

Alter, B. P. (1979). Fetal erythropoiesis in stress hematopoiesis. *Exp. Hematol.* **7,** 200–209.

Basu, P., Morris, P. E., Haar, J. L., Wani, M. A., Lingrel, J. B., Gaensler, K. M., and Lloyd, J. A. (2005). KLF2 is essential for primitive erythropoiesis and regulates the human and murine embryonic {beta}-like globin genes *in vivo*. *Blood* **106,** 2566–2571.

Belaousoff, M., Farrington, S. M., and Baron, M. H. (1999). Hematopoietic induction and respecification of A-P identity by visceral endoderm signaling in the mouse embryo. *Development* **125,** 5009–5018.

Bessis, M., Mize, C., and Prenant, M. (1978). Erythropoiesis: Comparison of *in vivo* and *in vitro* amplification. *Blood Cells* **4,** 155–174.

Bethlenfalvay, N. C., and Block, M. (1970). Fetal erythropoiesis. Maturation in megaloblastic (yolk sac) erythropoiesis in the C57Bl/6J mouse. *Acta Haematol.* **44,** 240–245.

Bielinska, M., Narita, N., Heikinheimo, M., Porter, S. B., and Wilson, D. B. (1996). Erythropoiesis and vasculogenesis in embryoid bodies lacking visceral yolk sac endoderm. *Blood* **88,** 3720–3730.

Boussios, T., Bertles, J. F., and Goldwasser, E. (1989). Erythropoietin-receptor characteristics during the ontogeny of hamster yolk sac erythroid cells. *J. Biol. Chem.* **264,** 16017–16021.

Brotherton, T. W., Chui, D. H. K., Gauldie, J., and Patterson, M. (1979). Hemoglobin ontogeny during normal mouse fetal development. *Proc. Natl. Acad. Sci. USA* **76,** 2853–2857.

Bulger, M., Schubeler, D., Bender, M. A., Hamilton, J., Farrell, C. M., Hardison, R. C., and Groudine, M. (2003). A complex chromatin landscape revealed by patterns of nuclease sensitivity and histone modification within the mouse beta-globin locus. *Mol. Cell. Biol.* **23,** 5234–5244.

Byrd, N., Becker, S., Maye, P., Narasimhaiah, R., St-Jacques, B., Zhang, X., McMahon, J., McMahon, A., and Grabel, L. (2002). Hedgehog is required for murine yolk sac angiogenesis. *Development* **129,** 361–372.

Chang, K. H., Nelson, A. M., Cao, H., Wang, L., Nakamoto, B., Ware, C. B., and Papayannopoulou, T. (2006). Definitive-like erythroid cells derived from human embryonic stem cells coexpress high levels of embryonic and fetal globins with little or no adult globin. *Blood* **108,** 1515–1523.

Chasis, J. A. (2006). Erythroblastic islands: Specialized microenvironmental niches for erythropoiesis. *Curr. Opin. Hematol.* **13,** 137–141.

Choi, K., Kennedy, M., Kazarov, A., Papadimitriou, J. C., and Keller, G. (1998). A common precursor for hematopoietic and endothelial cells. *Development* **125,** 725–732.

Cudennec, C. A., Thiery, J.-P., and Le Douarin, N. M. (1981). *In vitro* induction of adult erythropoiesis in early mouse yolk sac. *Proc. Natl. Acad. Sci. USA* **78,** 2412–2416.

Cumano, A., Furlonger, C., and Paige, C. J. (1993). Differentiation and characterization of B-cell precursors detected in the yolk sac and embryo body of embryos beginning at the 10- to 12-somite stage. *Proc. Natl. Acad. Sci. USA* **90,** 6429–6433.

Cumano, A., Dieterlen-Lievre, F., and Godin, I. (1996). Lymphoid potential, probed before circulation in mouse, is restricted to caudal intraembryonic splanchnopleura. *Cell* **86,** 907–916.

de Bruijn, M. F. T. R., Speck, N. A., Peeters, M. C. E., and Dzierzak, E. (2000). Definitive hematopoietic stem cells first develop within the major arterial regions of the mouse embryo. *EMBO J.* **19,** 2465–2474.

de la Chapelle, A., Fantoni, A., and Marks, P. (1969). Differentiation of mammalian somatic cells: DNA and hemoglobin synthesis in fetal mouse yolk sac erythroid cells. *Proc. Natl. Acad. Sci. USA* **63,** 812–819.

Debili, N., Coulombel, L., Croisille, L., Katz, J., Guichard, J., Breton-Gorius, J., and Vainchenker, W. (1996). Characterization of a bipotent erythro-megakaryocytic progenitor in human bone marrow. *Blood* **88,** 1284–1296.

Drake, C. J., and Fleming, P. A. (2000). Vasculogenesis in the day 6.5 to 9.5 mouse embryo. *Blood* **95,** 1671–1679.

Dyer, M. A., Farrington, S. M., Mohn, D., Munday, J. R., and Baron, M. H. (2001). Indian hedgehog activates hematopoiesis and vasculogenesis and can respecify neurectodermal cell fate in the mouse embryo. *Development* **128,** 1717–1730.

Ema, H., and Nakauchi, H. (2000). Expansion of hematopoietic stem cells in the developing liver of the mouse embryo. *Blood* **95,** 2284–2288.

Ema, H., Yokomizo, T., Wakamatsu, A., Terunuma, T., Yamamoto, M., and Takahashi, S. (2006). Primitive erythropoiesis from mesodermal precursors expressing VE-cadherin, PECAM-1, Tie2, endoglin, and CD34 in the mouse embryo. *Blood* **108,** 4018–4024.

Emerson, S. G., Shanti, T., Ferrara, J. L., and Greenstein, J. L. (1989). Developmental regulation of erythropoiesis by hematopoietic growth factors: Analysis on populations of BFU-E from bone marrow, peripheral blood, and fetal liver. *Blood* **74,** 49–55.

Enver, T., Raich, N., Ebens, A. J., Papayannopoulou, T., Costantini, F., and Stamatoyannopoulos, G. (1990). Developmental regulation of human fetal-to-adult globin gene switching in transgenic mice. *Nature* **344,** 309–313.

Fantoni, A., Bank, A., and Marks, P. A. (1967). Globin composition and synthesis of hemoglobins in developing fetal mice erythroid cells. *Science* **157,** 1327–1329.

Fantoni, A., de la Chapelle, A., Rifkind, R. A., and Marks, P. A. (1968). Erythroid cell development in fetal mice: Synthetic capacity for different proteins. *J. Mol. Biol.* **33,** 79–91.

Farace, M. G., Brown, B. A., Raschella, G., Alexander, J., Gambari, R., Fantoni, A., Hardies, S. C., Hutchison, C. A. I., and Edgell, M. H. (1984). The mouse bh1 gene codes for the z chain of embryonic hemoglobin. *J. Biol. Chem.* **259,** 7123–7128.

Ferkowicz, M. J., and Yoder, M. (2005). Blood island formation: Longstanding observations and modern interpretations. *Exp. Hematol.* **33,** 1041–1047.

Ferkowicz, M. J., Starr, M., Xie, X., Li, W., Johnson, S., Shelley, W. C., Morrison, P. R., and Yoder, M. (2003). CD41 expression defines the onset of primitive and definitive hematopoiesis in the murine embryo. *Development* **130,** 4393–4403.

Fraser, S. T., Isern, J., and Baron, M. H. (2007). Maturation and enucleation of primitive erythroblasts during mouse embryogenesis is accompanied by changes in cell-surface antigen expression. *Blood* **109,** 343–352.

Fujiwara, Y., Browne, C. P., Cuniff, K., Goff, S. C., and Orkin, S. H. (1996). Arrested development of embryonic red cell precursors in mouse embryos lacking transcription factor GATA-1. *Proc. Natl. Acad. Sci. USA* **93,** 12355–12358.

Furuta, C., Ema, H., Takayanagi, S.-I., Ogaeri, T., Okamura, D., Matsui, Y., and Nakauchi, H. (2006). Discordant developmental waves of angioblasts and hemangioblasts in the early gastrulating mouse embryo. *Development* **133,** 2771–2779.

Godin, I., Garcia-Porrero, J. A., Dieterlen-Lievre, F., and Cumano, A. (1999). Stem cell emergence and hematopoietic activity are incompatible in mouse intraembryonic sites. *J. Exp. Med.* **190,** 43–52.

Goldman, D. C., Berg, L. K., Heinrich, M. C., and Christian, J. L. (2006). Ectodermally derived steel/stem cell factor functions non-cell autonomously during primitive erythropoiesis in Xenopus. *Blood* **107,** 3114–3121.

Gulliver, G. (1875). Observations on the sizes and shapes of the red corpuscles of the blood of vertebrates, with drawings of them to a uniform scale, and extended and revised tables of measurements. *Proc. Zoolog. Soc. Lond.* **47,** 4–495.

Haar, J. L., and Ackerman, G. A. (1971). A phase and electron microscopic study of vasculogenesis and erythropoiesis in the yolk sac of the mouse. *Anat. Rec.* **170,** 199–224.

Hamada, K., Oike, Y., Takakura, N., Ito, Y., Jussila, L., Dumont, D. J., Alitalo, K., and Suda, T. (2000). VEGF-C signaling pathways through VEGFR-2 and VEGFR-3 in vasculoangiogenesis and hematopoiesis. *Blood* **96,** 3793–3800.

He, Z., and Russell, J. E. (2004). Antisickling effects of an endogenous human alpha-like globin. *Nat. Med.* **10,** 365–367.

Heath, D. S., Axelrod, A. A., McLeod, D. L., and Shreeve, M. M. (1976). Separation of the erythropoietin-responsive BFU-E and CFU-E in mouse bone marrow by unit gravity separation. *Blood* **47,** 777–792.

Hodge, D., Coghill, E., Keys, J., Maguire, T., Hartmann, B., McDowall, A., Weiss, M., Grimmond, S., and Perkins, A. (2006). A global role for EKLF in definitive and primitive erythropoiesis. *Blood* **107,** 3359–3370.

Houssaint, E. (1981). Differentiation of the mouse hepatic primordium. II. Extrinsic origin of the haemopoietic cell line. *Cell Differ.* **10,** 243–252.

Huber, K., Kouskoff, V., Fehling, H. J., Palis, J., and Keller, G. (2004). Haemangioblast commitment is initiated in the primitive streak of the mouse embryo. *Nature* **432,** 625–630.

Ji, R. P., Phoon, C. K. L., Aristizábal, O., McGrath, K. E., Palis, J., and Turnbull, D. H. (2003). Onset of cardiac function during early mouse embryogenesis coincides with entry of primitive erythroblasts into the embryo proper. *Circ. Res.* **92,** 133–135.

Johnson, G. R., and Moore, M. A. S. (1975). Role of stem cell migration in initiation of mouse foetal liver haemopoiesis. *Nature* **258,** 726–728.

Kieran, M. W., Perkins, A. C., Orkin, S. H., and Zon, L. Z. (1996). Thrombopoietin rescues *in vitro* erythroid colony formation from mouse embryos lacking the erythropoietin receptor. *Proc. Natl. Acad. Sci. USA* **93,** 9126–9131.

Kimura, T., Sonoda, Y., Iwai, N., Satoh, M., Yamaguchi-Tsukio, M., Izui, T., Suda, M., Sasaki, K., and Nakano, T. (2000). Proliferation and cell death of embryonic primitive erythrocytes. *Exp. Hematol.* **28,** 635–641.

Kinder, S. J., Tsang, T. E., Quinlan, G. A., Hadjantonakis, A.-K., Nagy, A., and Tam, P. P. L. (1999). The orderly allocation of mesodermal cells to the extraembryonic structures and the anteroposterior axis during gastrulation of the mouse embryo. *Development* **126,** 4691–4701.

Kingsley, P. D., Malik, J., Fantauzzo, K. A., and Palis, J. (2004). Yolk sac-derived primitive erythroblasts enucleate during mammalian embryogenesis. *Blood* **104,** 19–25.

Kingsley, P. D., Malik, J., Emerson, R. L., Bushnell, T. P., McGrath, K. E., Bloedorn, L. A., Bulger, M., and Palis, J. (2006). "Maturational" globin switching in primary primitive erythroid cells. *Blood* **104,** 1667–1672.

Koury, M. J. (2005). Erythropoietin: The story of hypoxia and a finely regulated hematopoietic hormone. *Exp. Hematol.* **33,** 1263–1270.

Kumano, G., and Smith, W. C. (2000). FGF signaling restricts the primary blood islands to ventral mesoderm. *Dev. Biol.* **228,** 304–314.

Kumaravelu, P., Hook, L., Morrison, A. M., Ure, J., Zhao, S., Zuyev, S., Ansell, J. D., and Medvinsky, A. (2002). Quantitative developmental anatomy of definitive haematopoietic stem cells/long-term repopulating units (HSC/RUs): Role of the aorta-gonad-mesonephros (AGM) region and the yolk sac in colonisation of the mouse embryonic liver. *Development* **129,** 4891–4899.

Kurata, H., Mancini, G. C., Alespieti, G., Migliaccio, A. R., and Migliaccio, G. (1998). Stem cell factor induces proliferation and differentiation of fetal progenitor cells in the mouse. *Br. J. Haematol.* **101,** 676–687.

Lenox, L. E., Perry, J. M., and Paulson, R. F. (2005). BMP4 and Madh5 regulate the erythroid response to acute anemia. *Blood* **105,** 2741–2748.

Li, W. L., Yamada, Y., Ueno, M., Nishikawa, S., Nishikawa, S., and Takakura, N. (2006). Platelet derived growth factor receptor alpha is essential for establishing a microenvironment that supports definitive erythropoiesis. *J. Biochem. (Tokyo)* **140,** 267–273.

Lin, C.-S., Lim, S.-K., D'Agati, V., and Constantini, F. (1996). Differential effects of an erythropoietin receptor gene disruption on primitive and definitive erythropoiesis. *Genes Dev.* **10,** 154–164.

Lucitti, J. L., Jones, E. A., Huang, C., Chen, J., Fraser, S. E., and Dickinson, M. E. (2007). Vascular remodeling of the mouse yolk sac requires hemodynamic force. *Development* **134,** 3317–3326.

Lugus, J. J., Chung, Y. S., Mills, J. C., Kim, S. I., Grass, J., Kyba, M., Doherty, J. M., Bresnick, E. H., and Choi, K. (2007). GATA2 functions at multiple steps in hemangioblast development and differentiation. *Development* **134,** 393–405.

Lux, C., Yoshimoto, M., McGrath, K. E., Conway, S., Palis, J., and Yoder, M. C. (2007). All primitive and definitive hematopoietic progenitor cells emerging prior to E10 in the mouse embryo are products of the yolk sac. *Blood* (in press).

Marks, P. A., and Rifkind, R. A. (1972). Protein synthesis: Its control in erythropoiesis. *Science* **175,** 955–961.

Marvin, M. J., Di Rocco, G., Gardiner, A., Bush, S. M., and Lassar, A. B. (2001). Inhibition of Wnt activity induces heart formation from posterior mesoderm. *Genes Dev.* **15,** 316–327.

Maximow, A. A. (1909). Untersuchungen uber blut und bindegewebe 1. Die fruhesten entwicklungsstadien der blut- und binde-gewebszellan bein saugetierembryo, bis zum anfang der blutbilding unden leber. *Arch. Mikroskop. Anat.* **73,** 444–561.

McCleod, D. L., Shreve, M. M., and Axelrod, A. A. (1976). Induction of megakaryocyte colonies with platelet formation *in vitro*. *Nature* **261,** 492–494.

McGann, J. K., Silver, L., Liesveld, J., and Palis, J. (1997). Erythropoietin-receptor expression and function during the initiation of murine yolk sac erythropoiesis. *Exp. Hematol.* **25,** 1149–1157.

McGrath, K. E., Koniski, A. D., Malik, J., and Palis, J. (2003). Circulation is established in a step-wise pattern in the mammalian embryo. *Blood* **101,** 1669–1676.

Migliaccio, A. R., and Migliaccio, G. (1988). Human embryonic hemopoiesis: Control mechanisms underlying progenitor differentiation *in vitro*. *Dev. Biol.* **125,** 127–134.

Migliaccio, G., Migliaccio, A. R., Petti, S., Mavilio, F., Russo, G., Lazzaro, D., Testa, U., Marinucci, M., and Peschle, C. (1986). Human embryonic hematopoiesis: Kinetics of progenitors and precursors underlying the yolk sac to liver transition. *J. Clin. Invest.* **78,** 51–60.

Mikkola, H. K., Gekas, C., Orkin, S. H., and Dieterlen-Lievre, F. (2005). Placenta as a site for hematopoietic stem cell development. *Exp. Hematol.* **33,** 1048–1054.

Morioka, K., and Minamikawa-Tachino, R. (1993). Temporal characteristics of the differentiation of embryonic erythroid cells in fetal peripheral blood of the Syrian hamster. *Dev. Growth Differ.* **35,** 569–582.

Motoyama, N., Kimura, T., Takahashi, T., Watanabe, T., and Nakano, T. (1999). bcl-x prevents apoptotic cell death of both primitive and definitive erythrocytes at the end of maturation. *J. Exp. Med.* **189,** 1691–1698.

Mucenski, M. L., McLain, K., Kier, A. B., Swerdlow, S. H., Schreiner, M. C., Miller, T. A., Pietryga, D. W., Scott, W. J., and Potter, S. S. (1991). A functional c-myb gene is required for normal murine fetal hepatic hematopoiesis. *Cell* **65,** 677–689.

Muench, M. O., and Namikawa, R. (2001). Disparate regulation of human fetal erythropoiesis by the microenvironments of the liver and bone marrow. *Blood Cells Mol. Dis.* **27,** 377–390.

Muller, A. M., Medvinsky, A., Strouboulis, J., Grosveld, F., and Dzierzak, E. (1994). Development of hematopoietic stem cell activity in the mouse embryo. *Immunity* **1,** 291–301.

Munugalavadla, V., and Kapur, R. (2005). Role of c-Kit and erythropoietin receptor in erythropoiesis. *Crit. Rev. Oncol. Hematol.* **54,** 63–75.

Nakazawa, F., Nagai, H., Shin, M., and Sheng, G. (2006). Negative regulation of primitive hematopoiesis by the FGF signaling pathway. *Blood* **108,** 3335–3343.

Nuez, B., Michalovich, D., Bygrave, A., Ploemacher, R., and Grosveld, F. (1995). Defective haematopoiesis in fetal liver resulting from inactivation of the EKLF gene. *Nature* **375,** 316–318.

Okuda, T., van Deursen, J., Hiebert, S. W., Grosveld, G., and Downing, J. R. (1996). AML-1, the target of multiple chromosomal translocations in human leukemia, is essential for normal fetal liver hematopoiesis. *Cell* **84,** 321–330.

Olivier, E. N., Qiu, C., Velho, M., Hirsch, R. E., and Bouhassira, E. E. (2006). Large-scale production of embryonic red blood cells from human embryonic stem cells. *Exp. Hematol.* **34,** 1635–1642.

Palis, J., McGrath, K. E., and Kingsley, P. D. (1995). Initiation of hematopoiesis and vasculogenesis in murine yolk sac explants. *Blood* **86,** 156–163.

Palis, J., Robertson, S., Kennedy, M., Wall, C., and Keller, G. (1999). Development of erythroid and myeloid progenitors in the yolk sac and embryo proper of the mouse. *Development* **126,** 5073–5084.

Palis, J., Chan, R. J., Koniski, A. D., Patel, R., Starr, M., and Yoder, M. C. (2001). Spatial and temporal emergence of high proliferative potential hematopoietic precursors during murine embryogenesis. *Proc. Natl. Acad. Sci. USA* **98,** 4528–4533.

Park, C., Ma, Y. D., and Choi, K. (2005). Evidence for the hemangioblast. *Exp. Hematol.* **33,** 965–970.

Perlingeiro, R. C. (2007). Endoglin is required for hemangioblast and early hematopoietic development. *Development* **134,** 3041–3048.

Perkins, A. C., Sharpe, A. H., and Orkin, S. H. (1995). Lethal beta-thalassaemia in mice lacking the erythroid CACCC-transcription factor EKLF. *Nature* **375,** 318–322.

Peschle, C., Mavilio, F., Care, A., Migliaccio, G., Migliaccio, A. R., Salvo, G., Samoggia, P., Petti, S., Guerriero, R., Marinucci, M., Lazzaro, D., Russo, G., *et al.* (1985). Haemoglobin switching in human embryos: Asynchrony of the $\zeta \rightarrow \alpha$ and $\epsilon \rightarrow \gamma$-globin switches in primitive and definitve erythropoietic lineage. *Nature* **313,** 235–238.

Pevny, L., Simon, M. C., Robertson, E., Klein, W. H., Tsai, S.-F., D'Agati, V., Orkin, S. H., and Constantini, F. (1991). Erythroid differentiation in chimaeric mice blocked by a targeted mutation in the gene for transcription factor GATA-1. *Nature* **349,** 257–260.

Porcher, C., Swat, W., Rockwell, K., Fujiwara, Y., Alt, F. W., and Orkin, S. H. (1996). The T cell leukemia oncoprotein SCL/tal-1 is essential for development of all hematopoietic lineages. *Cell* **86,** 47–57.

Rich, I. N., and Kubanek, B. (1976). Erythroid colony formation in foetal liver and adult bone marrow and spleen from the mouse. *Blood* **33,** 171–180.

Rich, I. N., and Kubanek, B. (1979). The ontogeny of erythropoiesis in the mouse detected by the erythroid colony-forming technique. *J. Embryol. Exp. Morphol.* **50,** 57–74.

Rifkind, R. A., Chui, D., and Epler, H. (1969). Ultrastructural study of early morphogenetic events during the establishment of fetal hepatic erythropoiesis. *J. Cell Biol.* **40,** 343–365.

Robb, L., Lyons, I., Li, R., Hartley, L., Kontgen, F., Harvey, R. P., Metcalf, D., and Begley, C. G. (1995). Absence of yolk sac hematopoiesis from mice with a targeted disruption of the scl gene. *Proc. Natl. Acad. Sci. USA* **92,** 7075–7079.

Russell, E., Thompson, M., and McFarland, E. (1968). Analysis of effects of W and f genic substitutions on fetal mouse hematology. *Genetics* **58,** 259–270.

Saleque, S., Cameron, S., and Orkin, S. H. (2002). The zinc-finger proto-oncogene Gfi-1b is essential for development of the erythroid and megakaryocyte lineages. *Genes Dev.* **16,** 301–306.

Samarut, J., Jurdic, P., and Nigon, V. (1979). Production of erythropoietic colony-forming units and erythrocytes during chick embryo development: An attempt at modelization of chick embryo erythropoiesis. *J. Embryol. Exp. Morphol.* **50,** 1–20.

Samokhvalov, I. M., Samokhvalova, N. I., and Nishikawa, S.-I. (2007). Cell tracing shows the contribution of the yolk sac to adult haematopoiesis. *Nature* **446,** 1056–1061.

Sangiorgi, F., Woods, C. M., and Lazarides, E. (1990). Vimentin downregulation is an inherent feature of murine erythropoiesis and occurs independently of lineage. *Development* **110,** 85–96.

Sasaki, K., and Matsumura, G. (1986). A karyometrical study of circulating erythroblasts of yolk sac origin in the mouse embryo. *Arch. Histol. Jpn.* **49,** 535–541.

Sasaki, K., and Sonoda, Y. (2000). Histometrical and three-dimensional analysis of liver hematopoiesis in the mouse embryo. *Arch. Histol. Jpn.* **63,** 137–146.

Sasaki, K., Yagi, H., Bronson, R. T., Tominaga, K., Matsunashi, T., Deguchi, K., Tani, Y., Kishimoto, T., and Komori, T. (1996). Absence of fetal liver hematopoiesis in mice deficient in transcriptional coactivator core binding factor *β*. *Proc. Natl. Acad. Sci. USA* **93,** 12359–12363.

Shalaby, F., Ho, J., Stanford, W. L., Fischer, K. D., Schuh, A. C., Schwartz, L., Bernstein, A., and Rossant, J. (1997). A requirement for Flk1 in primitive and definitive hematopoiesis and vasculogenesis. *Cell* **89,** 981–990.

Shivdasani, R. A., Mayer, E. L., and Orkin, S. H. (1995). Absence of blood formation in mice lacking T-cell leukemia oncoprotein tal-1/SCL. *Nature* **373,** 432–434.

Silver, L., and Palis, J. (1997). Initiation of murine embryonic erythropoiesis: A spatial analysis. *Blood* **89,** 1154–1164.

Socolovsky, M., Fallon, A. E. G., Wang, S. M., Brugnara, C., and Lodish, H. F. (1999). Fetal anemia and apoptosis of red cell progenitors in Sta5a−/−5b−/− mice: A direct role for Stat5 in bcl-XL induction. *Cell* **98,** 181–191.

Stamatoyannopoulos, G. (2005). Control of globin gene expression during development and erythroid differentiation. *Exp. Hematol.* **33,** 259–271.

Steiner, R., and Vogel, H. (1973). On the kinetics of erythroid cell differentiation in fetal mice: I. Microspectrophotometric determination of the hemoglobin content in erythroid cells during gestation. *J. Cell. Physiol.* **81,** 323–338.

Stephenson, J. R., Axelrad, A., McLeod, D., and Shreeve, M. (1971). Induction of colonies of hemoglobin-synthesizing cells by erythropoietin *in vitro*. *Proc. Natl. Acad. Sci. USA* **68,** 1542–1546.

Suda, T., Suda, J., and Ogawa, M. (1983). Single-cell origin of mouse hematopoietic colonies expressing multiple lineages in variable combinations. *Proc. Natl. Acad. Sci. USA* **80,** 6689–6693.

Sugiyama, D., Ogawa, M., Nakao, K., Osumi, N., Nishikawa, S., Nishikawa, S.-I., Arai, K.-I., Nakahata, T., and Tsuji, K. (2007). B cell potential can be obtained from pre-circulatory yolk sac, but with low frequency. *Dev. Biol.* **301,** 53–61.

Tavassoli, M. (1991). Embryonic and fetal hemopoiesis: An overview. *Blood Cells* **1,** 269–281.

Tober, J., Koniski, A., McGrath, K. E., Vemishetti, R., Emerson, R., de Mesy-Bentley, K. K., Waugh, R., and Palis, J. (2007). The megakaryocyte lineage originates from hemangioblast precursors and is an integral component both of primitive and of definitive hematopoiesis. *Blood* **109,** 1433–1441.

Trimborn, T., Bribnau, J., Grosveld, F., and Fraser, P. (1999). Mechanisms of developmental control of transcription in the murine a- and b-globin loci. *Genes Dev.* **13,** 112–124.

Tsai, F.-Y., Keller, G., Kuo, F. C., Weiss, M., Chen, J., Rosenblatt, M., Alt, F. W., and Orkin, S. H. (1994). An early haematopoietic defect in mice lacking the transcription factor GATA-2. *Nature* **371,** 221–226.

Ueno, H., and Weissman, I. L. (2006). Clonal analysis of mouse development reveals a polyclonal origin for yolk sac blood islands. *Dev. Cell* **11,** 519–533.

Valtieri, M., Gabbianelli, M., Bassano, E., Petti, S., Russo, G., Testa, U., and Peschle, C. (1989). Erythropoietin alone induces erythroid burst formation by human embryonic but not adult BFU-E in unicellular serum-free culture. *Blood* **74,** 460–470.

Wang, H., Gilner, J. B., Bautch, V. L., Wang, D. Z., Wainwright, B. J., Kirby, S. L., and Patterson, C. (2007). Wnt2 coordinates the commitment of mesoderm to hematopoietic, endothelial, and cardiac lineages in embryoid bodies. *J. Biol. Chem.* **282,** 782–791.

Wang, Q., Stacey, T., Binder, M., Marin-Padilla, M., Sharpe, A. H., and Speck, N. A. (1996). Disruption of the Cbfa2 gene causes necrosis and hemorrhaging in the central nervous system and blocks definitive hematopoiesis. *Proc. Natl. Acad. Sci. USA* **93,** 3444–3449.

Warren, A. J., Colledge, W. H., Carlton, M. B. L., Evans, M., Smith, A. J. H., and Rabbitts, T. H. (1994). The oncogenic cysteine-rich LIM domain protein is essential for erythroid development. *Cell* **78,** 45–57.

Whitelaw, E., Tsai, S.-F., Hogben, P., and Orkin, S. H. (1990). Regulated expression of globin chains and the erythroid transcription factor GATA-1 during erythropoiesis in the developing mouse. *Mol. Cell. Biol.* **10,** 6596–6606.

Wilt, F. H. (1965). Erythropoiesis in the chick embryo: The role of endoderm. *Science* **147,** 1588–1590.

Wong, P. M. C., Chung, S.-W., White, J. S., Reicheld, S. M., Patterson, M., Clarke, B. J., and Chui, D. H. K. (1983). Adult hemoglobins are synthesized in murine fetal liver erythropoietic cells. *Blood* **62,** 1280–1288.

Wong, P. M. C., Chung, S. W., Chui, D. H. K., and Eaves, C. J. (1986). Properties of the earliest clonogenic hematopoietic precursors to appear in the developing murine yolk sac. *Proc. Natl. Acad. Sci. USA* **83,** 3851–3854.

Wu, H., Liu, X., Jaenisch, R., and Lodish, H. F. (1995). Generation of committed erythroid BFU-E and CFU-E progenitors does not require erythropoietin or the erythropoietin receptor. *Cell* **83,** 59–67.

Xie, X., Chan, R. J., Johnson, S. A., Starr, M., McCarthy, J., Kapur, R., and Yoder, M. C. (2003). Thrombopoietin promotes mixed lineage and megakaryocyte cell growth but inhibits primitive and definitive erythropoiesis in cells isolated from early murine yolk sacs. *Blood* **101,** 1329–1335.

Xu, M.-J., Matsuoka, S., Yang, F.-C., Ebihara, Y., Manabe, A., Tanaka, R., Eguchi, M., Asano, S., Nakahata, T., and Tsuji, K. (2001). Evidence for the presence of murine primitive megakarycytopoiesis in the early yolk sac. *Blood* **97,** 2016–2022.

Yamada, Y., Warren, A. J., Dobson, C., Forster, A., Pannell, R., and Rabbitts, T. H. (1998). The T cell leukemia LIM protein Lmo2 is necessary for adult mouse hematopoiesis. *Proc. Natl. Acad. Sci. USA* **95,** 3890–3895.

Yoder, M. C., Hiatt, K., and Mukherjee, P. (1997). In vivo repopulating hematopoietic stem cells are present in the murine yolk sac at day 9.0 postcoitus. *Proc. Natl. Acad. Sci. USA* **94,** 6776–6780.

Yokota, T., Huang, J., Tavian, M., Nagai, Y., Hirose, J., Zuniga-Pflucker, J.-C., Peault, B., and Kincade, P. W. (2006). Tracing the first waves of lymphopoiesis in mice. *Development* **133,** 2041–2051.

Zambidis, E. T., Peault, B., Park, T. S., Bunz, F., and Civin, C. I. (2005). Hematopoietic differentiation of human embryonic stem cells progresses through sequential hematoendothelial, primitive, and definitive stages resembling human yolk sac development. *Blood* **106,** 860–870.

Zeigler, B. M., Sugiyama, D., Chen, M., Guo, Y., Downs, K. M., and Speck, N. A. (2006). The allantois and chorion, when isolated before circulation or chorio-allantoic fusion, have hematopoietic potential. *Development* **133,** 4183–4192.

CHAPTER TWO

The Erythroblastic Island

Deepa Manwani*,† *and* James J. Bieker†

Contents

* Schneider Children's Hospital, New York 11040
† The Brookdale Department of Molecular, Cell and Developmental Biology, Mount Sinai School of Medicine, New York 10029

Current Topics in Developmental Biology, Volume 82
ISSN 0070-2153, DOI: 10.1016/S0070-2153(07)00002-6

Abstract

Erythroblastic islands are specialized microenvironmental compartments within which definitive mammalian erythroblasts proliferate and differentiate. These islands consist of a central macrophage that extends cytoplasmic protrusions to a ring of surrounding erythroblasts. The interaction of cells within the erythroblastic island is essential for both early and late stages of erythroid maturation. It has been proposed that early in erythroid maturation the macrophages provide nutrients, proliferative and survival signals to the erythroblasts, and phagocytose extruded erythroblast nuclei at the conclusion of erythroid maturation. There is also accumulating evidence for the role of macrophages in promoting enucleation itself. The central macrophages are identified by their unique immunophenotypic signature. Their pronounced adhesive properties, ability for avid endocytosis, lack of respiratory bursts, and consequent release of toxic oxidative species, make them perfectly adapted to function as nurse cells. Both macrophages and erythroblasts display adhesive interactions that maintain island integrity, and elucidating these details is an area of intense interest and investigation. Such interactions enable regulatory feedback within islands via cross talk between cells and also trigger intracellular signaling pathways that regulate gene expression. An additional control mechanism for cellular growth within the erythroblastic islands is through the modulation of apoptosis via feedback loops between mature and immature erythroblasts and between macrophages and immature erythroblasts.

The focus of this chapter is to outline the mechanisms by which erythroblastic islands aid erythropoiesis, review the historical data surrounding their discovery, and highlight important unanswered questions.

1. Introduction

Hematopoiesis is the process by which a self-renewing population of stem cells provides a continuous replenishment of differentiated blood cells by generating progeny with sequentially altered gene expression patterns (Kondo *et al.*, 2003; Orkin, 2000). As this process proceeds, there is progressive restriction in potential, first generating lineage-restricted progenitors, then morphologically identifiable precursors, and finally the mature blood cells. The very first site of hematopoiesis occurs in the yolk sac and provides primitive erythrocytes that are essential for the survival of the embryo until the next, definitive wave of hematopoiesis is established (Tavassoli, 1991). The sites of definitive erythropoiesis are the fetal liver and postnatal bone marrow, and occur in three distinct stages. The first stage is the evolution of lineage-committed progenitors that are microscopically invisible (Emerson *et al.*, 1985; Rosse, 1976). Identification relies on a selective enrichment via cell surface markers combined with culture and *in vivo* cellular assays. The earliest recognizable erythroid-specific progenitor is the burst-forming unit erythroid (BFU-E) that in semisolid media gives rise to large colonies of red blood cells (RBCs), identifiable between 7 and 10 days after plating of murine-derived cells. BFU-E generates colony-forming unit erythroid (CFU-E). The later progenitors can also give rise to colonies of RBCs; however, they are smaller and tighter and arise within 2–3 days of culture. CFU-E expresses the erythropoietin receptor and can give rise to the characteristic colonies in the presence of erythropoietin alone (Axelrad *et al.*, 1974).

The second stage of erythroid differentiation consists of morphologically identifiable nucleated precursors that progress from the proerythroblast to basophilic, polychromatophilic, and orthochromatic forms (Granick and Levere, 1964). In mammals, four distinctive processes characterize the progression through these stages: accumulation of hemoglobin contributing to the change from basophilic to acidophilic cytoplasm in more mature forms, expansion of erythroblast numbers through a limited number of cell divisions, a progressive decrease in cell size, and progressive nuclear condensation and enucleation (Granick and Levere, 1964). This second stage occurs when the erythroblasts are in physical contact with a macrophage.

The third and final stage of erythroid differentiation involves the maturation of the reticulocytes into circulating erythrocytes. The reticulocytes dismantle their ribosomal machinery, expel organelles, and assume a biconcave discoid shape. These mature erythrocytes then circulate in the blood stream until senescent, when they are removed by the macrophages within the reticuloendothelial system (Gifford *et al.*, 2006).

In contrast to definitive erythropoiesis in the fetal liver, primitive erythroblasts arising from the yolk sac were thought to retain their nuclei and diverge in the second and third stages of erythroblast maturation. More recently, evidence for enucleation of primitive erythroblasts has been uncovered and for the persistence of these enucleated forms later into gestation than previously described (Fraser *et al.*, 2007; Kingsley *et al.*, 2004). Thus, there are many more parallels between the stages of maturation of primitive and definitive erythroblasts than previously conceived.

Marcel Bessis first described erythroblastic islands, the specialized microenvironmental compartments within which mammalian erythroblasts proliferate and differentiate during their second stage of maturation (Bessis, 1958). These islands consist of a central macrophage that extends cytoplasmic protrusions (Gifford *et al.*, 2006) to a ring of surrounding erythroblasts. Twenty years later, after extensive studies to delineate the biological significance of these structures (Bessis and Breton-Gorius, 1961, 1962; Keyhani and Bessis, 1969; Policard and Bessis, 1962), he concluded that "the anatomic existence of the erythroblastic island consisting of the central histiocyte and rings of erythroblasts surrounding it, the fact that the histiocyte does phagocytose nuclei extruded by late erythroid cells, and the close contact between the two, strongly suggest that the erythroblastic island may constitute an example of an 'ecologic niche', a sociologic notion applied to cytology, which contributes to the maturation of RBCs. This is as much as can be said at the moment'' (Bessis *et al.*, 1978). Another two decades later, our current understanding has substantially advanced and the appeal for sociological analogies has withheld the passage of time. As noted by James Palis, ''no red cell is an island'' (Palis, 2004) and indeed extensive macrophage–erythroblast and erythroblast–erythroblast adhesive interactions are necessary for a thriving definitive erythropoietic community. Whether maturation of primitive erythroblasts also occurs in proximity with macrophages and the precise location of these units is currently an area of intense research interest.

2. Composition and Sites of Formation of Erythroblastic Islands

Erythroblastic islands have been demonstrated *in vivo* in the fetal liver, bone marrow, and splenic red pulp, all sites of mammalian definitive erythropoiesis, as well as in long-term bone marrow cultures *in vitro* (Allen and Dexter, 1982). Thus, erythroblastic islands are essential for the maturation of erythroblasts that are destined to enucleate. Mature avian RBCs do not enucleate and differentiate in bone marrow that does not contain islands. It is unclear if primitive erythroblastic islands exist or whether primitive erythroblasts are nomads migrating to the flourishing community of definitive erythroblastic islands for assistance in their terminal stages of maturation. Erythroblastic islands have not been demonstrated in the yolk sac and should not be confused with yolk sac blood islands, which have long been recognized to be the first site for blood cell emergence during embryonic development (Ferkowicz and Yoder, 2005). These blood islands are a cluster of primitive erythroblasts surrounded by an endothelial covering and nestled between the outer visceral endoderm and inner mesothelial cell layers comprising the yolk sac. These cells arise from a common precursor called the hemangioblast that exists at the primitive streak, followed by the migration of committed progenitors to the proximal yolk sac. Further evidence for this model is provided by the emergence of the first wave of embryonic hematopoiesis well before any morphological indication of blood island development. While this may have no relevance to the subject at hand, it does raise parallel questions for the erythroblastic island such as: Do the erythroblastic cells and macrophage arise from a common precursor, and if so where does this precursor reside and what inductive signals are responsible for the assembly of the erythroblastic island? Is the island a true colony and do the macrophage and erythroblasts arise from the same cell, for example a granulocyte-erythroid-macrophage-monocyte (GEMM) precursor or even more speculative, a dedicated erythroid-macrophage precursor? Support for this possibility is provided by recent studies revealing the presence of primitive megakaryocyte-erythroid progenitors that give rise to yolk sac-derived megakaryocytes in addition to the previously identified primitive erythroid cells (Tober *et al.*, 2007). Whether the primitive megakaryocyte-erythroid progenitors derive from primitive GEMM progenitors has not been investigated and needs to be studied specifically.

2.1. Localization of erythroblastic islands within the bone marrow

Erythroblastic islands are uniformly distributed throughout the bone marrow (Mohandas and Prenant, 1978). In order to demonstrate their presence and study their spatial distribution, Mohandas and Prenant performed a three-dimensional reconstruction of rat bone marrow based on 0.5 μM serial sections (Fig. 2.1). These sections were stained and studied by light and electron microscopy. They were also photographed and the individual photographs were transferred to plastic sheets that were stacked to provide the three-dimensional distribution within the bone marrow. This formed the basis of a model built with cork balls scaled to the size of the cells and flat

Figure 2.1 A three-dimensional scale model of a small volume of bone marrow from normal rat is shown. In this model, the large sphere is a megakaryocyte (m), and also shown are sinuses (S), macrophages (M), and the clusters of small spheres are the erythroblasts. This research was originally published in *Blood* (Mohandas and Prenant, 1978). © The American Society of Hematology.

sheets of cork cut out to represent the sinusoids. A similar model was constructed for both normal and hypertransfused animals.

It was speculated that the erythroblastic islands would be most abundant adjacent to the sinusoids, allowing for the egress of reticulocytes into the vasculature. However, the equal predominance of islands in other locations led to the question of whether islands migrate toward the sinusoids as they mature. Quantitative light and electron microscopy of rat bone marrow did indeed show a difference in the composition of islands adjacent and nonadjacent to the sinusoids. Nonadjacent islands contain more proerythroblasts, and adjacent islands are rich in orthochromatophilic erythroblasts, the numbers of basophilic and polychromatophilic erythroblasts being comparable in both (Yokoyama *et al.*, 2003). This remarkable finding suggests a migration of islands to the sinusoids as the cells within the island become more mature. Confirmation of such observations might uncover the role of proteases, secreted either by the mature erythroblasts or by central macrophage, that remodel underlying extracellular matrix making movement possible.

2.2. Structure of erythroblastic islands

Erythroblastic islands, consisting of a central macrophage surrounded by a ring of developing erythroblasts (Fig. 2.2 Bernard, 1991), had not been described prior to the pioneering studies by Marcel Bessis. The reason for

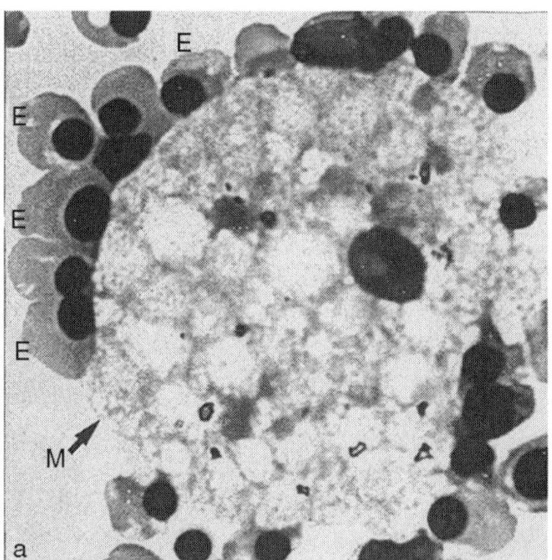

Figure 2.2 Characteristic appearance of human erythroblastic island. A Wright-Giemsa stained, cytocentrifuged preparation of cells obtained from erythroblasts generated by *in vitro* culture of peripheral blood-derived mononuclear cells using two-phase liquid culture system was analyzed. An erythroblastic island consisting of a central macrophage (M) surrounded by a ring of late erythroblasts (E) on day 12 of the second phase of a macrophage-containing culture is shown. This research was originally published in *Blood* (Hanspal *et al.*, 1998). © The American Society of Hematology. (See Color Insert.)

this is that they are easily disrupted when bone marrow smears are prepared for examination. However, Bessis found that "it is fairly easy to see the erythroblastic island in the living state in phase-contrast microscopy. To do so, one need only remove, very carefully, a very small fragment of bone marrow from a freshly opened bone and dissociate it in a drop of plasma. In the resulting preparations, which admittedly are not very fine, the crown of erythroblasts surrounding the reticular cell is clearly visible." Howard Mel, one of Bessis' students, was the first to accomplish this task while spending a sabbatical year in Bessis' laboratory in 1970. He used an apparatus that permitted live cells to be isolated simply by the action of gravity. Following this, Le Charpentier isolated and examined these structures by treating the bone marrow with gentle physical or enzymatic disruption (Le Charpentier and Prenant, 1975), and techniques based on similar principles are currently in use (Crocker and Gordon, 1985; Lee *et al.*, 2006; Sadahira *et al.*, 1990). Erythroblastic islands can also be reconstituted when erythroblasts are coincubated with macrophage cells (Fig. 2.3; Iavarone *et al.*, 2004; Lee *et al.*, 2006) and isolated islands can also be maintained in culture (Hanspal *et al.*, 1998; Le Charpentier and Prenant, 1975). The establishment of these techniques has been critical for the advancement in our understanding of the biological mechanisms governing the erythroblastic islands.

In steady-state erythropoiesis, erythroblastic islands are composed of erythroid cells in various stages of differentiation, ranging from CFU-E to young reticulocytes (Bessis *et al.*, 1978; Le Charpentier and Prenant, 1975; Lee *et al.*, 1988; Sadahira *et al.*, 1999; Yokoyama *et al.*, 2002). There is variation in the number of erythroblasts per island. Tissue sections from rat femur reveal about 10 cells per island (Yokoyama *et al.*, 2002) whereas islands harvested from human bone marrow contain 5–30 erythroblasts per island (Lee *et al.*, 1988).

Scanning electron microscopy of these islands shows cytoplasmic extensions arising from the macrophage that surround peripheral erythroid cells providing intimate contact between the macrophage and the developing erythroblasts (Fig. 2.4; Allen and Dexter, 1982). The formation and integrity of this island structure involves multiple adhesive interactions between adjacent erythroblasts and between cells of the island and the extracellular matrix (Arkin *et al.*, 1991; Armeanu *et al.*, 1995; Coulombel *et al.*, 1991; El Nemer *et al.*, 1998; Kansas *et al.*, 1990; Rosemblatt *et al.*, 1991).

2.3. Unique immunophenotypic signature of the central macrophage

The central macrophage arises from a resident monocyte precursor. Mouse central macrophages can be distinguished from other stromal cells in hematopoietic tissues by the expression of F4/80 antigen and Forssman glycosphingolipid. F4/80 is a cell surface glycoprotein with homology to the

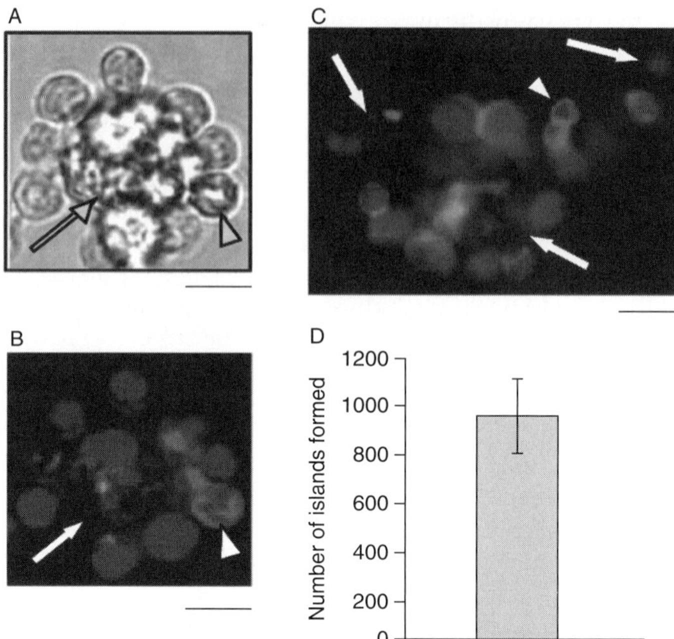

Figure 2.3 Reconstituted erythroblastic islands. Bright field (A), immunofluorescent standard (B), and confocal (C) micrographs of typical erythroblastic islands formed from single-cell suspensions of MacGreen mouse bone marrow. Macrophages from MacGreen mice express the macrophage colony-stimulating factor (M-CSF) receptor-green fluorescent protein transgene, thereby providing a useful macrophage identifier. Results from an assay for reforming islands from single-cell suspensions of freshly harvested mouse bone marrow are depicted. Adults 3–5 months of age were used. A single-cell suspension was prepared, then cells were incubated for carefully controlled times in media containing manganese. Islands and their cellular components were identified by three-color immunofluorescence microscopy. Immunofluorescent micrographs of islands show cells stained for erythroid-specific marker GPA (Ter119; red), macrophage marker M-CSF receptor GFP transgene expression (green), and DNA (Hoechst 33342; blue). Because surface expression of glycophorin A increases during terminal differentiation, the intensity of Ter119 staining served as an effective indicator of erythroblast stage. A faint blush of Ter119 fluorescence was present in early erythroblasts and increasing degrees of staining were observed in progressively more differentiated cells. The fluorescence intensity of Ter119 label varied among erythroblasts in an individual island, indicating that islands were composed of erythroblasts at various stages of differentiation. Young, multilobulated reticulocytes were present in many islands, again consistent with prior descriptions of erythroblastic islands formed *in vivo*. In the confocal image, some of the cells appear blurred because they are not in the plane of focus. However, macrophage staining is apparent in various regions of the island. Reticulocytes, arrowheads; macrophage, arrows; bars represent 10 μm. (D) Histogram shows number of erythroblastic islands formed from 1×105 single cells; $n=10$. Results are shown as mean $\pm$ SD. This research was originally published in *Blood* (Lee *et al.*, 2006). © The American Society of Hematology. (See Color Insert.)

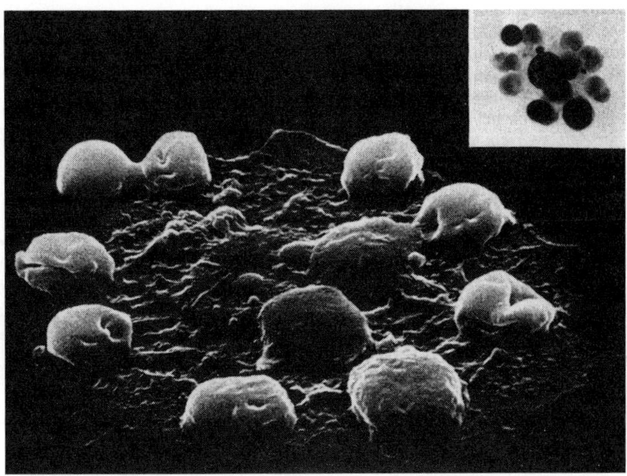

Figure 2.4 Scanning electron micrography of an erythroblastic island from rat bone marrow. Erythroblastic island after 2 hours *in vitro* culture examined by scanning electron microscopy (×5000). Top left insert: the same island seen with light microscopy (Giemsa ×1026). Note the two nuclear extrusions, one early (bottom) and the other almost complete (top left). This research was originally published in *Blood Cells* (Marcel Bessis *et al.*, 1978). © Springer-Verlag.

G–protein-linked transmembrane 7-hormone receptor family (Austyn and Gordon, 1981; Hirsch *et al.*, 1981; Hume and Gordon, 1983; Hume *et al.*, 1983; McKnight *et al.*, 1996). In early studies by Crocker and Gordon (1985), macrophages were isolated from hematopoietic cell clusters in mouse bone marrow (Fig. 2.5). These were closely compared to resident macrophages from the peritoneal cavity and to circulating monocytes (Table 2.1). The erythroblastic island macrophages are very large with diameters frequently exceeding 15 μm and have a nuclear/cytoplasmic ratio of much less than 1. These cells were intensely F4/80 avid and possessed delicate plasma membrane processes. The elaborately branched processes appear to cradle the attached hematopoietic cells (Fig. 2.5C). The F4/80 weakly stained cells were comparatively smaller, rounded, and without plasma membrane extensions. They resembled promonocytes and monocytes with a reniform nuclei and a nuclear/cytoplasmic ratio greater than 1. Resident bone marrow and peritoneal macrophages were similar in that both stained uniformly with F4/80 and 2.4G2 (IgG1/2b FcR), although the intensity of staining for both antigens was considerably greater in the resident bone marrow macrophages. In contrast to peritoneal macrophages however, resident bone marrow macrophages had no detectable Mac-I antigen expression, the presumed ligand-binding site of C3. Complement receptors were readily identified on peritoneal macrophages, monocytes, and neutrophils. The central macrophages showed high

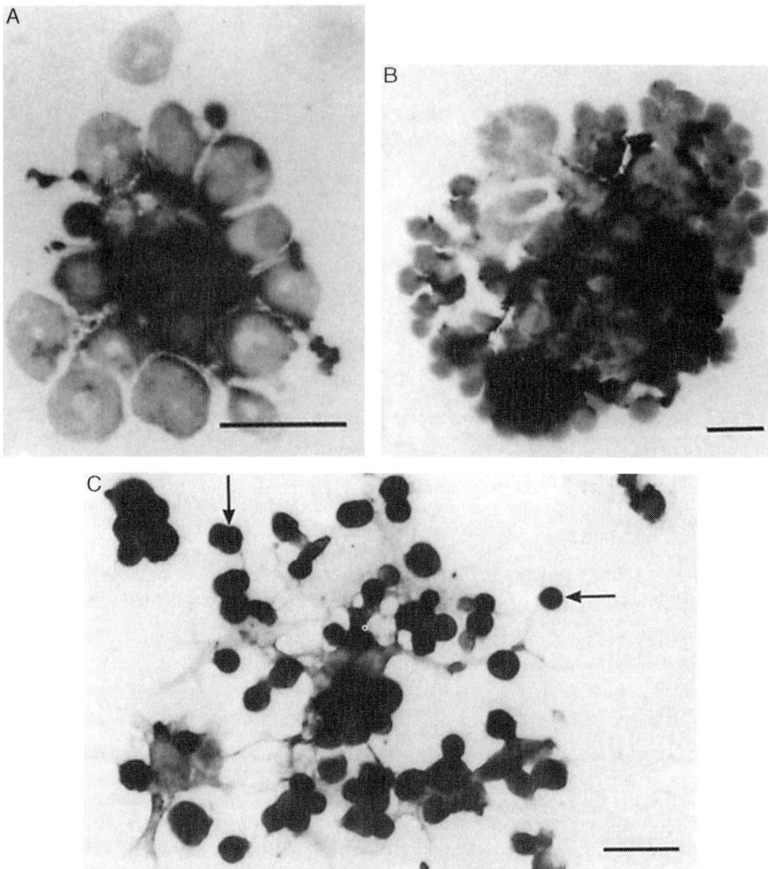

Figure 2.5 Analysis of erythroblastic islands after collagenase digestion and cluster purification. Direct cytocentrifuge preparations (A and B) or following adherence of cells to glass coverslip (C) are shown. (A) Small cluster stained with Ab F4/80, counterstained with hematoxylin, showing central macrophage surrounded by erythroblasts. Microscopic analysis at different depths of focus indicated that a single macrophage was present in such clusters. (B) Large cluster stained with Ab F4/80, counterstained with hematoxylin, showing four to five macrophages with processes ramifying among clustering cells. It has been proposed that this may represent several erythroblastic islands clumped together. (C) Central macrophage stained with Ab F4/80 after adhesion to glass coverslip. Note that delicate macrophage processes ramify extensively, establishing intimate contact with hematopoietic cells distal from the macrophage cell body (arrow) and appear to cradle hematopoietic cells. Counterstained with hematoxylin. Bar, 10 μm. This research was originally published in *J. Exp Med.* (Crocker and Gordon, 1985). © The Rockefeller University Press.

phagocytic activity and absence of a respiratory burst with a potent stimulator. They bound strongly to unopsonized sheep erythrocytes. In addition, the cells that the central macrophages associated with were actively cycling,

Table 2.1 Comparison of phenotype of central island macrophages and inflammatory peritoneal macrophages from C57BL/6 mice

	Number of experiments	Bone marrow island macrophage	Peritoneal macrophage
Surface antigens:			
F4/80	>20	+++	++
Mac-1 (CR3)	6	−	++
FcR IgG1/IIb	2	+++	++
Surface receptors:			
Zymosan (potent stimulator of phagocytosis and respiratory burst)			
• Phagocytosis	3	+++	++
• Respiratory burst	2	−	+/−
Complement	3	−	++
Sheep Erythrocytes	>20	+++	−
FcR IgG2a/2b	3	+++	+

This research is taken with permission from Crocker and Gordon(1985), © The Rockefeller University Press.

suggesting their trophic roles as opposed to simply being scavengers of dead cells. With their pronounced adhesive properties and ability for avid endocytosis, they are perfectly adapted to function as nurse cells. The lack of respiratory burst and the consequent release of toxic oxidative species are perhaps protective to the surrounding erythroblasts. Macrophages obtained from splenic (Sadahira *et al.*, 1990, 2000) and hepatic erythroblastic islands (Naito *et al.*, 1997) share a similar profile to bone marrow-derived central macrophages. In contrast, the monocytes and peritoneal macrophages with their complement receptors and respiratory burst activity play an important role in inflammatory states.

Forssman glycosphingolipid distinguishes central macrophages from monocytes and inflammatory macrophages in hematopoietic tissues (Sadahira *et al.*, 1988). This antigen has a structure of GalNAcα1–3GalNAcβ1–3Galα1–4 Galβ1–4Glcβ1-Cer and is synthesized by glycoside α1–1,3-*N*-acetylgalactosaminyl-transferase (Haslam and Baenziger, 1996). Using a combination of F4/80 and Forssman GSL, hematopoietic tissue macrophages can be classified as F4/80+Forssman− (immature) and F4/80+Forssman+ (mature) subpopulations. Using allogeneic bone marrow transplantation, Forssman GSL was not expressed in macrophages soon after lodgment in hematopoietic tissues but was expressed as they matured in the tissues (Sadahira *et al.*, 1991).

Human bone marrow-derived central macrophages also express FcRI, FcRII, FcRIII, CD4, CD31, CD11a, CD 11c, CD 18, CD 31, and HLA-DR.

2.4. Enucleation of primitive and definitive erythrocytes

It was recognized more than 125 years ago that the mature RBCs of adult vertebrates circulate either in nucleated or in enucleated forms. RBCs of birds, fish, reptiles, and amphibians retain their nucleus and contain three filamentous systems: an actin–spectrin-based cytoskeleton, intermediate filaments that attach the cytoskeleton to the nuclear membrane, and a group of microtubules organized into a circumferential marginal band (Cohen, 1991). In contrast, the definitive RBCs of mammals lose intermediate filaments and microtubules during terminal differentiation and enucleate prior to entering the bloodstream; that is, the cells mature extravascularly. Enucleation of definitive erythrocytes is accompanied by several cell divisions and a progression of morphologically identifiable forms, resulting from the accumulation of hemoglobin in the cytoplasm and condensation of the nucleus. The loss of the intermediate filaments allows the nucleus to move freely to the periphery of the cell and occupy an acentric position prior to enucleation. A number of studies have proposed that enucleation may be a process similar to cytokinesis (Campbell, 1968; Repasky and Eckert, 1981a,b,c; Simpson and Kling, 1967; Skutelsky and Danon, 1967, 1970) and have suggested that F-actin is present in the form of a ring in the constriction, similar to the cleavage furrow of mitotic cells (Perry *et al.*, 1971; Schroeder, 1973). Using splenic erythroblasts from mice infected with the anemia-inducing strain of Friend virus, Koury *et al.* (1989) showed that F-actin is concentrated between the extruding nucleus and incipient reticulocyte in enucleating erythroblasts. The combination of the cleavage furrow formed by coalescing vacuoles with new membranes and the constriction of the actin ring results in the nucleus being pinched off with a thin rim of cytoplasm and the surrounding plasma membrane from the incipient reticulocyte. The disruption of F-actin bundles with cytochalasin D treatment inhibits enucleation (Koury *et al.*, 1988), while Emp, B1-integrin, and glycoconjugates are recognized by concanavalin A partition to the nucleus (Geiduschek and Singer, 1979; Koury *et al.*, 1989; Lee *et al.*, 2004). The furrow also contains several mitochondria, which are also seen in a similar location during cytokinesis, supporting the concept that erythroid enucleation is a specialized form of cell division. The process of enucleation involves segregation of cytoskeletal and cell surface proteins between the plasma membrane of the nucleus and the incipient reticulocyte. Most of the major cytoskeletal proteins including band 3, spectrin, ankyrin, and 4.1 protein segregate to the newly formed reticulocyte. The processes described above occur within the definitive erythroblastic islands, allowing the central macrophage easy access to the expelled nucleus for phagocytosis and digestion. Ongoing studies suggest that the disproportionately greater segregation of adhesive receptors to the cytoplasmic membrane surrounding the expelled nucleus (Lee *et al.*, 2004) favors its engulfment by the macrophage and the reciprocal decrease in adhesive proteins on the newly formed reticulocyte, its separation from the nurse cell.

The long-standing perception had been that primitive mammalian erythrocytes retain their nuclei and are thus more similar to avian, fish, and reptile RBCs. However, Kingsley *et al.* (2004) have shown that primitive erythroblasts progressively enucleate in circulation between embryonic day 12.5 (E12.5) and E16.5. Using antibodies to specific regions of murine embryonic βH1-globin and adult βmajor-globin, they were able to differentiate yolk sac-derived primitive RBCs. These antibodies, in combination with nuclear staining, identified three distinct populations in the peripheral blood of E13.5–E15.5 fetuses—nucleated primitive erythroblasts, enucleated primitive erythrocytes that express βH1-globin, and the definitive cells that are enucleated and express βmajor-globin. There are a number of similarities in the erythropoietic program of primitive and definitive cells; for one, in both programs there is progressive maturation with enucleation (De la Chapelle *et al.*, 1969; Steiner and Vogel, 1973). In addition, prior to enucleation, the cells lose intermediate filaments (Sangiorgi *et al.*, 1990) and the nuclei condense and move to the plasma membrane. Despite these similarities, the one remarkable difference is that there is no evidence yet that the primitive erythroblasts require contact with macrophages to enucleate. Does this occur while the cells are in circulation and if so, what are the unique molecular mechanisms that govern enucleation in these cells? One might alternately speculate that the primitive erythroblasts migrate to sites, for instance, in the fetal liver, where reticuloendothelial cells play the role of the macrophage "nurse" cells.

3. Cell–Cell Adhesive Interactions Within Erythroblastic Islands

A number of cell adhesion molecules and their interactions within the erythroblastic islands have been described and have been proven to be critical for island integrity. The precise role of these adhesive interactions is still unclear. Given that the integrin–actin cytoskeleton interactions regulate intracellular signaling, they may coordinate adhesion and gene expression in the erythroblastic islands (Table 2.2).

Table 2.2 Adhesion molecules and their interactions within the erythroblastic island

Erythroblast adhesion molecule	Macrophage adhesion molecule	Erythroblast–erythroblast interaction	Erythroblast–macrophage interaction	References
Emp	Emp	Yes	Yes	Bala *et al.*, 2006; Hanspal *et al.*, 1998; Soni *et al.*, 2006
$\alpha 4\beta 1$ integrin/ very late antigen-4 (VLA-4)	VCAM-1		Yes	Hamamura *et al.*, 1996; Sadahira *et al.*, 1995
ICAM-4	αv		Yes	Lee *et al.*, 2006; Spring *et al.*, 2001; Telen, 2005
?	SER or sialoadhesion			Crocker *et al.*, 1990; Crocker and Gordon, 1985, 1986
?	ED2 antigen			Barbe *et al.*, 1996
?	EbR			Barbe *et al.*, 1996

EbR, Erythroblast receptor; Emp, Erythroblast macrophage protein; ICAM-4, intercellular adhesion molecule; SER, sheep erythrocyte receptor.

3.1. Erythroblast macrophage protein

Erythroblast macrophage protein is a 36-kD transmembrane protein expressed on the surface of both erythroblasts and macrophages (Hanspal *et al.*, 1998). The cytoplasmic domain contains several binding sites for SH2 domains and a potential binding site for phosphotyrosine-binding domains, suggesting a signaling function. Erythroblasts cultured in the presence of anti-Emp show the same phenotype as those cultured in the absence of macrophages—a sixfold increase in apoptosis is observed, accompanied by a marked decrease in erythroid proliferation, maturation, and enucleation (Hanspal *et al.*, 1998). Furthermore, Emp-null murine fetuses are severely anemic and die *in utero* at E19.5, that is perinatally, strongly supporting a crucial role for Emp in definitive erythropoiesis (Soni *et al.*, 2006). Emp-null fetal liver macrophages exhibit both a quantitative and a qualitative defect. The number of F4/80 positive cells is reduced to 25–30% of wild-type values and in appearance they are smaller, round in shape, and lack cytoplasmic extensions, suggesting an immature morphology. Consistent with this, ER-MP12, an antigen expressed by immature macrophage precursors, is expressed normally. This suggests that loss of Emp impairs terminal maturation but not commitment to macrophage lineage. *In vitro* erythroblast reconstitution assays using Emp-null erythroblasts or Emp-null macrophages with their wild-type counterparts indicate that Emp function is both RBC and macrophage autonomous. Interestingly, recent studies using transfected mammalian cells showed that Emp is localized in the contractile ring during cytokinesis and that it exists in a complex with actin *in vivo* (Bala *et al.*, 2006). Immunofluorescent labeling of wild-type and mutant erythroblasts with phalloidin detected a striking difference in the localization pattern of F-actin: in the wild-type erythroblasts, actin staining, which completely colocalized with Emp staining, was present throughout the cytoplasm as well as on the plasma membrane. In the mutant erythroblasts, however, actin staining was detected largely near the plasma membrane. Almost no cytoplasmic actin was detected in mutant cells, suggesting that in the absence of Emp, the actin distribution is impaired in erythroblasts. Emp also colocalizes with concentrated F-actin bundles that are detected in wild-type erythroblasts undergoing enucleation. Thus, in addition to being involved in nuclear expulsion, Emp–actin association may function to regulate the actin cytoskeleton in reticulocytes.

The potential role of Emp in regulating actin cytoskeleton has implications in macrophage function as well. Emp-deficient macrophages display condensed, less organized actin filaments, and are therefore unable to efficiently develop long cytoplasmic extensions. Detailed characterization of Emp-null macrophages in terms of their capacity to migrate, invade, and phagocytose, all of which are actin-based cellular events, will shed further light on these processes.

3.2. $\alpha 4\beta 1$(vla-4)/VCAM-1

Adhesive interactions occur between $\alpha 4\beta 1$ integrin expressed on erythroblasts and its counter–receptor—vascular adhesion molecule-1 (VCAM-1) expressed on macrophage cells. Monoclonal antibodies to either receptor disrupt island integrity *in vitro* (Sadahira *et al.*, 1995). Treatment of mice with very late antigen-4 antibodies *in utero* specifically induces anemia (Hamamura *et al.*, 1996).

3.3. Intercellular adhesion molecule-4/αv

The erythroid-specific isoform of intercellular adhesion molecule-4 (ICAM-4) expressed on erythroid cells (LW blood group glycoprotein, CD 242) interacts with αv integrin on macrophage cells (Spring *et al.*, 2001) as well as the leukocyte β1 integrin and platelet integrin αIIβ3 (Telen, 2005). Blocking ICAM-4-αv binding with αv synthetic peptides produces a 70% decrease in islands reconstituted *in vitro* (Lee *et al.*, 2006). ICAM-4-null mice have a 50% decrease in island formation in the bone marrow; however, steady-state erythropoiesis is not adversely affected in the adult animal. A secreted isoform of mouse ICAM-4, ICAM-4S, that is upregulated in terminal differentiation has been described. ICAM-4S may compete with membrane-bound integrin counter-receptors, thereby blocking the interaction of ICAM-4 with αv. This might facilitate the detachment of young reticulocytes from the islands, thereby enabling their egress into the vasculature.

Three other macrophage receptors that could be involved in adhesion to erythroblasts include lectin-like sheep erythrocyte receptor (Crocker and Gordon, 1985, 1986; Crocker *et al.*, 1990), erythroblast receptor (EbR), and ED2 antigen (Barbe *et al.*, 1996). Neither these receptors nor their corresponding ligands have been fully characterized.

A recent study shows palladin, an actin cytoskeleton-associated protein, is an important regulator of fetal liver definitive erythropoiesis (Liu *et al.*, 2007). Its disruption results in significant fetal anemia caused by impaired fetal liver definitive erythropoiesis resulting from disrupted erythroblastic island formation, suggesting a role for paladin in cell–cell interaction. The mutant HSCs in fetal liver can reconstitute lethally irradiated mice and differentiate into different lineages in methylcellulose culture system. These results suggest that palladin may regulate definitive erythropoiesis through a noncell autonomous manner. Erythroblastic island reconstitution assays show that intrinsic defects in palladin−/− macrophages but not erythroblasts are responsible for impaired erythroblastic island formation, and consequently definitive erythropoiesis deficiency. Many known adhesive proteins were characterized within the palladin-null islands and were found to be unaffected, further underscoring the possibility of novel adhesive pathways operant within the erythroblastic islands (Liu *et al.*, 2007).

4. Erythroblastic Island Functions

Erythroblasts can proliferate, mature, and enucleate *in vitro* in the absence of other cell types; however, this process is typically very inefficient at all stages (Hanspal *et al.*, 1998) and strikingly only a minority of *in vitro* differentiated erythroblasts complete the final step of enucleation. The generation of fully mature, enucleated erythrocytes is enhanced by the coculture with macrophages (Qiu *et al.*, 1995) or other accessory cells such as murine stromal cell lines or human mesenchymal cells (Giarratana *et al.*, 2005). The interaction of cells within the erythroblastic island is essential for both early and late stages of erythroid maturation. It has been proposed that early in erythroid maturation, the macrophages provide nutrients, and proliferative and survival signals to the erythroblasts. It has long been recognized that macrophages phagocytose extruded erythroblast nuclei at the conclusion of erythroid maturation. There is also accumulating evidence for the role of macrophages in promoting enucleation itself. Furthermore, macrophages provide adhesive interactions that maintain island integrity. The role of the adhesive interactions has yet to be clearly delineated—it is likely that they allow for regulatory feedback within islands via cross talk between cells and also trigger intracellular signaling pathways that regulate gene expression

4.1. Positive and negative growth regulatory effects on developing erythroblasts within the erythroblastic island

The concept of a control mechanism for cellular growth through modulation of apoptosis has recently come to include a wide variety of tissue systems, including hematopoietic cells. Changes in the balance between cell survival and death are clear signs of development of hematologic disorders such as the myelodysplastic syndromes (Greenberg, 1998; Parker and Mufti, 1998) and chronic myelogenous leukemia (Clarkson *et al.*, 1997). Therefore, tight regulation of apoptosis is needed to maintain hematopoietic homeostasis. The apoptosis of hematopoietic progenitor cells is regulated both positively and negatively by an interacting network of various cytokines and adhesive molecules (Wickremasinghe and Hoffbrand, 1999). It is therefore not surprising that such mechanisms are operant within the erythroblastic island.

Macrophages secrete cytokines that promote erythroblast proliferation and maturation. These include burst-promoting activity and insulin-like growth factor-1 (Kurtz *et al.*, 1985; Sawada *et al.*, 1989), factors that can stimulate the growth of BFU-E and CFU-E. Although erythropoietin mRNA expression has been demonstrated in mouse bone marrow macrophages (Rich *et al.*, 1988a,b), the precise role of macrophages in either synthesizing erythropoietin or presenting erythropoietin synthesized elsewhere remains to be elucidated. It has, however, been shown that coculture of erythroblasts with macrophages prevents erythroblast apoptosis (Hanspal *et al.*, 1998).

Erythroblasts express Fas throughout differentiation; however, only immature erythroblasts are susceptible to the death signal resulting from Fas/Fas ligand cross-linking. Late differentiating erythroblasts exhibit a Fas-based cytotoxicity against the immature erythroblasts via Fas ligand induction (De Maria *et al.*, 1999). It has been speculated that high levels of erythropoietin within the island protect the immature erythroblasts from this signaling pathway, promoting increased erythroid survival. This mechanism would upregulate erythropoiesis in anemic individuals that exhibit increased erythropoietin levels. Besides a negative feedback loop between mature and immature erythroblasts, there is one described between macrophages and immature erythroblasts that involves RCAS1 (receptor binding cancer antigen expressed in Siso cells) and its receptor (Matsushima *et al.*, 2001). Immature erythroblasts express RCAS1 receptor. Soluble RCAS1, secreted by bone marrow macrophages, activates proapoptotic caspases-8 and -3 in immature erythroblasts. These observations indicate that during erythropoiesis the level of apoptotic cell death is finely modulated in the erythroblastic islands by positive and negative regulatory factors, mainly at specific stages of cellular maturation that are predominantly at a high proliferative capacity.

4.2. The role of macrophages in supplying iron for hemoglobin synthesis

Another important unanswered question is whether macrophages supply iron for hemoglobin synthesis as was originally thought. In 1962, Bessis reported that in humans (but not rats or mice), ferritin molecules always occur in the space between erythroblasts and the histiocyte cell membrane and/or attached to the erythroblast membrane. It is unclear what the source of the ferritin is, whether it is derived from plasma or from the central histiocyte that extrudes it when it is in contact with an erythroblast. Whatever the origin of the ferritin, it is incorporated into the erythroblast by "rhopheocytosis," a process now known to be micropinocytosis (Policard and Bessis, 1962). These vesicles are filled with ferritin and form as a result of invagination of the surface membrane to which ferritin is attached utilizing a specific acid ferritin receptor (Konijn *et al.*, 1994). The vesicles disappear in the interior of the cytoplasm and the ferritin molecules are liberated. Over the last 10 years, our understanding of mammalian iron transport and homeostasis has advanced dramatically and it is likely that these questions will be addressed in the near future in a more meaningful manner. We now know some of the key players in iron homeostasis, such as hepcidin, a circulating peptide hormone primarily produced by hepatocytes (Nicolas *et al.*, 2001; Park *et al.*, 2001; Pigeon *et al.*, 2001) and ferroportin, the major transmembrane transporter transferring iron out of enterocytes, macrophages, and, to a lesser extent, hepatocytes (Abboud and Haile, 2000; Donovan *et al.*, 2005; Fraenkel *et al.*, 2005) (13–16Nar). Studies in cultured cells have established that that hepcidin binds directly to ferroportin, triggering its internalization and degradation within lysosomes (Nemeth *et al.*, 2004). Hepcidin expression is altered in response to each of the stimuli known to affect iron homeostasis: in conditions of stress erythropoiesis, for example in hypoxia or iron deficiency anemia, levels of hepcidin are decreased resulting in decreased ferroportin inactivation (Andrews and Schmidt, 2006). While the increase in ferroportin at the basilar surface of enterocytes results in increased absorption of dietary iron to provide additional substrate in face of an increased demand, it is intriguing to speculate that the ferroportin on the surface of central macrophages is also upregulated. This could result in local export of iron to the erythroblasts within the island.

4.3. Engulfment and breakdown of extruded nuclei play an important role in the regulation of late stage erythropoiesis

It has been shown by time lapse videography that the macrophages actively phagocytose extruded definitive erythroblast nuclei at the end of terminal differentiation (reviewed by Bessis *et al.*, 1978). This macrophage function has been documented *in vitro* in long-term bone marrow cultures (Allen and Dexter, 1982) and *in vivo* in mice lacking DNase II, where fetal liver macrophages become engorged with multiple ingested, but not digested, nuclei (Kawane *et al.*, 2001). It has further been demonstrated that the macrophage function of engulfment and digestion of extruded nuclei is vital for continued erythropoiesis. In the mice lacking DNase II, the lack of nuclear digestion resulted in a severe defect in erythropoiesis and embryonic lethal anemia (Yoshida *et al.*, 2005b). DNase II is an endonuclease present in the lysosomes of macrophages that cleaves DNA after macrophages engulf apoptotic cells or the nuclei that are expelled from erythroid precursor cells. Yoshida *et al.* have shown that the gene encoding IFN-β (Ifnb1) was activated in DNase II−/− fetal liver and that lack of signals from the interferon type I receptor "rescued" the anemia and the lethality.

DNase II−/− embryos produced not only IFN-β but also IFN-γ. But, in contrast to IFN-β, IFN-γ had little cytotoxic effect on erythroid cells. RNA hybridization analysis indicated that Ifnb1 was activated in DNase II−/− fetal liver, but expression was low and no interferon activity was detected in the serum of the embryos. In contrast, *in situ* hybridization detected IFN-β mRNA in the macrophages carrying undigested DNA in the blood islands of DNase II−/− fetal liver. The authors concluded that IFN-β produced in the fetal liver is responsible for inhibiting erythropoiesis that occurs in association with macrophages at the blood islands. Many cytokines have more potent activity in a membrane-associated form than in a soluble form. It is likely that IFN-β expression by macrophages in the blood islands, even in low concentrations, has a deleterious effect on erythropoiesis.

Analysis of erythroblast plasma membrane protein (Emp) that partitions between the nucleus, as it is being expelled, and the reticulocyte shows that Emp partitions predominantly to the plasma membrane surrounding the extruded nucleus (Lee *et al.*, 2004). This process would result in effective macrophage binding at the site, thereby facilitating phagocytosis. Furthermore, the extruded nuclei contain markedly decreased ATP and increased calcium (Yoshida *et al.*, 2005a). It has been speculated that these alterations inactivate the aminotranslocase and activate the scramblase, resulting in movement of the phosphatidylserine from the inner to the outer leaflet of the lipid bilayer. This rapid exposure of phosphatidylserine and engulfment of the nuclei by macrophages has been demonstrated *in vitro* (Yoshida *et al.*, 2005a). Furthermore, engulfment of nuclei is blocked by a dominant negative mutant of milk fat globule epidermal growth factor EGF-8, known to inhibit phagocytosis of apoptotic cells by blocking surface phosphatidylserine (Hanayama *et al.*, 2002).

The defect of severe anemia and *in utero* death in late gestation seen in DNase II-null mice is recapitulated in one of the three phosphatidylserine receptor knockout mouse models (Kunisaki *et al.*, 2004) that presumably leads to the inability of macrophage cells to ingest extruded nuclei. The conserved mitogen-activated protein kinase family members c-Jun N-terminal kinase (JNK) and p38 have been implicated in stress and proinflammatory signal transduction and recently in erythropoiesis. Homozygous deletion of either JNK1 or JNK2 in mice has no apparent effect on hematopoiesis (Bonnesen *et al.*, 2005), which might be explained by redundancy between the JNK isoforms. In contrast, the Jnk1−/− Jnk2−/− embryos die early in embryonic development (E11.5–12.5) due to dysregulation of apoptosis in brain development, making the contribution of the JNK in definitive erythropoiesis *in vivo* difficult to study further. Bonnesen *et al.* (2005) investigated the role of the JNK-activating kinase mitogen-activated protein kinase/extracellular signal to regulated kinase (MEK), kinase 1 (MEKK1), in development. Mice deficient in MEKK1 kinase activity (Mekk1KD mice) were generated and are alive and fertile on a C57/BL6×129 background. Unexpectedly, after backcrossing the Mekk1KD/+ heterozygotes into the C57/BL6 background, a dramatic decrease in the frequency of live Mekk1KD embryos that develop past E14.5 was observed. At E14.5, all mutant embryos studied, although morphologically normal, were anemic and showed defective definitive erythropoiesis with accumulation of nucleated late erythroblasts. These studies strongly suggest that MEKK1-JNK signaling is required for degradation of nuclear DNA extruded from erythroid precursors during the late stages of definitive erythropoiesis in the fetal liver. Crossing the Mekk1KD/+ C57/BL/6×129 hybrids with Jnk1−/− or Jnk2−/− mice resulted in embryonic lethality of all Mekk1KD Jnk1−/− and Mekk1KD Jnk2−/− embryos (regardless of background), suggesting that an intact MEKK1-JNK signaling pathway is required for normal embryonic development. Mekk1KD, Mekk1KD Jnk1−/−, and Mekk1KD Jnk2−/− embryos have identical phenotypes, survive up to midgestation, and display normal morphology but are highly anemic. Interestingly, a similar phenotype has been observed in p38-deficient mice. However, while the anemia in p38−/− mice is attributed to defective erythropoietin gene expression, normal levels of erythropoietin at both mRNA and protein level were observed in Mekk1KD and Mekk1KD Jnk2−/− embryos. The fetal livers of Mekk1KD and Mekk1KD Jnk2−/− also contain BFU-Es and CFU-Es formed at normal frequencies, and can reconstitute erythropoiesis in lethally irradiated hosts. It is therefore clear that MEKK1 is not required for the production of, or the response to, cytokines and/or growth factors required for the expansion and differentiation of erythroid progenitors up to the stage where the orthochromatic erythroblasts lose their nuclei and become reticulocytes. The phenotype of these mice is almost identical to those lacking DNaseII (described above). However, fetal livers from Mekk1KD and Mekk1KD Jnk2−/− embryos contain reduced levels of macrophages and TUNEL stains showed extensive accumulation of apoptotic bodies. These apoptotic bodies did not colocalize with Ter119-positive erythroblasts but rather appeared free in the extracellular space, suggesting that they represent nuclear DNA extruded from erythroid precursor cells. The fetal liver macrophages from the mutant mice exhibit normal phagocytic function, implicating the decrease in the number of macrophages in the development of defective erythropoiesis.

All together these results suggest that engulfment and breakdown of extruded nuclei play an important role in the regulation of late stage erythropoiesis.

4.4. Macrophages promote enucleation

Another important question is whether the macrophages also promote enucleation or whether this is mediated via retinoblastoma tumor suppressor (Rb) protein. Rb is present in the nucleus of all cells and plays a critical role in cell cycle decisions. Mice lacking Rb, in addition to other defects, have defective erythropoiesis and erythroblasts in the fetal liver that fail to mature and enucleate (Clarke *et al.*, 1992; Jacks *et al.*, 1992; Lee *et al.*, 1992).

It is yet unclear whether the role of Rb in RBC enucleation and maturation is intrinsic (Clark *et al.*, 2004; Spike *et al.*, 2007) to the RBCs or extrinsic to it (Iavarone *et al.*, 2004; Palis, 2004).

When Rb-deficient stem cells are transplanted into normal mice, the cells fail to enucleate and the animals are anemic. However, chimeric mice composed of both normal and mutant cells give rise to erythrocytes that are Rb− mutant in origin, raising the possibility that the Rb protein in macrophages may be essential for the process (Williams *et al.*, 1994). Evidence for this hypothesis is provided by Iavarone *et al.* (2004). They reported that these animals have a marked reduction of mature macrophages in the liver as well as the presence of immature macrophages that are unable to bind erythroblast. They suggest that the effects of Rb on macrophage maturation are mediated by Id2 and PU.1. Id2, a nuclear protein, is bound by Rb. Id2 can bind to transcription factor PU.1, a critical regulator of genes expressed in macrophages, and this interaction inhibits the transcription of macrophage specific genes. The macrophage defect in Rb-deficient mice can be reversed by loss of Id2. Their results are consistent with a model in which Rb binds to Id2 in the nucleus of immature macrophages, thus freeing PU.1 to transcribe genes involved in maturation of macrophages into erythroblast nursing cells, playing a critical role in RBC enucleation. In complete contrast to the studies described above, analysis of erythroblast maturation *in vitro* (Clark *et al.*, 2004; Spike and Macleod, 2005) or in Rb-null chimeric mice that are challenged *in vivo* with phenylhydrazine to induce stress erythropoiesis recapitulates the erythroid maturation defects seen in the developing Rb-null fetal liver, including the failure to enucleate and upregulate TER119 (Spike and Macleod, 2005), raising the possibility that Rb is required in a cell intrinsic manner to regulate erythropoiesis. This is specifically true under conditions of oxidative or proliferative stress. Spike *et al.* (2007) argue that it is unlikely that Rb plays a specific role in maintaining erythroblastic island integrity as was originally concluded from the examination Rb-null fetal livers. They suggest that the disruption in erythroblastic island formation is a direct result of hypoxia. To test this hypothesis, they counted the total number of erythroblastic islands and the number of erythroblasts per macrophage in cultures of native islands grown at either regular tissue culture or in hypoxic conditions. For these assays, they used fetal liver from control mice and conditionally targeted mice in which Rb was deleted in the embryo but not in the placenta (Wu *et al.*, 2003) to control for differential effects of ischemia experienced *in vivo*. At 21% oxygen, they failed to observe any significant difference in numbers of islands or in erythroblasts per macrophage formed with Rb-deficient fetal liver cells relative to control. However, when they cultured fetal liver from controls or from Rb-deficient mice under hypoxic conditions, they observed a dramatic reduction in both the number of islands and the number of erythroblasts per macrophage suggesting that hypoxia, as opposed to Rb gene status, was instrumental in disrupting erythroblastic islands. Histological examination of fetal livers of Rb-deficient mice revealed a preponderance of F4/80 positive macrophages adjacent to the ischemic areas, suggesting that macrophages in the Rb-null fetal liver may be diverted from their role in erythropoietic island formation to promote clearance of dying cells. Furthermore, they demonstrate the absence of any reduction in proportionate representation of F4/80-positive macrophages between wild-type and Rb-null fetal livers and a similar quality of erythroblastic islands in the intact proximal regions of the Rb-null fetal liver is similar to wild type. The same authors showed that when the erythroblastic islands were cultured, Rb-null macrophages and wild-type macrophages were both competent to bind erythroblasts. They also used microarray data and quantitative real-time PCR to examine how loss of Rb affected expression of macrophage-specific genes that have been implicated in the erythroblastic island defect in Rb-null fetal livers, including c-Fms, myeloperoxidase, cathepsin S, complement components, and lysozyme. In contrast to the effect of Rb loss on hypoxia-inducible gene expression, they failed to detect any significant change in the expression levels of most macrophage-specific genes in Rb-null fetal liver relative to wild type at E12.5 (Spike *et al.*, 2007). Expression of c-Fms (a PU1 target gene) was previously shown to be deregulated in Rb-null MEFs, and together with data implicating deregulated PU1 in erythroblastic island defects in Rb-null fetal liver was taken as evidence that macrophages were not functioning normally in Rb-null fetal liver (Iavarone *et al.*, 2004). However, Spike *et al.* were unable to detect any reduction in c-Fms levels by quantitative real-time PCR in Rb-null fetal liver relative to wild type. While further studies will be required to resolve the role of Rb within macrophages, the observation that *in vitro* erythroid enucleation progresses more efficiently in the presence of stromal cells (Giarratana *et al.*, 2005) and requires physical force (Yoshida *et al.*, 2005a) further supports the possibility that tethering to macrophages facilitates enucleation.

4.5. Erythroblasts within the island are a source of angiogenic factors that exert paracrine effects

Erythroblasts secrete two angiogenic factors, vascular endothelial growth factor A (VEGF-A) and placental growth factor (PIGF) (Tordjman *et al.*, 2001). Media from cultured erythroblasts induces migration of monocytes and endothelial cell permeability, both inhibited by VEGF-A- and PIGF-A-specific antibodies. Erythroid progenitors do not express receptors for either of these angiogenic factors whereas central macrophages do, and thus the secreted molecules may have paracrine effects regulating island integrity. In addition, the angiogenic factors may enable egress of the reticulocytes into the vasculature by modulating endothelial cell junctional integrity.

4.6. Erythroblast-mediated regulation of erythropoiesis via cell–cell interaction

Erythroblast intercellular signaling seems to regulate the activity and gene expression of GATA-1, a transcription factor crucial for erythropoiesis. While the absence of GATA-1 results in proerythroblast apoptosis (Pevny *et al.*, 1991, 1995; Weiss *et al.*, 1994) and embryonic lethality in knockout mice (Fujiwara *et al.*, 1996), its overexpression blocks terminal differentiation (Whyatt *et al.*, 1997; Lindeboom *et al.*, 2000). Strikingly, the presence of normally expressing cells enables overexpressing cells to complete terminal differentiation. Using a mouse model in which only half the RBCs overexpress GATA-1, it has been observed that the normal late stage erythroblasts produce a signal termed (RBC differentiation signal (REDS) that corrects the flaw in overexpressing cells (Gutierrez *et al.*, 2004). Mechanistically this appears to involve cell–cell interaction rather than a soluble factor, underscoring the importance of island integrity in the execution of this regulatory function.

5. Conclusion

The biological significance of islands has frequently been questioned. Unraveling the processes that lead to effective erythropoiesis in these niches is in an early stage and future work will contribute to the understanding of the true relevance of this system to normal and pathological hematopoiesis *in vivo*.

ACKNOWLEDGMENTS

Studies in the authors' laboratories have been supported by the National Institutes of Health (K08 DK02871, R01 DK46865, DK48721, and HL73437).

REFERENCES

Abboud, S., and Haile, D. J. (2000). A novel mammalian iron-regulated protein involved in intracellular iron metabolism. *J. Biol. Chem.* **275,** 19906–19912.

Allen, T. D., and Dexter, T. M. (1982). Ultrastructural aspects of erythropoietic differentiation in long-term bone marrow culture. *Differentiation* **21,** 86–94.

Andrews, N. C., and Schmidt, P. J. (2006). Iron homeostasis. *Annu. Rev. Physiol.* **69,** 69–85.

Arkin, S., Naprstek, B., Guarini, L., Ferrone, S., and Lipton, J. M. (1991). Expression of intercellular adhesion molecule-1 (CD54) on hematopoietic progenitors. *Blood* **77,** 948–953.

Armeanu, S., Buhring, H. J., Reuss-Borst, M., Muller, C. A., and Klein, G. (1995). E-cadherin is functionally involved in the maturation of the erythroid lineage. *J. Cell Biol.* **131,** 243–249.

Austyn, J. M., and Gordon, S. (1981). F4/80, a monoclonal antibody directed specifically against the mouse macrophage. *Eur. J. Immunol.* **11,** 805–815.

Axelrad, A. A., McLeod, D. L., Shreeve, M. M., and Heath, D. S. (1974). Properties of cells that produce erythrocytic colonies *in vitro*. *In* "Hemopoiesis in Culture" (W. A. Robinson, Ed.), p. 226. DHEW publication, Washington, DC.

Bala, S., Kumar, A., Soni, S., Sinha, S., and Hanspal, M. (2006). Emp is a component of the nuclear matrix of mammalian cells and undergoes dynamic rearrangements during cell division. *Biochem. Biophys. Res. Commun.* **342,** 1040–1048.

Barbe, E., Huitinga, I., Dopp, E. A., Bauer, J., and Dijkstra, C. D. (1996). A novel bone marrow frozen section assay for studying hematopoietic interactions in situ: The role of stromal bone marrow macrophages in erythroblast binding. *J. Cell Sci.* **109,** (Pt 12), 2937–2945.

Bernard, J. (1991). The erythroblastic island: Past and future. *Blood Cells* **17,** 5–10; discussion 10–14.

Bessis, M. (1958). Erythroblastic island, functional unity of bone marrow. *Rev. Hematol.* **13,** 8–11.

Bessis, M., and Breton-Gorius, J. (1961). The erythroblastic islet and the rhopheocytosis of ferritin in inflammation. *Nouv. Rev. Fr. Hematol.* **1,** 569–582.

Bessis, M. C., and Breton-Gorius, J. (1962). Iron metabolism in the bone marrow as seen by electron microscopy: A critical review. *Blood* **19,** 635–663.

Bessis, M., Mize, C., and Prenant, M. (1978). Erythropoiesis: Comparison of *in vivo* and *in vitro* amplification. *Blood Cells* **4,** 155–174.

Bonnesen, B., Orskov, C., Rasmussen, S., Holst, P. J., Christensen, J. P., Eriksen, K. W., Qvortrup, K., Odum, N., and Labuda, T. (2005). MEK kinase 1 activity is required for definitive erythropoiesis in the mouse fetal liver. *Blood* **106,** 3396–3404.

Campbell, F. R. (1968). Nuclear elimination from the normoblast of fetal guinea pig liver as studied with electron microscopy and serial sectioning techniques. *Anat. Rec.* **160,** 539–554.

Clark, A. J., Doyle, K. M., and Humbert, P. O. (2004). Cell-intrinsic requirement for pRb in erythropoiesis. *Blood* **104,** 1324–1326.

Clarke, A. R., Maandag, E. R., van Roon, M., van der Lugt, N. M., van der Valk, M., Hooper, M. L., Berns, A., and te Riele, H. (1992). Requirement for a functional Rb-1 gene in murine development. *Nature* **359,** 328–330.

Clarkson, B. D., Strife, A., Wisniewski, D., Lambek, C., and Carpino, N. (1997). New understanding of the pathogenesis of CML: A prototype of early neoplasia. *Leukemia* **11,** 1404–1428.

Cohen, W. D. (1991). The cytoskeletal system of nucleated erythrocytes. *Int. Rev. Cytol.* **130,** 37–84.

Coulombel, L., Vuillet-Gaugler, M. H., Leroy, C., Rosemblatt, M., and Breton-Gorius, J. (1991). Adhesive properties of human erythroblastic precursor cells. *Blood Cells* **17,** 65–78; discussion 79–81.

Crocker, P. R., and Gordon, S. (1985). Isolation and characterization of resident stromal macrophages and hematopoietic cell clusters from mouse bone marrow. *J. Exp. Med.* **162,** 993–1014.

Crocker, P. R., and Gordon, S. (1986). Properties and distribution of a lectin-like hemagglutinin differentially expressed by murine stromal tissue macrophages. *J. Exp. Med.* **164,** 1862–1875.

Crocker, P. R., Werb, Z., Gordon, S., and Bainton, D. F. (1990). Ultrastructural localization of a macrophage-restricted sialic acid binding hemagglutinin, SER, in macrophage-hematopoietic cell clusters. *Blood* **76,** 1131–1138.

De la Chapelle, A., Fantoni, A., and Marks, P. A. (1969). Differentiation of mammalian somatic cells: DNA and hemoglobin synthesis in fetal mouse yolk sac erythroid cells. *Proc. Natl. Acad. Sci. USA* **63,** 812–819.

De Maria, R., Testa, U., Luchetti, L., Zeuner, A., Stassi, G., Pelosi, E., Riccioni, R., Felli, N., Samoggia, P., and Peschle, C. (1999). Apoptotic role of Fas/Fas ligand system in the regulation of erythropoiesis. *Blood* **93,** 796–803.

Donovan, A., Lima, C. A., Pinkus, J. L., Pinkus, G. S., Zon, L. I., Robine, S., and Andrews, N. C. (2005). The iron exporter ferroportin/Slc40a1 is essential for iron homeostasis. *Cell Metab.* **1,** 191–200.

El Nemer, W., Gane, P., Colin, Y., Bony, V., Rahuel, C., Galacteros, F., Cartron, J. P., and Le Van Kim, C. (1998). The Lutheran blood group glycoproteins, the erythroid receptors for laminin, are adhesion molecules. *J. Biol. Chem.* **273,** 16686–16693.

Emerson, S. G., Sieff, C. A., Wang, E. A., Wong, G. G., Clark, S. C., and Nathan, D. G. (1985). Purification of fetal hematopoietic progenitors and demonstration of recombinant multipotential colony-stimulating activity. *J. Clin. Invest.* **76,** 1286–1290.

Ferkowicz, M. J., and Yoder, M. C. (2005). Blood island formation: Longstanding observations and modern interpretations. *Exp. Hematol.* **33,** 1041–1047.

Fraenkel, P. G., Traver, D., Donovan, A., Zahrieh, D., and Zon, L. I. (2005). Ferroportin1 is required for normal iron cycling in zebrafish. *J. Clin. Invest.* **115,** 1532–1541.

Fraser, S. T., Isern, J., and Baron, M. H. (2007). Maturation and enucleation of primitive erythroblasts during mouse embryogenesis is accompanied by changes in cell-surface antigen expression. *Blood* **109,** 343–352.

Fujiwara, Y., Browne, C. P., Cunniff, K., Goff, S. C., and Orkin, S. H. (1996). Arrested development of embryonic red cell precursors in mouse embryos lacking transcription factor GATA-1. *Proc. Natl. Acad. Sci. USA* **93,** 12355–12358.

Geiduschek, J. B., and Singer, S. J. (1979). Molecular changes in the membranes of mouse erythroid cells accompanying differentiation. *Cell* **16,** 149–163.

Giarratana, M. C., Kobari, L., Lapillonne, H., Chalmers, D., Kiger, L., Cynober, T., Marden, M. C., Wajcman, H., and Douay, L. (2005). *Ex vivo* generation of fully mature human red blood cells from hematopoietic stem cells. *Nat. Biotechnol.* **23,** 69–74.

Gifford, S. C., Derganc, J., Shevkoplyas, S. S., Yoshida, T., and Bitensky, M. W. (2006). A detailed study of time-dependent changes in human red blood cells: From reticulocyte maturation to erythrocyte senescence. *Br. J. Haematol.* **135,** 395–404.

Granick, S., and Levere, R. D. (1964). Heme synthesis in erythroid cells. *Prog. Hematol.* **27,** 1–47.

Greenberg, P. L. (1998). Apoptosis and its role in the myelodysplastic syndromes: Implications for disease natural history and treatment. *Leuk. Res.* **22,** 1123–1136.

Gutierrez, L., Lindeboom, F., Langeveld, A., Grosveld, F., Philipsen, S., and Whyatt, D. (2004). Homotypic signalling regulates Gata1 activity in the erythroblastic island. *Development* **131,** 3183–3193.

Hamamura, K., Matsuda, H., Takeuchi, Y., Habu, S., Yagita, H., and Okumura, K. (1996). A critical role of VLA-4 in erythropoiesis *in vivo*. *Blood* **87,** 2513–2517.

Hanayama, R., Tanaka, M., Miwa, K., Shinohara, A., Iwamatsu, A., and Nagata, S. (2002). Identification of a factor that links apoptotic cells to phagocytes. *Nature* **417,** 182–187.

Hanspal, M., Smockova, Y., and Uong, Q. (1998). Molecular identification and functional characterization of a novel protein that mediates the attachment of erythroblasts to macrophages. *Blood* **92,** 2940–2950.

Haslam, D. B., and Baenziger, J. U. (1996). Expression cloning of Forssman glycolipid synthetase: A novel member of the histo-blood group ABO gene family. *Proc. Natl. Acad. Sci. USA* **93,** 10697–10702.

Hirsch, S., Austyn, J. M., and Gordon, S. (1981). Expression of the macrophage-specific antigen F4/80 during differentiation of mouse bone marrow cells in culture. *J. Exp. Med.* **154,** 713–725.

Hume, D. A., and Gordon, S. (1983). Mononuclear phagocyte system of the mouse defined by immunohistochemical localization of antigen F4/80. Identification of resident macrophages in renal medullary and cortical interstitium and the juxtaglomerular complex. *J. Exp. Med.* **157,** 1704–1709.

Hume, D. A., Robinson, A. P., MacPherson, G. G., and Gordon, S. (1983). The mononuclear phagocyte system of the mouse defined by immunohistochemical localization of antigen F4/80. Relationship between macrophages, Langerhans cells, reticular cells, and dendritic cells in lymphoid and hematopoietic organs. *J. Exp. Med.* **158,** 1522–1536.

Iavarone, A., King, E. R., Dai, X. M., Leone, G., Stanley, E. R., and Lasorella, A. (2004). Retinoblastoma promotes definitive erythropoiesis by repressing Id2 in fetal liver macrophages. *Nature* **432,** 1040–1045.

Jacks, T., Fazeli, A., Schmitt, E. M., Bronson, R. T., Goodell, M. A., and Weinberg, R. A. (1992). Effects of an Rb mutation in the mouse. *Nature* **359,** 295–300.

Kansas, G. S., Muirhead, M. J., and Dailey, M. O. (1990). Expression of the CD11/CD18, leukocyte adhesion molecule 1, and CD44 adhesion molecules during normal myeloid and erythroid differentiation in humans. *Blood* **76,** 2483–2492.

Kawane, K., Fukuyama, H., Kondoh, G., Takeda, J., Ohsawa, Y., Uchiyama, Y., and Nagata, S. (2001). Requirement of DNase II for definitive erythropoiesis in the mouse fetal liver. *Science* **292,** 1546–1549.

Keyhani, E., and Bessis, M. (1969). Electron microscope study of the erythroblastic island using the cryo-scouring method. *Nouv. Rev. Fr. Hematol.* **9,** 803–816.

Kingsley, P. D., Malik, J., Fantauzzo, K. A., and Palis, J. (2004). Yolk sac-derived primitive erythroblasts enucleate during mammalian embryogenesis. *Blood* **104,** 19–25.

Kondo, M., Wagers, A. J., Manz, M. G., Prohaska, S. S., Scherer, D. C., Beilhack, G. F., Shizuru, J. A., and Weissman, I. L. (2003). Biology of hematopoietic stem cells and progenitors: Implications for clinical application. *Annu. Rev. Immunol.* **21,** 759–806.

Konijn, A. M., Meyron-Holtz, E. G., Fibach, E., and Gelvan, D. (1994). Cellular ferritin uptake: A highly regulated pathway for iron assimilation in human erythroid precursor cells. *Adv. Exp. Med. Biol.* **356,** 189–197.

Koury, S. T., Koury, M. J., and Bondurant, M. C. (1988). Morphological changes in erythroblasts during erythropoietin-induced terminal differentiation *in vitro*. *Exp. Hematol.* **16,** 758–763.

Koury, S. T., Koury, M. J., and Bondurant, M. C. (1989). Cytoskeletal distribution and function during the maturation and enucleation of mammalian erythroblasts. *J. Cell Biol.* **109,** 3005–3013.

Kunisaki, Y., Masuko, S., Noda, M., Inayoshi, A., Sanui, T., Harada, M., Sasazuki, T., and Fukui, Y. (2004). Defective fetal liver erythropoiesis and T lymphopoiesis in mice lacking the phosphatidylserine receptor. *Blood* **103,** 3362–3364.

Kurtz, A., Hartl, W., Jelkmann, W., Zapf, J., and Bauer, C. (1985). Activity in fetal bovine serum that stimulates erythroid colony formation in fetal mouse livers is insulin-like growth factor I. *J. Clin. Invest.* **76,** 1643–1648.

Le Charpentier, Y., and Prenant, M. (1975). [Isolation of erythroblastic islands. Study by optical and scanning electron microscopy (author's transl)]. *Nouv. Rev. Fr. Hematol.* **15,** 119–140.

Lee, E. Y., Chang, C. Y., Hu, N., Wang, Y. C., Lai, C. C., Herrup, K., Lee, W. H., and Bradley, A. (1992). Mice deficient for Rb are nonviable and show defects in neurogenesis and haematopoiesis. *Nature* **359,** 288–294.

Lee, G., Lo, A., Short, S. A., Mankelow, T. J., Spring, F., Parsons, S. F., Yazdanbakhsh, K., Mohandas, N., Anstee, D. J., and Chasis, J. A. (2006). Targeted gene deletion demonstrates that the cell adhesion molecule ICAM-4 is critical for erythroblastic island formation. *Blood* **108,** 2064–2071.

Lee, J. C., Gimm, J. A., Lo, A. J., Koury, M. J., Krauss, S. W., Mohandas, N., and Chasis, J. A. (2004). Mechanism of protein sorting during erythroblast enucleation: Role of cytoskeletal connectivity. *Blood* **103,** 1912–1919.

Lee, S. H., Crocker, P. R., Westaby, S., Key, N., Mason, D. Y., Gordon, S., and Weatherall, D. J. (1988). Isolation and immunocytochemical characterization of human bone marrow stromal macrophages in hemopoietic clusters. *J. Exp. Med.* **168,** 1193–1198.

Liu, X. S., Luo, H. J., Yang, H., Wang, L., Kong, H., Jin, Y. E., Wang, F., Gu, M. M., Chen, Z., Lu, Z. Y., and Wang, Z. G. (2007). Palladin regulates cell and extracellular matrix interaction through maintaining normal actin cytoskeleton architecture and stabilizing beta1-integrin. *J. Cell Biochem.* **100,** 1288–1300.

Matsushima, T., Nakashima, M., Oshima, K., Abe, Y., Nishimura, J., Nawata, H., Watanabe, T., and Muta, K. (2001). Receptor binding cancer antigen expressed on SiSo cells, a novel regulator of apoptosis of erythroid progenitor cells. *Blood* **98,** 313–321.

McKnight, A. J., Macfarlane, A. J., Dri, P., Turley, L., Willis, A. C., and Gordon, S. (1996). Molecular cloning of F4/80, a murine macrophage-restricted cell surface glycoprotein with homology to the G-protein-linked transmembrane 7 hormone receptor family. *J. Biol. Chem.* **271,** 486–489.

Mohandas, N., and Prenant, M. (1978). Three-dimensional model of bone marrow. *Blood* **51,** 633–643.

Naito, M., Hasegawa, G., and Takahashi, K. (1997). Development, differentiation, and maturation of Kupffer cells. *Microsc. Res. Tech.* **39,** 350–364.

Nemeth, E., Tuttle, M. S., Powelson, J., Vaughn, M. B., Donovan, A., Ward, D. M., Ganz, T., and Kaplan, J. (2004). Hepcidin regulates cellular iron efflux by binding to ferroportin and inducing its internalization. *Science* **306,** 2090–2093.

Nicolas, G., Bennoun, M., Devaux, I., Beaumont, C., Grandchamp, B., Kahn, A., and Vaulont, S. (2001). Lack of hepcidin gene expression and severe tissue iron overload in upstream stimulatory factor 2 (USF2) knockout mice. *Proc. Natl. Acad. Sci. USA* **98,** 8780–8785.

Orkin, S. H. (2000). Diversification of haematopoietic stem cells to specific lineages. *Nat. Rev. Genet.* **1,** 57–64.

Palis, J. (2004). Developmental biology: No red cell is an island. *Nature* **432,** 964–965.

Park, C. H., Valore, E. V., Waring, A. J., and Ganz, T. (2001). Hepcidin, a urinary antimicrobial peptide synthesized in the liver. *J. Biol. Chem.* **276,** 7806–7810.

Parker, J. E., and Mufti, G. J. (1998). Ineffective haemopoiesis and apoptosis in myelodysplastic syndromes. *Br. J. Haematol.* **101,** 220–230.

Perry, M. M., John, H. A., and Thomas, N. S. (1971). Actin-like filaments in the cleavage furrow of newt egg. *Exp. Cell Res.* **65,** 249–253.

Pevny, L., Simon, M. C., Robertson, E., Klein, W. H., Tsai, S. F., D'Agati, V., Orkin, S. H., and Costantini, F. (1991). Erythroid differentiation in chimaeric mice blocked by a targeted mutation in the gene for transcription factor GATA-1. *Nature* **349,** 257–260.

Pevny, L., Lin, C. S., D'Agati, V., Simon, M. C., Orkin, S. H., and Costantini, F. (1995). Development of hematopoietic cells lacking transcription factor GATA-1. *Development* **121,** 163–172.

Pigeon, C., Ilyin, G., Courselaud, B., Leroyer, P., Turlin, B., Brissot, P., and Loreal, O. (2001). A new mouse liver-specific gene, encoding a protein homologous to human antimicrobial peptide hepcidin, is overexpressed during iron overload. *J. Biol. Chem.* **276,** 7811–7819.

Policard, A., and Bessis, M. (1962). Micropinocytosis and rhopheocytosis. *Nature* **194,** 110–111.

Qiu, L. B., Dickson, H., Hajibagheri, N., and Crocker, P. R. (1995). Extruded erythroblast nuclei are bound and phagocytosed by a novel macrophage receptor. *Blood* **85,** 1630–1639.

Repasky, E. A., and Eckert, B. S. (1981a). A reevaluation of the process of enucleation in mammalian erythroid cells. *Prog. Clin. Biol. Res.* **55,** 679–692.

Repasky, E. A., and Eckert, B. S. (1981b). Microtubules in mammalian erythroblasts. Are marginal bands present? *Anat. Embryol. (Berl)* **162,** 419–424.

Repasky, E. A., and Eckert, B. S. (1981c). The effect of cytochalasin B on the enucleation of erythroid cells *in vitro*. *Cell Tissue Res.* **221,** 85–91.

Rich, I. N., Vogt, C., and Pentz, S. (1988a). Erythropoietin gene expression in macrophages detected by *in situ* hybridization. *Behring Inst. Mitt.* **83,** 202–206.

Rich, I. N., Vogt, C., and Pentz, S. (1988b). Erythropoietin gene expression *in vitro* and *in vivo* detected by in situ hybridization. *Blood Cells* **14,** 505–520.

Rosemblatt, M., Vuillet-Gaugler, M. H., Leroy, C., and Coulombel, L. (1991). Coexpression of two fibronectin receptors, VLA-4 and VLA-5, by immature human erythroblastic precursor cells. *J. Clin. Invest.* **87,** 6–11.

Rosse, C. (1976). Small lymphocyte and transitional cell populations of the bone marrow; their role in the mediation of immune and hemopoietic progenitor cell functions. *Int. Rev. Cytol.* **45,** 155–290.

Sadahira, Y., Mori, M., Awai, M., Watarai, S., and Yasuda, T. (1988). Forssman glycosphingolipid as an immunohistochemical marker for mouse stromal macrophages in hematopoietic foci. *Blood* **72,** 42–48.

Sadahira, Y., Mori, M., and Kimoto, T. (1990). Isolation and short-term culture of mouse splenic erythroblastic islands. *Cell Struct. Funct.* **15,** 59–65.

Sadahira, Y., Yasuda, T., and Kimoto, T. (1991). Regulation of Forssman antigen expression during maturation of mouse stromal macrophages in haematopoietic foci. *Immunology* **73,** 498–504.

Sadahira, Y., Yoshino, T., and Monobe, Y. (1995). Very late activation antigen 4-vascular cell adhesion molecule 1 interaction is involved in the formation of erythroblastic islands. *J. Exp. Med.* **181,** 411–415.

Sadahira, Y., Wada, H., Manabe, T., and Yawata, Y. (1999). Immunohistochemical assessment of human bone marrow macrophages in hematologic disorders. *Pathol. Int.* **49,** 626–632.

Sadahira, Y., Yasuda, T., Yoshino, T., Manabe, T., Takeishi, T., Kobayashi, Y., Ebe, Y., and Naito, M. (2000). Impaired splenic erythropoiesis in phlebotomized mice injected with CL2MDP-liposome: An experimental model for studying the role of stromal macrophages in erythropoiesis. *J. Leukoc. Biol.* **68,** 464–470.

Sangiorgi, F., Woods, C. M., and Lazarides, E. (1990). Vimentin downregulation is an inherent feature of murine erythropoiesis and occurs independently of lineage. *Development* **110,** 85–96.

Sawada, K., Krantz, S. B., Dessypris, E. N., Koury, S. T., and Sawyer, S. T. (1989). Human colony-forming units-erythroid do not require accessory cells, but do require direct interaction with insulin-like growth factor I and/or insulin for erythroid development. *J. Clin. Invest.* **83,** 1701–1709.

Schroeder, T. E. (1973). Actin in dividing cells: Contractile ring filaments bind heavy meromyosin. *Proc. Natl. Acad. Sci. USA* **70,** 1688–1692.

Simpson, C. F., and Kling, J. M. (1967). The mechanism of denucleation in circulating erythroblasts. *J. Cell Biol.* **35,** 237–245.

Skutelsky, E., and Danon, D. (1967). An electron microscopic study of nuclear elimination from the late erythroblast. *J. Cell Biol.* **33,** 625–635.

Skutelsky, E., and Danon, D. (1970). Comparative study of nuclear expulsion from the late erythroblast and cytokinesis. *Exp. Cell Res.* **60,** 427–436.

Soni, S., Bala, S., Gwynn, B., Sahr, K. E., Peters, L. L., and Hanspal, M. (2006). Absence of erythroblast macrophage protein (Emp) leads to failure of erythroblast nuclear extrusion. *J. Biol. Chem.* **281,** 20181–20189.

Spike, B. T., and Macleod, K. F. (2005). The Rb tumor suppressor in stress responses and hematopoietic homeostasis. *Cell Cycle* **4,** 42–45.

Spike, B. T., Dibling, B. C., and Macleod, K. F. (2007). Hypoxic stress underlies defects in erythroblast islands in the Rb null mouse. *Blood* **110,** (6), 2173–2181.

Spring, F. A., Parsons, S. F., Ortlepp, S., Olsson, M. L., Sessions, R., Brady, R. L., and Anstee, D. J. (2001). Intercellular adhesion molecule-4 binds alpha(4)beta(1) and alpha (V)-family integrins through novel integrin-binding mechanisms. *Blood* **98,** 458–466.

Steiner, R., and Vogel, H. (1973). On the kinetics of erythroid cell differentiation in fetal mice. I. Microspectrophotometric determination of the hemoglobin content in erythroid cells during gestation. *J. Cell Physiol.* **81,** 323–338.

Tavassoli, M. (1991). Embryonic and fetal hemopoiesis: An overview. *Blood Cells* **17,** 269–281; discussion 282–266.

Telen, M. J. (2005). Erythrocyte adhesion receptors: Blood group antigens and related molecules. *Transfus. Med. Rev.* **19,** 32–44.

Tober, J., Koniski, A., McGrath, K. E., Vemishetti, R., Emerson, R., de Mesy-Bentley, K. K., Waugh, R., and Palis, J. (2007). The megakaryocyte lineage originates from hemangioblast precursors and is an integral component both of primitive and of definitive hematopoiesis. *Blood* **109,** 1433–1441.

Tordjman, R., Delaire, S., Plouet, J., Ting, S., Gaulard, P., Fichelson, S., Romeo, P. H., and Lemarchandel, V. (2001). Erythroblasts are a source of angiogenic factors. *Blood* **97,** 1968–1974.

Weiss, M. J., Keller, G., and Orkin, S. H. (1994). Novel insights into erythroid development revealed through *in vitro* differentiation of GATA-1 embryonic stem cells. *Genes Dev.* **8,** 1184–1197.

Whyatt, D. J., Karis, A., Harkes, I. C., Verkerk, A., Gillemans, N., Elefanty, A. G., Vairo, G., Ploemacher, R., Grosveld, F., and Philipsen, S. (1997). The level of the tissue-specific factor GATA-1 affects the cell-cycle machinery. *Genes Funct.* **1,** 11–24.

Whyatt, D., Lindeboom, F., Karis, A., Ferreira, R., Milot, E., Hendriks, R., de Bruijn, M., Langeveld, A., Gribnau, J., Grosveld, F., and Philipsen, S. (2000). An intrinsic but cell-nonautonomous defect in GATA-1-overexpressing mouse erythroid cells. *Nature* **406,** 519–524.

Wickremasinghe, R. G., and Hoffbrand, A. V. (1999). Biochemical and genetic control of apoptosis: Relevance to normal hematopoiesis and hematological malignancies. *Blood* **93,** 3587–3600.

Williams, B. O., Schmitt, E. M., Remington, L., Bronson, R. T., Albert, D. M., Weinberg, R. A., and Jacks, T. (1994). Extensive contribution of Rb-deficient cells to adult chimeric mice with limited histopathological consequences. *EMBO J.* **13,** 4251–4259.

Wu, L., de Bruin, A., Saavedra, H. I., Starovic, M., Trimboli, A., Yang, Y., Opavska, J., Wilson, P., Thompson, J. C., Ostrowski, M. C., Rosol, T. J., Woollett, L. A., *et al.* (2003). Extra-embryonic function of Rb is essential for embryonic development and viability. *Nature* **421,** 942–947.

Yokoyama, T., Kitagawa, H., Takeuchi, T., Tsukahara, S., and Kannan, Y. (2002). No apoptotic cell death of erythroid cells of erythroblastic islands in bone marrow of healthy rats. *J. Vet. Med. Sci.* **64,** 913–919.

Yokoyama, T., Etoh, T., Kitagawa, H., Tsukahara, S., and Kannan, Y. (2003). Migration of erythroblastic islands toward the sinusoid as erythroid maturation proceeds in rat bone marrow. *J. Vet. Med. Sci.* **65,** 449–452.

Yoshida, H., Kawane, K., Koike, M., Mori, Y., Uchiyama, Y., and Nagata, S. (2005a). Phosphatidylserine-dependent engulfment by macrophages of nuclei from erythroid precursor cells. *Nature* **437,** 754–758.

Yoshida, H., Okabe, Y., Kawane, K., Fukuyama, H., and Nagata, S. (2005b). Lethal anemia caused by interferon-beta produced in mouse embryos carrying undigested DNA. *Nat. Immunol.* **6,** 49–56.

CHAPTER THREE

Epigenetic Control of Complex Loci During Erythropoiesis

Ryan J. Wozniak* *and* Emery H. Bresnick*

Contents

Abstract

Epigenetic mechanisms involving dynamic changes in posttranslational histone modifications commonly control gene transcription and therefore the execution of all cellular differentiation programs. The differentiation of hematopoietic stem cells into specific progenitor cells and the diverse blood cell types represents a particularly powerful system for the study of epigenetic mechanisms. The hematopoietic system allows one to define mechanisms underlying the establishment and regulation of histone modification patterns covering entire genes and/or chromosomes at distinct stages of differentiation. This chapter

* Department of Pharmacology, University of Wisconsin School of Medicine and Public Health, 385 Medical Sciences Center, Madison, Wisconsin 53706

Current Topics in Developmental Biology, Volume 82
ISSN 0070-2153, DOI: 10.1016/S0070-2153(07)00003-8

reviews progress in elucidating principles underlying epigenetic control of complex loci, specifically focusing on genes differentially expressed during hematopoiesis.

1. Introduction

The organization of DNA into chromatin constitutes a prominent mode of regulating fundamental nuclear processes including transcription, replication, recombination, and DNA repair. Such organization is highly dynamic, especially at the level of the fundamental repeating unit of chromatin, the nucleosome, which consists of DNA wrapped 1.6 times around a core histone octamer. Core histones are subjected to extensive posttranslational modifications, including acetylation, methylation, phosphorylation, sumoylation, and ubiquitination (Kouzarides, 2007; Strahl and Allis, 2000). These modifications serve as a tractable set of switches, governing the accessibility of chromatin to DNA- and histone-binding factors (Lee *et al.*, 1993; Vettese-Dadey *et al.*, 1996). Once bound to DNA, transcription factors recruit additional coregulators (coactivators or corepressors) that catalyze histone posttranslational modifications (chromatin-modifying enzymes) (Brownell *et al.*, 1996; Ogryzko *et al.*, 1996; Yang *et al.*, 1996) and remodel chromatin (chromatin-remodeling complexes) via directly regulating nucleosome structure and positioning (Saha *et al.*, 2006). This process by which phenotype is modified without alterations in genotype is referred to as epigenetic regulation, and accordingly, histone modifications are designated as epigenetic marks (Bernstein *et al.*, 2007). Epigenetic marks can influence gene activity by directly modulating chromatin structure (Lee *et al.*, 1993; Tse *et al.*, 1998) or via functioning as ligands that attract regulatory factors to the template (Bannister *et al.*, 2001; Jacobs *et al.*, 2001; Lachner *et al.*, 2001; Nakayama *et al.*, 2001).

As open, accessible chromatin (euchromatin) is generally transcriptionally permissive and condensed, inaccessible chromatin (heterochromatin) is often transcriptionally repressive (Felsenfeld and Groudine, 2003; van Holde, 1989; Wolffe, 1998), epigenetic regulation constitutes a fundamentally important mode of transcriptional control. The plasticity of epigenetic marks allows for the exquisite fine-tuning of gene expression necessary for proper embryonic development and critical processes that continue into adulthood, such as the development of red blood cells or erythropoiesis.

Epigenetic marks are commonly classified based on their links to activation or repression. Although individual epigenetic marks can function independently, it has been hypothesized that ensembles of epigenetic marks function combinatorially, thereby constituting a "histone code" (Jenuwein and Allis, 2001; Strahl and Allis, 2000). The numerous activating and repressing epigenetic marks in mammalian systems and the enzymes responsible for their deposition or removal have been recently reviewed (Kouzarides, 2007). Given the plethora of epigenetic marks and chromatin-modifying enzymes identified and the common occurrence in the mammalian genome of complex loci governed by multiple, dispersed regulatory elements (Dean, 2006), understanding how chromatin dynamics establishes and regulates expression states during development is a formidable task. Inherent complexities involve the need to make concurrent measurements of epigenetic marks and chromatin structure, as well as transcription factor and coregulator occupancy at conserved *cis*-elements across large chromatin domains containing genes or clusters of genes and their surrounding intergenic regions. Furthermore, unlike studies to dissect mechanisms underlying basal transcription or the biochemical functions of chromatin-modifying and -remodeling enzymes, which commonly use simple organisms such as *Saccharomyces cerevisiae* and cancer cell lines such as 293, HeLa, and 3T3, determining how endogenous histone modification patterns are established, regulated, and function during mammalian development requires the use of physiologically relevant cells and tissues. Such systems are typically less amenable to biochemical and molecular analyses than cancer cell lines, which are easy to grow and manipulate.

This chapter reviews progress in elucidating principles underlying epigenetic control of complex loci, specifically highlighting studies of two chromatin domains containing genes differentially expressed during hematopoiesis—the *β-globin* locus containing *β*-like globin genes that are differentially expressed in embryonic (primitive) and adult (definitive) erythroid cells, and the *Gata2* locus expressed in hematopoietic stem cells and erythroid precursor cells and repressed during erythropoiesis.

2. Utility of Analyzing Erythropoiesis to Dissect Epigenetic Mechanisms

Cellular differentiation requires coordinated changes in gene expression to alter phenotype. As pluripotent stem cells make lineage-determining decisions in the path toward terminal differentiation, certain genes are activated, while other genes are silenced. A major determinant of these sequential changes in cellular type and function involves modifications in broad histone modification patterns, leading to chromatin reorganization, altered transcription factor access, and changes in transcription. Elucidating molecular mechanisms underlying the establishment and regulation of chromatin domains during differentiation will lead to an understanding of how genetic networks emerge to orchestrate differentiation. Such mechanisms will provide the requisite conceptual foundation for devising approaches to manipulate these processes should they go awry.

Terminally differentiated cells have relatively static expression profiles, limiting their utility to dissect mechanisms that orchestrate chromatin modification and remodeling during cellular differentiation. By contrast, given the dynamic conversion of erythroid precursor cells into increasingly differentiated erythroid cells during erythropoiesis (Cantor and Orkin, 2002; Orkin and Zon, 1997; Richmond *et al.*, 2005; Kim and Bresnick, 2007), this system can be exploited to analyze how chromatin modification and remodeling function in the context of normal cellular differentiation. Commonly used systems to study erythropoiesis include primary cells isolated from murine hematopoietic tissues, namely bone marrow, fetal liver, and spleen (Orkin and Zon, 1997); human or mouse hematopoietic precursor cells differentiated *ex vivo* (Fibach *et al.*, 1993; Klingmuller *et al.*, 1997; Uddin *et al.*, 2004; Wojda *et al.*, 2002); physiologically validated cell lines established by murine embryonic stem cell differentiation into hematopoietic cells (Cantor *et al.*, 2002; Weiss *et al.*, 1997); and transformed erythroleukemia cell lines that retain some capacity to mature in response to chemical inducers (Dean *et al.*, 1981; Nudel *et al.*, 1977). An emerging system with considerable promise for addressing questions relevant to human hematopoieisis involves the differentiation of human ES cells into erythroid cells (Olivier *et al.*, 2006). While each system has advantages and disadvantages, it is obviously desirable to use systems that recapitulate normal erythropoiesis. In this regard, considerable progress has been made in defining epigenetic mechanisms that regulate the endogenous *β*-globin and *Gata2* loci.

3. Epigenetic Control of the *β*-like Globin Genes During Erythropoiesis

3.1. Locus organization and regulation

The *β*-like globin genes encode *β*-globin polypeptides that combine with *α*-globin polypeptides to form the hemoglobin tetramer (Stamatoyannopoulos, 1991). The murine and human *β*-globin loci consist of four and five genes, respectively (Fig. 3.1A), which were believed for many years to be expressed during erythropoiesis in the order in which they are arranged on the chromosome. The murine *βH1* and *Eγ* genes are expressed in embryonic/fetal erythroid cells and *βmajor* and *βminor* are expressed in the adult (Whitelaw *et al.*, 1990). Analysis of hemoglobin switching in primitive erythroid cells revealed an exception to the rule in which *β*-like globin genes are expressed in their order on the chromosome. *β*H1 is expressed in the early yolk sac prior to a "maturational switch" in which *Eγ* is activated, followed by *βmajor* and *βminor* (Kingsley *et al.*, 2006). Human ε is active in early embryogenesis,

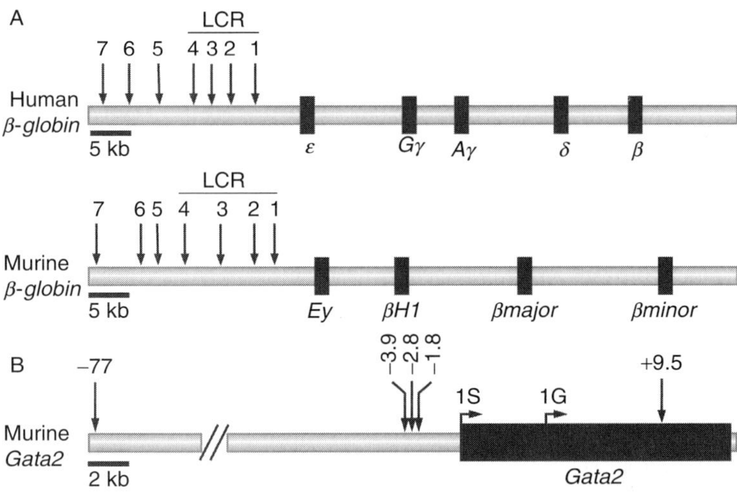

Figure 3.1 (A) Organization of human and murine *β-globin* loci. The black boxes depict the human and murine *β-globin* genes, while the arrows denote the position of upstream DNaseI hypersensitive sites (HS). Functional studies have identified HS1–HS4 as the locus control region (LCR) in both species. (B) Organization of the murine *Gata2* locus. Bent arrows indicate the locations of the 1S and 1G promoters, while vertical arrows denote the position of GATA switch sites in kilobases from the 1S promoter.

Gγ and *Aγ* are fetally transcribed, and *δ* and *β* are expressed after birth (Stamatoyannopoulos, 1991).

Both the murine and human *β*-globin loci are preceded upstream by DNaseI hypersensitive sites (HS1–HS4) scattered over an approximately 12 kb region (Forrester *et al.*, 1986; Tuan and London, 1984). Functional analyses in transgenic mice led to the designation of these sequences as a locus control region (LCR) (Forrester *et al.*, 1987; Grosveld *et al.*, 1987). The concept of an LCR, defined by its activity to confer copy number-dependent and chromosome position site-independent expression of transgenes, was derived from these *β-globin* LCR studies (Li *et al.*, 2002). Although additional HSs exist upstream of the *β-globin* LCR and promoter (Bulger *et al.*, 2003; Xiang *et al.*, 2006), many questions remain unanswered regarding their structure and function, and therefore they will not be discussed in this chapter.

3.2. Nucleoprotein structure of the endogenous locus

A major experimental approach that has greatly facilitated mechanistic analysis at complex loci, such as the *β*-globin locus, is chromatin immunoprecipitation (ChIP) (Im *et al.*, 2004; Johnson and Bresnick, 2002; Orlando *et al.*, 1997), which allows one to obtain molecular snapshots of a specific chromosomal region in living cells. ChIP involves cross-linking proteins to chromatin at endogenous sites and measuring the levels and distribution of proteins bound directly to DNA and those tethered to DNA via protein–protein interactions. Based on the availability of antibodies against specific posttranslationally modified histones, ChIP is also routinely used to measure the levels and distribution of epigenetic marks.

Analyses of the histone modification pattern of the murine *β*-globin locus in adult erythroid cells revealed that the LCR and adult *β-globin* genes reside in chromatin enriched with transcriptionally permissive epigenetic marks—acetylated histones H3 and H4 and dimethylated K4 of histone H3 (H3dimeK4), while the embryonic/fetal *β-globin* genes reside in chromatin containing low levels of these modifications (Bulger *et al.*, 2003; Forsberg *et al.*, 2000b; Kiekhaefer *et al.*, 2002). The discontinuous histone acetylation pattern was entirely unexpected, as studies on the chicken *β*-globin locus suggested that active chromatin domains are broadly enriched in acetylated histones (Hebbes *et al.*, 1988; Litt *et al.*, 2001a,b), which correlates with general DNaseI sensitivity, an approximately two-fold increased sensitivity to DNaseI cleavage in isolated nuclei (Groudine and Weintraub, 1981; Groudine *et al.*, 1978; Hebbes *et al.*, 1988).

By contrast to the hypoacetylated chromatin near the murine embryonic/fetal *β-like globin* genes in definitive erythroid cells, these genes and flanking sequences are hyperacetylated in primitive erythroid cells from the yolk sac of E10.5 mice (Forsberg *et al.*, 2000b; Kingsley *et al.*, 2006). Repression of *Ey* and *βH1* on embryonic development from E10.5 to E12.5 is not accompanied by reduced acetylation at or near these genes in primitive erythroid cells (Kingsley *et al.*, 2006). However, as the adult *β*-like globin genes are activated during the E10.5–E12.5 progression, histone acetylation at the respective regions increases considerably, although acetylation at the promoters is high at both stages (Kingsley *et al.*, 2006). The human embryonic (ε) and fetal *β-globin* genes (*G*γ and *A*γ) in K562 erythroleukemia cells, but not in nonerythroid HeLa cells, exist in hyperacetylated chromatin, consistent with their transcriptional activity in these cells (Kim and Dean, 2004). In aggregate, these results support a model in which the *β-globin* locus assembles into a discontinuous histone modification pattern, which is dynamically reconfigured during erythropoiesis.

What mechanisms might establish a broad region of low-level acetylation within a hyperacetylated chromatin domain? Because histone deacetylase (HDAC) inhibition by trichostatin A or butyrate restores acetylated H4, but not H3, levels within the hypoacetylated subdomain in adult erythroid cells, low levels of acetylated H4 appear to reflect an HDAC scanning mechanism in which HDACs continuously act within the subdomain (Im *et al.*, 2002). By contrast, low levels of acetylated H3 could result from inability of the respective histone acetyltransferases (HATs) to gain access to the subdomain or from the actions of inhibitor-resistant HDACs to maintain low-level acetylated H3, as a subset of HDACs are resistant to trichostatin A and butyrate (Marks *et al.*, 2004).

3.3. LCR function

Based on the erythroid cell-specific histone modification pattern of the β-globin locus, an obvious possibility is that the LCR functions to establish and/or modulate this pattern. Initial evidence implicating the *β-globin* LCR as a long-range regulator of chromatin structure emerged from analysis of a naturally occurring LCR deletion in a patient with severe hispanic β-thalassemia (Forrester *et al.*, 1990). This deletion resulted in loss of general DNaseI sensitivity across the *β-globin* locus and transcriptional repression (Forrester *et al.*, 1990). However, in addition to removing the LCR, approximately 20 kb of sequence upstream of the LCR was deleted in this patient. Subsequent work in which the endogenous murine *β-globin* LCR was deleted by homologous recombination supported the notion that the LCR confers high-level expression to the β-like globin genes at all developmental stages (Bender *et al.*, 2001; Epner *et al.*, 1998; Fiering *et al.*, 1995; Hug *et al.*, 1996; Reik *et al.*, 1998). Surprisingly, however, the LCR deletion does not abrogate general DNaseI sensitivity (Bender *et al.*, 2000) nor the occupancy of certain transcription factors at the *βmajor* promoter (Sawado *et al.*, 2003; Vakoc *et al.*, 2005a; Zhou *et al.*, 2006). Whereas histone acetylation at the adult β-like globin genes (*βmajor* and *βminor*) persists in the LCR-deleted allele, it is reduced in sequences immediately upstream and downstream of the LCR (Schubeler *et al.*, 2001). As long-range activation is abrogated, despite persistence of epigenetic marks associated with permissive or active chromatin at *βmajor* and *βminor*, the LCR activity to confer high-level transcription does not appear to include a broad reconfiguration of epigenetic marks throughout the locus.

Deleting the LCR results in an approximately two-fold reduction in polymerase II (Pol II) at the *βmajor* promoter (Sawado *et al.*, 2003), indicating that promoter-bound factors cannot recruit maximal Pol II levels without the LCR. Either LCR-mediated establishment of maximal Pol II loading is crucial to achieve high-level expression, or the LCR has additional functions post-Pol II recruitment that confer high-level expression. As the LCR directly recruits Pol II (Johnson *et al.*, 2001, 2003), and given the proximity of the LCR to the β-like globin genes (Carter *et al.*, 2002; Drissen *et al.*, 2004; Tolhuis *et al.*, 2002), LCR-bound Pol II might transfer to the β-like globin promoters, thereby maintaining ample levels for high-level transcription (Johnson *et al.*, 2001). Alternatively, because deleting the LCR is associated with relocalization of the β-globin locus within the three-dimensional confines of the nucleus (Ragoczy *et al.*, 2003, 2006), this LCR-driven subnuclear localization might underlie LCR-mediated Pol II recruitment. Nevertheless, it is unclear whether relocalization is a consequence of, or an intrinsic step in, repression. Considerable additional work is required to determine how the LCR functions over a long distance of the chromosome to confer high-level expression.

Given the apparent lack of a role for the LCR in establishing epigenetic marks at and near the adult β-like globin genes, sequences upstream of the LCR, within the locus, or downstream of β*minor* might confer such a function. HS5 has properties of a chromatin insulator (Li and Stamatoyannopoulos, 1994), analogous to chicken HS4 (Chung *et al.*, 1993), but deletion of this element from a human β-globin locus BAC (Wai *et al.*, 2003) and from the endogenous locus (Bender *et al.*, 1998) does not affect β-like globin gene expression. Similarly, targeted deletion of an HS at the 3′ end of the endogenous locus (3′ HS1) also lacks functional consequences (Bender *et al.*, 2006). Given the apparent lack of a role for the LCR, HS5, and 3′ HS1 in establishing general DNaseI sensitivity and the broad histone modification pattern, it seems reasonable that sequences establishing and regulating these parameters reside within the locus. Nevertheless, the targeted deletion studies provided unequivocal evidence that the LCR functions over a long distance on the chromosome to confer high-level transcription to the β-like globin genes at all stages of development.

3.4. Establishment of the histone modification pattern: *Trans*-acting factor requirements

Although *cis*-elements mediating establishment and/or maintenance of the histone modification pattern have not been defined, considerable progress has been made in defining the contributions of *trans*-acting factors. The GATA family of proteins (GATA-1–6) are zinc-finger transcription factors that both activate and repress target genes by binding the consensus sequence WGATAR (Ko and Engel, 1993; Merika and Orkin, 1993). The founding member of this family, GATA-1, was discovered as a β-globin locus-binding protein (Evans and Felsenfeld, 1989; Tsai *et al.*, 1989). Gene-targeting experiments revealed that GATA-1 is required for definitive erythropoiesis (Pevny *et al.*, 1995), and GATA-1 functions redundantly with GATA-2 to support the generation and/or survival of primitive erythroblasts (Fujiwara *et al.*, 2004).

GATA-1 structure/function analyses have been greatly facilitated by the development of a mouse ES cell-derived GATA-1-null cell line (G1E) in which *Gata1* was disrupted via homologous recombination (Weiss *et al.*, 1997). Differentiation of G1E cells into erythroid cells yielded an immortalized proerythroblast-like cell line that recapitulates a normal window of erythropoiesis (Welch *et al.*, 2004). β-Estradiol- or tamoxifen-mediated activation of a stably expressed estrogen receptor ligand-binding domain fusion to GATA-1 (ER-GATA-1) rescues erythroid differentiation (Gregory *et al.*, 1999).

ER-GATA-1-mediated activation of *βmajor* transcription is associated with ER-GATA-1 occupancy at HS1–HS4 and the *βmajor* promoter in G1E–ER-GATA-1 cells (Im *et al.*, 2005; Johnson *et al.*, 2003). Endogenous GATA-1 has a similar pattern of occupancy in mouse erythroleukemia (MEL) cells and murine fetal liver, a major site of erythropoiesis (Johnson *et al.*, 2002). Importantly, ER-GATA-1 and GATA-1 do not occupy the vast majority of conserved GATA motifs, demonstrating an exquisite discrimination among such motifs in a chromosomal context (Bresnick *et al.*, 2006).

At least one determinant of GATA-1 chromatin occupancy is friend of GATA-1 (FOG-1), the first coregulator identified that mediates GATA-1 activity (Tsang *et al.*, 1997, 1998). GATA-1 target genes can be either FOG-1-dependent or -independent (Fig. 3.2) (Crispino *et al.*, 1999; Johnson *et al.*, 2007). FOG-1 facilitates GATA-1 chromatin occupancy at certain target sites (Letting *et al.*, 2004; Pal *et al.*, 2004), the GATA-1–FOG-1 interaction is required for looping to bring the LCR in proximity of the adult *β*-like globin genes (Vakoc *et al.*, 2005a), and FOG-1 directly binds the NuRD corepressor complex (Hong *et al.*, 2005). GATA-1 can be isolated in multiple complexes containing corepressors, such as NuRD (Hong *et al.*, 2005; Rodriquez *et al.*, 2005), or coactivators such as the HAT CBP/p300 (Blobel *et al.*, 1998; Hung *et al.*, 1999; Ogryzko

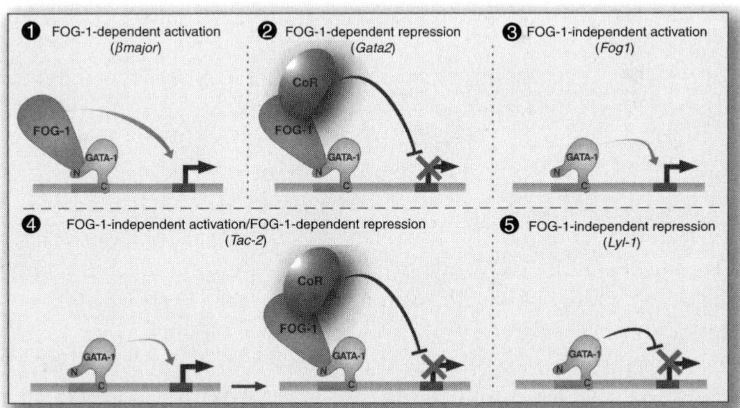

Figure 3.2 Multiple modes of GATA-1 function. Five modes of GATA-1 function are depicted, including FOG-1-dependent activation, FOG-1-dependent repression, FOG-1-independent activation, coupled FOG-1-independent activation followed by FOG-1-dependent repression, and FOG-1-independent repression. These mechanisms were identified via comparison of GATA-1 function in wild-type and FOG-1-null hematopoietic precursor cells, as well as comparison of the activities of wild-type GATA-1 and GATA-1 mutants impaired in FOG-1 binding.

et al., 1996). As CBP/p300 can interact with other HATs, such as the p300/CBP-associated factor (P/CAF) complex (Ogryzko *et al.*, 1998; Yang *et al.*, 1996), CBP recruitment to the β-globin locus might attract additional regulatory factors via protein–protein interactions. Many questions remain unanswered regarding how GATA-1 utilizes these coregulators, including whether multiple coregulators are utilized simultaneously, sequentially, or if the coregulator requirements are target site-specific.

The hematopoietic transcription factor p45/NF-E2, which heterodimerizes with the Maf protein p18 to form the heterodimeric transcription factor NF-E2, is also implicated in β-like globin gene transcriptional regulation (Andrews *et al.*, 1993a,b; Mignotte *et al.*, 1989; Ney *et al.*, 1993). p45/NF-E2 binds a tandem motif in HS2 *in vitro* that confers exceptionally strong erythroid cell-specific enhancer activity in transfection assays (Moon and Ley, 1991; Ney *et al.*, 1990a,b; Talbot and Grosveld, 1991). However, targeted deletion of the endogenous mouse p45/NF-E2 locus did not reveal a nonredundant role in erythropoiesis and/or β-like globin transcription (Shivdasani and Orkin, 1995; Shivdasani *et al.*, 1995). p45/NF-E2-null mice die shortly after birth due to bleeding resulting from defective megakaryopoiesis. Because other transcription factors bind the same motif as NF-E2 (Chan *et al.*, 1993, 1996; Farmer *et al.*, 1997; Igarashi *et al.*, 1998; Johnsen *et al.*, 1996), presumably other factors function redundantly with p45/NF-E2 *in vivo*. However, efforts to validate such redundancy have been unsuccessful (Martin *et al.*, 1998).

By contrast to the apparent redundant function of p45/NF-E2 at the β-globin locus *in vivo*, retroviral insertion into the p45/NF-E2 locus in CB3 mouse erythroleukemia cells abrogates β*major* expression (Lu *et al.*, 1994), and stable expression of p45/NF-E2 in CB3 cells strongly activates β*major* transcription (Bean and Ney, 1997; Kotkow and Orkin, 1995; Lu *et al.*, 1994; Mosser *et al.*, 1998). ChIP analyses in murine fetal liver, MEL cells, CB3 cells reconstituted with recombinant p45/NF-E2, and the G1E system demonstrated p45/NF-E2 occupancy at the LCR and the β*major* promoter (Daftari *et al.*, 1999; Forsberg *et al.*, 2000a; Johnson *et al.*, 2001, 2002; Sawado *et al.*, 2001). While p45/NF-E2 occupies HS2 independently of GATA-1, its occupancy at other HSs of the LCR and the β*major* promoter requires GATA-1 (Johnson *et al.*, 2002). The GATA-1-dependent cross-linking of p45/NF-E2 to certain HSs might be related to the finding that LCR HSs adopt a higher-order structure termed an active chromatin hub (Tolhuis *et al.*, 2002).

Analogous to GATA-1, p45/NF-E2 also interacts with CBP/p300 (Forsberg *et al.*, 1999; Hung *et al.*, 2001), but the transactivation domain of p45/NF-E2 uniquely contains two PPXY sequences that bind certain WW domains (Mosser *et al.*, 1998). As mutation of the PPXY motifs strongly reduces p45/NF-E2-mediated activation of endogenous β-globin expression in CB3 cells (Kiekhaefer *et al.*, 2004; Mosser *et al.*, 1998), it appears that a WW domain-containing coregulator is an important mediator of NF-E2 function. Although the endogenous WW domain protein(s) that binds p45/NF-E2 is unknown, transcriptional coregulators containing WW domains [yes-associated protein (YAP) (Yagi *et al.*, 1999) and transcriptional coactivator with PDZ-binding motif (TAZ) (Kanai *et al.*, 2000)] have been described.

Unlike GATA-1 and NF-E2, erythroid Kruppel-like factor (EKLF) is an established regulator of hemoglobin switching. EKLF was cloned from MEL cells and demonstrated to bind the functionally important *β*-globin CACCC motif (Miller and Bieker, 1993). Targeted deletion of EKLF established a crucial role in regulating definitive erythropoiesis (Perkins *et al.*, 1995), and subsequent work provided evidence for EKLF regulation of primitive erythropoiesis (Hodge *et al.*, 2006). EKLF occupies HS2, HS3, *βmajor* promoter and to a lesser extent HS1 in G1E–ER-GATA-1 cells (Im *et al.*, 2005) and HS1-HS4 in murine E14.5 fetal liver, Ter119+ bone marrow cells, and E10.5 yolk sac (Zhou *et al.*, 2006). Besides regulating *β*-globin expression, studies on EKLF-null mice revealed deregulated expression of multiple genes encoding erythroid membrane components (Nilson *et al.*, 2006).

EKLF physically and functionally interacts with the chromatin remodeler BRG1 (Armstrong *et al.*, 1998; Brown *et al.*, 2002), a component of the SWI–SNF complex (Khavari *et al.*, 1993) that occupies the HSs of the LCR and the *βmajor* promoter (Im *et al.*, 2005). However, the BRG1 distribution at the *β*-globin locus does not correlate precisely with the EKLF occupancy pattern (Im *et al.*, 2005), indicating that EKLF is not the sole determinant of BRG1 occupancy at the locus. CBP/p300 binds and acetylates EKLF, thereby regulating its transactivation activity (Zhang *et al.*, 2001). EKLF is required for hypersensitivity at HS3 (Tewari *et al.*, 1998) that presumably involves EKLF-mediated recruitment of chromatin-modifying and/or -remodeling factors at this site. Finally, EKLF levels differ in primitive versus definitive murine erythroid cells, and such a difference might constitute a mechanism underlying hemoglobin switching (Zhou *et al.*, 2006).

GATA-1, NF-E2, and EKLF are prime candidates for regulators of the broad histone modification pattern at the *β*-globin locus. GATA-1 deficiency in G1E cells does not abolish the erythroid cell-specific histone modification pattern, but rather results in modestly reduced acetylated histones H3 and H4 at the LCR, *βmajor* promoter, and ORF (Im *et al.*, 2005; Kiekhaefer *et al.*, 2002; Letting *et al.*, 2003). ER-GATA-1 activation rescues acetylation at these sites (Im *et al.*, 2005; Kiekhaefer *et al.*, 2002; Letting *et al.*, 2003), increases permissive H4K20 monomethylation, and decreases repressive H3meK27 at the *βmajor* ORF (Vakoc *et al.*, 2006). GATA-1 is not required to establish and/or maintain H3dimeK4 at the LCR and *βmajor* promoter (Im *et al.*, 2005; Kiekhaefer *et al.*, 2002), but H3dimeK4 at the *βmajor* ORF is reduced in G1E cells, and ER-GATA-1 rescues H3dimeK4 at this site (Im *et al.*, 2005; Kiekhaefer *et al.*, 2002). This GATA-1 dependence might be secondary to GATA-1-mediated transcriptional activation, as H3dimeK4 can be established in an elongation-dependent manner in other contexts (Krogan *et al.*, 2003).

By contrast to G1E cells, CB3 cells lacking p45/NF-E2 have reduced H3dimeK4 at the *βmajor* promoter and ORF (Kiekhaefer *et al.*, 2002). Because *βmajor* transcription is repressed in both cell types, and H3dimeK4 at the promoter is relatively unaffected in G1E cells, the decreased H3dimeK4 in CB3 cells is not likely a consequence of repression. Resembling G1E cells, acetylated H3 levels at the *βmajor* promoter and ORF are reduced in CB3 cells, and stable expression of p45/NF-E2 rescues these defects, although p45/NF-E2 deficiency has a considerably lesser effect on H4 versus H3 acetylation (Im *et al.*, 2005; Johnson *et al.*, 2001; Kiekhaefer *et al.*, 2002). Another similarity between GATA-1 and p45/NF-E2 is that both establish H3meK79 at the *βmajor* promoter and ORF (Im *et al.*, 2005; Im *et al.*, 2003); H3meK79 synthesis, catalyzed by disruptor of telomeric silencing 1, is associated with the generation of permissive or active chromatin (Okada *et al.*, 2005; van Leeuwen *et al.*, 2002).

H3dimeK4 persists at the LCR in G1E and CB3 cells, indicating that factors other than GATA-1 and NF-E2 establish and maintain this epigenetic mark at the LCR in erythroid cells. Identifying such factors will yield novel insights into *β*-like globin gene regulation and perhaps the regulation of other complex loci during development. Intriguingly, studies with multipotent hematopoietic precursor cells from EKLF and p45/NF-E2 knockout mice revealed defective epigenetic marks and factor occupancy at a *β*-globin locus BAC (Bottardi *et al.*, 2003, 2006). As lineage-specific factors, such as EKLF and p45/NF-E2, can be expressed "promiscuously" at the onset of hematopoiesis (Miyamoto *et al.*, 2002), such factors might function in the earliest stages of chromatin domain activation, perhaps via differentiation stage-specific mechanisms. Moreover, other studies have provided evidence for establishment of permissive chromatin structural features prior to transcriptional activation (Bottardi *et al.*, 2003; Forsberg *et al.*, 2000b; Jimenez *et al.*, 1992; Kiekhaefer *et al.*, 2002).

Much of the work described above involved the analysis of epigenetic marks in cells containing or lacking a given factor, but it is essential to consider the dynamics of epigenetic regulatory mechanisms. As ER-GATA-1 activity can be titrated with increasing concentrations of agonist, and kinetic analyses can be conducted following the addition of a maximally effective agonist concentration, G1E cells represent a powerful system for delineating mechanisms underlying GATA-1 function at endogenous loci. Kinetic studies revealed ER-GATA-1 occupancy at the LCR, followed by occupancy at the *βmajor* promoter 6–10 h later (Im *et al.*, 2005). Moreover, titration of ER-GATA-1 activity revealed GATA-1 occupancy at the LCR prior to the promoter (Im *et al.*, 2005). Analysis of p45/NF-E2, EKLF, BRG1, and RNA polymerase II (Pol II) occupancy, as well as epigenetic marks associated with active chromatin (acetylated H3 and H4, H3dimeK4, H3trimeK4, and H3K79 methylation), led to the development of a three-phase model of GATA-1-mediated *βmajor* activation (Im *et al.*, 2005; Kim *et al.*, 2007c). The three phases (Fig. 3.3) are as follows:

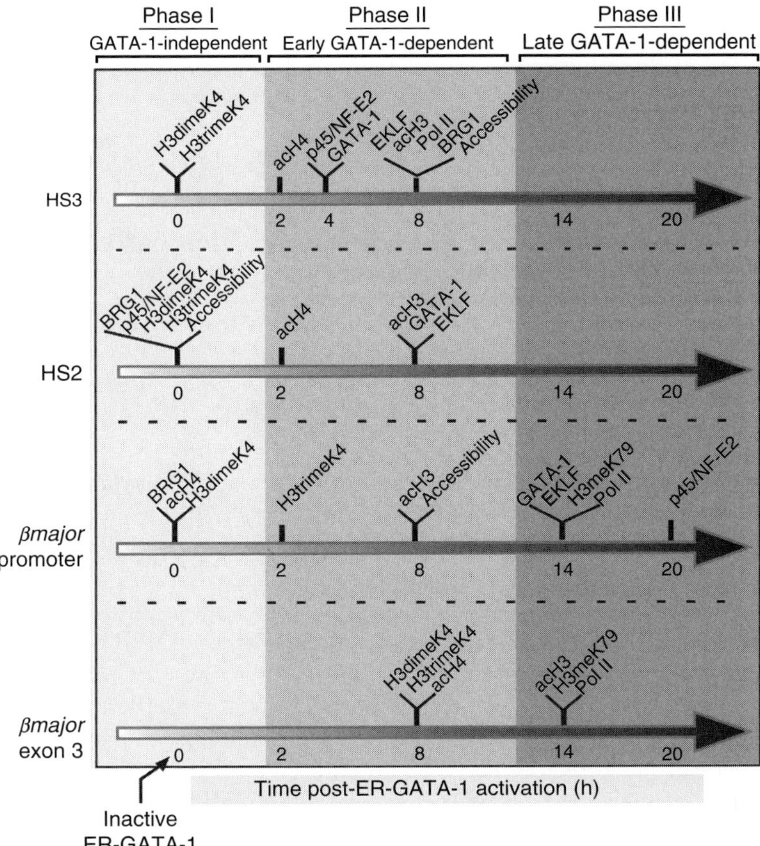

Figure 3.3 Multistep model of *β-globin* locus activation. The model [modified from Im *et al.* (2005)] depicts molecular events measured at the endogenous murine *β-globin* locus various times after tamoxifen-mediated activation of ER-GATA-1 stably expressed in G1E cells. Molecular events are indicated at the time in which their values equal 50±20% of the maximal value achieved during the kinetic analysis. Three phases in the establishment of *β-globin* domain activation are indicated at the top. In the GATA-1-independent phase, certain molecular events have already occurred, indicating that other cellular factors mediate their establishment. Note that ER-GATA-1 occupancy at the LCR precedes ER-GATA-1 occupancy at the promoter. A central aspect of the model is that differential utilization of GATA motifs within the chromatin domain underlies the multistep activation mechanism.

(1) GATA-1-independent establishment of certain epigenetic features and assembly of an LCR subcomplex, (2) GATA-1-dependent maturation of the LCR complex and epigenetic modification at specific sites within the locus, and (3) GATA-1-dependent establishment of epigenetic marks at the promoter and promoter complex assembly (Im *et al.*, 2005). While this model provides a solid foundation for understanding how GATA-1 activates the complex β-globin locus in adult erythroid cells, considerable additional work is required to identify additional steps in the multistep mechanism, to define interrelationships among individual steps, to determine the importance of individual steps, and to establish how this mechanism operates during hemoglobin switching *in vivo*.

3.5. Establishment of the histone modification pattern: A role for intergenic transcription?

Intergenic transcription occurs at multiple sites within the locus (Ashe *et al.*, 1997; Gribnau *et al.*, 2000; Haussecker and Proudfoot, 2005; Johnson *et al.*, 2003; Kim *et al.*, 2007a,b; Kong *et al.*, 1997; Ling *et al.*, 2004a) and represents another prospective determinant of the histone modification pattern. In principle, Pol II can mediate long-range effects by carrying coactivator cargo along a chromosome or by physically modifying chromatin structure as a consequence of transcription itself. The elongation inhibitor 5,6-dichloro-1-beta-D-ribofuranosylbenzimidazole (DRB) does not abrogate the β-globin locus histone modification pattern in G1E–ER-GATA-1 cells (Im *et al.*, 2005), suggesting that ongoing elongation is not required for maintenance of the pattern. However, the DRB result does not rule out an elongation requirement for establishing the pattern, which might only require a single round of elongation. Intergenic transcripts and acetylated H4 are reciprocally distributed across the human *β-globin* locus in K562 erythroleukemia cells, and knockdown of Dicer upregulates intergenic transcripts as well as increases epigenetic marks associated with permissive or active chromatin at the β-globin locus (Haussecker and Proudfoot, 2005). Accordingly, it was suggested that RNAi-dependent mechanisms regulate intergenic transcription at the β-globin locus, analogous to mechanisms functioning at centromeres (Fukugawa *et al.*, 2004; Kanellopoulou *et al.*, 2005; Verdel *et al.*, 2004). Nonetheless, further dissection of mechanisms underlying establishment and regulation of the dynamic histone modification pattern at the β-globin locus should be facilitated by the development of systems that allow one to analyze nucleoprotein structure at distinct stages of differentiation, rather than piecing together the components of a dynamic mechanism with data derived from multiple systems.

4. Epigenetic Control of the *Gata2* Locus During Erythropoiesis

4.1. *Gata2* transcriptional regulation via GATA factor interplay

Through both unique and overlapping functions, GATA-1 and GATA-2 regulate hematopoiesis (Bresnick *et al.*, 2005; Cantor and Orkin, 2002). GATA-2 is expressed in hematopoietic stem cells and erythroid progenitors and functions to maintain the multipotent hematopoietic stem cell population (Ling *et al.*, 2004b; Tsai and Orkin, 1997; Tsai *et al.*, 1994). As GATA-1 levels rise during erythropoiesis, GATA-2 expression declines (Grass *et al.*, 2003; Weiss *et al.*, 1994, 1997).

Quantitative ChIP and ChIP coupled with genomic microarray analysis (ChIP-chip) studies established a direct link between GATA-1 and GATA-2 function that explains the reciprocal expression pattern. GATA-2 occupies five conserved GATA motif-containing regions spanning approximately 100 kb of the repressed murine *Gata2* locus in erythroid precursor cells (Grass *et al.*, 2003, 2006; Martowicz *et al.*, 2005). These regions are located at −77, −3.9, −2.8, −1.8, and +9.5 kb in relation to the *Gata2* hematopoietic-specific 1S promoter, with the +9.5 kb site located in the intron between exons four and five (Fig. 3.1B). *Gata2* has an additional promoter, 1G, which appears to be active more broadly in *Gata2*-expressing cells (Menegishi *et al.*, 1998). GATA-1 displaces GATA-2 from these sites, thereby instigating repression (Grass *et al.*, 2003, 2006; Lugus *et al.*, 2007). Thus, we designated these GATA-binding regions as "GATA switch sites." GATA-2 occupancy at the GATA switch sites is consistent with positive autoregulation.

Although GATA-1 and GATA-2 have nearly identical DNA-binding domains, additional conserved sequences, a similar DNA binding specificity, and common coregulators, occupancy at the *Gata2* GATA switch sites elicits distinct transcriptional outputs. It is instructive therefore to consider how these factors function in a context-dependent manner and whether they differentially affect the *Gata2* histone modification pattern.

4.2. Context-dependent molecular actions

In untreated G1E cells expressing endogenous GATA-2, acetylated H3 and H4 and H3dimeK4 are enriched throughout the *Gata2* locus, extending approximately 4 kb upstream of the 1S promoter and downstream to the 3′ end of *Gata2* (Grass *et al.*, 2003, 2006). ER-GATA-1 activation reduces acetylated H3 and H4 at the *Gata2* 1S promoter, ORF, and all GATA switch sites except the distant −77 kb site (Fig. 3.4; Grass *et al.*, 2003, 2006).

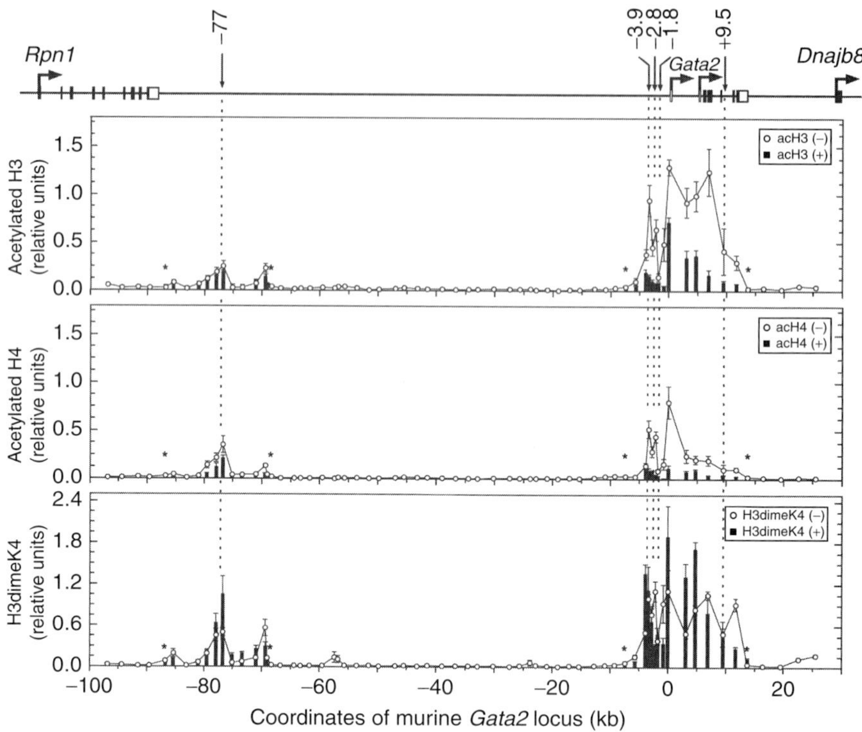

Figure 3.4 Quantitative ChIP analysis of acH3, acH4, and H3dimeK4 levels at endogenous *Gata2* locus, adapted from Grass *et al.* (2006). Organization of the *Gata2* locus and neighboring genes is shown at the top. Open and filled boxes depict noncoding and coding exons, respectively. Arrows pointing down depict GATA switch sites. Coordinate 1 reflects the first nucleotide of the *Gata2* 1S exon. ER-GATA-1 induces deacetylation at and near the *Gata2* promoter and open reading frame but not at the −77 kb region. The graphs below illustrate quantitative ChIP analysis of acH3, acH4, and H3dimeK4 levels across the locus with an average distance of 2 kb between primer sets in untreated (open circles) and β-estradiol-treated (48 h; black bars) G1E–ER-GATA-1 cells (mean standard error, two to five independent experiments). Asterisks indicate the limit of signal detection in the treated condition. DNA immunoprecipitated with preimmune (PI) antibody was analyzed with all primer sets, and the average signal from all experiments is 0.0035.

The deacetylation is not complete, however, as approximately 25–50% of the acetylation persists upon repression; it is unclear whether this residual acetylation reflects a certain percentage of persistent active templates or the incomplete deacetylation of repressed templates. Although 1S promoter-bound Pol II decreases upon repression, the basal transcription factor TFIIB persists, indicating that repression does not overtly abrogate chromatin accessibility—at least not to TFIIB (Martowicz *et al.*, 2006). The GATA-1-dependent deacetylation is likely to be intrinsic to, and not a consequence of, repression, as repression instigated by the elongation inhibitor DRB is not accompanied by deacetylation.

Context-dependent GATA-1 activity is exemplified by the GATA-1-mediated deacetylation and decreased Pol II occupancy at *Gata2* promoters and GATA-1-induced acetylation and Pol II recruitment at the *β*-globin locus (Grass *et al.*, 2003, 2006; Kiekhaefer *et al.*, 2002). Consistent with ER-GATA-1-mediated histone deacetylation of the *Gata2* locus, CBP/p300 occupancy is reduced at certain but not all *Gata2* GATA switch sites. ER-GATA-1 also induces repressive H3K27 methylation at the *Gata2* promoter and extending into the ORF post-ER-GATA-1 activation (Vakoc *et al.*, 2006). Despite the GATA-1-mediated repression of *Gata2*, H3dimeK4 levels persist (Fig. 3.4; Grass *et al.*, 2003, 2006), implying that the factors responsible for maintaining H3dimeK4 at *Gata2* are stable and not displaced or negated by ER-GATA-1 and its associated corepressors. In terms of chromatin accessibility, the −77, −3.9, −2.8, and −1.8 GATA switch sites coincide with DNaseI HSs, and hypersensitivity is differentially affected by ER-GATA-1 activation. Hypersensitivity at the −1.8 kb site is lost upon repression and unchanged at the −77 and −3.9 kb sites (Grass *et al.*, 2006; Martowicz *et al.*, 2005). The −2.8 kb hypersensitivity is weak and therefore whether this is sustained upon repression is unclear. These results as well as reporter gene assays in cultured cells and transgenic mouse embryos demonstrating distinct cell type-specific enhancer activities of the GATA switch sites (Grass *et al.*, 2006; Martowicz *et al.*, 2005; Wozniak *et al.*, 2007) indicate that the GATA switch sites have both shared and distinct properties. Such differences might reflect unique, nonredundant functions *in vivo*, or analogous to *β*-globin locus HSs, major differences might only be apparent when HSs are analyzed in isolation, away from the endogenous locus. Studies on the underlying mechanisms hold enormous promise for revealing novel insights into how complexes dispersed within complex chromatin domains coordinately establish epigenetic regulation.

5. Principles of Epigenetic Control Emerging from Studies of Erythropoiesis

Studies of the regulation of complex loci during erythropoiesis have revealed important insights into epigenetic mechanisms operational during cellular differentiation. Although histone modification patterns at complex loci are a composite of multiple epigenetic marks, the establishment and regulation of individual marks are distinct and can be dissociated via the use of mutant cells, such as those discussed that lack GATA-1 or p45/NF-E2. Intriguingly, even within an "active" domain, regions of relative hypoacetylation can exist, indicating that complex loci can consist of distinct subdomains, each having unique chromatin structural features (Bresnick *et al.*, 2006; Forsberg and Bresnick, 2001). Furthermore, uniform enrichments in epigenetic marks associated with permissive or active chromatin are apparently not required to ensure that chromatin domains do not succomb to the prevalent transcriptional repression machinery.

Currently, it is not possible to make high fidelity predictions of the functional state of a given chromatin region based on its hallmark histone modification pattern. In this regard, certain epigenetic marks associated with permissive or active chromatin, for example H3dimeK4 and acetylated H3 and H4, can persist at repressed loci and can precede differentiation-associated activation. Thus, these epigenetic marks are not indicative of active transcription. The apparent complexity in which a given epigenetic mark is not diagnostic of a specific functional state of chromatin is further exemplified by H3dimeK9 and H3trimeK9, which are paradoxically enriched in repressed chromatin (Bannister *et al.*, 2001; Cheutin *et al.*, 2003; Lachner *et al.*, 2001; Litt *et al.*, 2001a,b; Nakayama *et al.*, 2001; Nielsen *et al.*, 2001) and induced on activation (Vakoc *et al.*, 2005b). As proposed by Allis and coworkers, it might be necessary to consider the combinatorial consequences of a potential histone code (Fischle *et al.*, 2003; Jenuwein and Allis, 2001), but in other contexts defined biochemical activities of one or a limited number of epigenetic marks might dominantly establish the regulatory mechanism.

Investigations of epigenetic regulation during erythropoiesis have revealed dramatic context-dependent activities of GATA factors vis-à-vis histone modification patterns, nucleoprotein complex assembly, and target gene expression. One can assume that a delicate balance exists between coactivator and corepressor utilization, analogous to that proposed for nuclear receptor superfamily members (Perissi *et al.*, 2004). Importantly, however, almost nothing is known about the molecular parameters underlying this balance in the context of GATA factors.

As GATA-1 and GATA-2 undergo diverse posttranslational modifications (Boyes *et al.*, 1998; Chun *et al.*, 2003; Collavin *et al.*, 2004; Crossley and Orkin, 1994; Hung *et al.*, 1999; Partington and Patient, 1999; Towatari *et al.*, 1995, 2004; Zhao *et al.*, 2006), it is reasonable to assume that such modifications control coactivator/corepressor switches. However, initial studies to examine the functional consequences of such modifications have yet to yield definitive conclusions. Although multiple amino acid residues of GATA-1 are phosphorylated in response to cell signaling, knocking in a gene encoding a triple phosphorylation site-mutant yields essentially normal mice, albeit colony forming assays with bone marrow suggest modest decreases in Burst Forming Unit (BFU-E) and Colony Forming Unit (CFU-E) activity (Rooke and Orkin, 2006). This analysis demonstrated definitively that three phosphorylation sites are not required for GATA-1 function *in vivo*, even under conditions of stress erythropoiesis, and therefore raise significant questions regarding the importance of these sites. Nevertheless, multiple additional posttranslationally modified sites exist, and it remains an attractive possibility that signal-dependent GATA factor modifications are a major determinant of coactivator/corepressor utilization. Analogous to the proposed histone code, it would not be surprising if signals conferred by multiple modifications must be integrated, and importantly not a single modification has been shown to be absolutely crucial for GATA factor function. Understanding how signaling systems converge upon GATA factors to establish GATA factor-dependent genetic networks during differentiation promises to be an exciting and productive area of research, which should yield fundamental insights into epigenetic mechanisms that control complex loci.

ACKNOWLEDGMENTS

The authors acknowledge support from DK50107, DK55700, and DK68634 and thank colleagues in the Bresnick laboratory for critical comments.

REFERENCES

Andrews, N. C., Erdjument-Bromage, H., Davidson, M. B., Tempst, P., and Orkin, S. H. (1993a). Erythroid transcription factor NF-E2 is a haematopoietic-specific basic-leucine zipper protein. *Nature* **362,** 722–728.

Andrews, N. C., Kotkow, K. J., Ney, P. A., Erdjument-Bromage, H., Tempst, P., and Orkin, S. H. (1993b). The ubiquitous subunit of erythroid transcription factor NF-E2 is a small basic-leucine zipper protein related to the v-maf oncogene. *Proc. Natl. Acad. Sci. USA* **90,** 11488–11492.

Armstrong, J. A., Bieker, J. J., and Emerson, B. M. (1998). A SWI/SNF-related chromatin remodeling complex, E-RC1, is required for tissue-specific transcriptional regulation by EKLF *in vitro*. *Cell* **95,** 93–104.

Ashe, H. L., Monks, J., Wijgerde, M., Fraser, P., and Proudfoot, N. J. (1997). Intergenic transcription and transinduction of the human beta-globin locus. *Genes Dev.* **11,** 2494–2509.

Bannister, A. J., Zegerman, P., Partridge, J. F., Miska, E. A., Thomas, J. O., Allshire, R. C., and Kouzarides, T. (2001). Selective recognition of methylated lysine 9 on histone H3 by the HP1 chromo domain. *Nature* **410,** 120–124.

Bean, T. L., and Ney, P. A. (1997). Multiple regions of p45 NF-E2 are required for beta-globin gene expression in erythroid cells. *Nucleic Acids Res.* **25,** 2509–2515.

Bender, M. A., Reik, A., Close, J., Telling, A., Epner, E., Fiering, S., Hardison, R., and Groudine, M. (1998). Description and targeted deletion of 5″ hypersensitive site 5 and 6 of the mouse beta-globin locus control region. *Blood* **92,** 4394–4403.

Bender, M. A., Bulger, M., Close, J., and Groudine, M. (2000). Beta-globin gene switching and DNaseI sensitivity of the endogenous beta-globin locus in mice do not require the locus control region. *Mol. Cell* **5,** 387–393.

Bender, M. A., Roach, J. N., Halow, J., Close, J., Alami, R., Bouhassira, E. E., Groudine, M., and Fiering, S. N. (2001). Targeted deletion of 5′HΣ1 and 5′HΣ4 of the beta-globin locus control region reveals additive activity of the DNaseI hypersensitive sites. *Blood* **98,** 2022–2027.

Bender, M. A., Byron, R., Ragoczy, T., Telling, A., Bulger, M., and Groudine, M. (2006). Flanking HS-62.5 and 3′HΣ1, and regions upstream of the LCR, are not required for beta-globin transcription. *Blood* **108,** 1395–1401.

Bernstein, B. E., Meissner, A., and Lander, E. S. (2007). The mammalian epigenome. *Cell* **128,** 669–681.

Blobel, G. A., Nakajima, T., Eckner, R., Montminy, M., and Orkin, S. H. (1998). CREB-binding protein cooperates with transcription factor GATA-1 and is required for erythroid differentiation. *Proc. Natl. Acad. Sci. USA* **95,** 2061–2066.

Bottardi, S., Aumont, A., Grosveld, F., and Milot, E. (2003). Developmental stage-specific epigenetic control of human beta-globin gene expression is potentiated in hematopoietic progenitor cells prior to their transcriptional activation. *Blood* **102,** 3989–3997.

Bottardi, S., Ross, J., Pierre-Charles, N., Blank, V., and Milot, E. (2006). Lineage-specific activators affect beta-globin locus chromatin in multipotent hematopoietic progenitors. *EMBO J.* **25,** 3586–3595.

Boyes, J., Byfield, P., Nakatani, Y., and Ogryzko, V. (1998). Regulation of activity of the transcription factor GATA-1 by acetylation. *Nature* **396,** 594–598.

Bresnick, E. H., Martowicz, M. L., Pal, S., and Johnson, K. D. (2005). Developmental control via GATA factor interplay at chromatin domains. *J. Cell. Physiol.* **205,** 1–9.

Bresnick, E. H., Johnson, K. D., Kim, S.-I., and Im, H. (2006). Establishment and regulation of chromatin domains: Mechanistic insights from studies of hemoglobin synthesis. *Prog. Nucleic Acids Res. Mol. Biol.* **81,** 435–471.

Brown, R. C., Pattison, S., van Ree, J., Coghill, E., Perkins, A., Jane, S. M., and Cunningham, J. M. (2002). Distinct domains of erythroid Kruppel-like factor modulate chromatin remodeling and transactivation at the endogenous beta-globin gene promoter. *Mol. Cell. Biol.* **22,** 161–170.

Brownell, J. E., Zhou, J., Ranalli, T., Kobayashi, R., Edmondson, D. G., Roth, S. Y., and Allis, C. D. (1996). Tetrahymena histone acetyltransferase A: A homolog to yeast Gcn5p linking histone acetylation to gene activation. *Cell* **84,** 843–851.

Bulger, M., Schubeler, D., Bender, M. A., Hamilton, J., Farrell, C. M., Hardison, R. C., and Groudine, M. (2003). A complex chromatin landscape revealed by patterns of nuclease sensitivity and histone modification within the mouse beta-globin locus. *Mol. Cell. Biol.* **23,** 5234–5244.

Cantor, A. B., and Orkin, S. H. (2002). Transcriptional regulation of erythropoiesis: An affair involving multiple partners. *Oncogene* **21,** 3368–3376.

Cantor, A. B., Katz, S. G., and Orkin, S. H. (2002). Distinct domains of the GATA-1 cofactor FOG-1 differentially influence erythroid versus megakaryocytic maturation. *Mol. Cell. Biol.* **22,** 4268–4279.

Carter, D., Chakalova, L., Osborne, C. S., Dai, Y. F., and Fraser, P. (2002). Long-range chromatin regulatory interactions *in vivo*. *Nat. Genet.* **32,** 623–626.

Chan, K., Lu, R., Chang, J. C., and Kan, Y. W. (1996). NRF2, a member of the NFE2 family of transcription factors, is not essential for murine erythropoiesis, growth, and development. *Proc. Natl. Acad. Sci. USA* **93,** 13943–13948.

Chan, J. Y., Han, X. L., and Kan, Y. W. (1993). Cloning of Nrf1, an NF-E2-related transcription factor, by genetic selection in yeast. *Proc. Natl. Acad. Sci. USA* **90,** 11371–13375.

Cheutin, T., McNairn, A. J., Jenuwein, T., Gilbert, D. M., Singh, P. B., and Misteli, T. (2003). Maintenance of stable heterochromatin domains by dynamic HP1 binding. *Science* **299,** 721–725.

Chun, T. H., Itoh, H., Subramanian, L., Ingiguez-Lluhi, J. A., and Nakao, K. (2003). Modification of GATA-2 transcriptional activity in endothelial cells by the SUMO E3 ligase PIASgamma. *Circ. Res.* **92,** 1201–1208.

Chung, J. H., Whiteley, M., and Felsenfeld, G. (1993). A 5″ element of the chicken beta-globin domain serves as an insulator in human erythroid cells and protects against position effect in *Drosophila*. *Cell* **74,** 505–514.

Collavin, L., Gostissa, M., Avolio, F., Secco, P., Ronchi, A., Santoro, C., and Del Sal, G. (2004). Modification of the erythroid transcription factor GATA-1 by SUMO-1. *Proc. Natl. Acad. Sci. USA* **101,** 8870–8875.

Crispino, J. D., Lodish, M. B., MacKay, J. P., and Orkin, S. H. (1999). Use of altered specificity mutants to probe a specific protein-protein interaction in differentiation: The GATA-1:FOG complex. *Mol. Cell* **3,** 219–228.

Crossley, M., and Orkin, S. H. (1994). Phosphorylation of the erythroid transcription factor GATA-1. *J. Biol. Chem.* **269,** 16589–16596.

Daftari, P., Gavva, N. R., and Shen, C. K. (1999). Distinction between AP1 and NF-E2 factor-binding at specific chromatin regions in mammalian cells. *Oncogene* **18,** 5482–5486.

Dean, A. (2006). On a chromosome far, far away: LCRs and gene expression. *Trends Genet.* **22,** 38–45.

Dean, A., Erard, F., Schneider, A. P., and Schechter, A. N. (1981). Induction of hemoglobin accumulation in human K562 cells by hemin is reversible. *Science* **212,** 459–461.

Drissen, R., Palstra, R. J., Gillemans, N., Splinter, E., Grosveld, F., Philipsen, S., and de Laat, W. (2004). The active spatial organization of the beta-globin locus requires the transcription factor EKLF. *Genes Dev.* **18,** 2485–2490.

Epner, E., Reik, A., Cimbora, D., Telling, A., Bender, M. A., Fiering, S., Enver, T., Martin, D. I., Kennedy, M., Keller, G., and Groudine, M. (1998). The beta-globin LCR is not necessary for an open chromatin structure or developmentally regulated transcription of the native mouse beta-globin locus. *Mol. Cell* **2,** 447–455.

Evans, T., and Felsenfeld, G. (1989). The erythroid-specific transcription factor Eryf1: A new finger protein. *Cell* **58,** 877–885.

Farmer, S. C., Sun, C. W., Winnier, G. E., Hogan, B. L., and Townes, T. M. (1997). The bZIP transcription factor LCR-F1 is essential for mesoderm formation in mouse development. *Genes Dev.* **11,** 786–798.

Felsenfeld, G., and Groudine, M. (2003). Controlling the double helix. *Nature* **421,** 448–453.

Fibach, E., Burke, K. P., Schechter, A. N., Noguchi, C. T., and Rodgers, G. P. (1993). Hydroxyurea increases fetal hemoglobin in cultured erythroid cells derived from normal individuals and patients with sickle cell anemia or beta-thalassemia. *Blood* **81,** 1630–1635.

Fiering, S., Epner, E., Robinson, K., Zhuang, Y., Telling, A., Hu, M., Martin, D. I., Enver, T., Ley, T. J., and Groudine, M. (1995). Targeted deletion of 5′HΣ2 of the murine beta-globin LCR reveals that it is not essential for proper regulation of the beta-globin locus. *Genes Dev.* **9,** 2203–2213.

Fischle, W., Wang, Y., and Allis, C. D. (2003). Histone and chromatin crosstalk. *Curr. Opin. Cell Biol.* **15,** 172–183.

Forrester, W. C., Thompson, C., Elder, J. T., and Groudine, M. (1986). A developmentally stable chromatin structure in the human beta-globin gene cluster. *Proc. Natl. Acad. Sci. USA* **83,** 1359–1363.

Forrester, W. C., Takegawa, S., Papayannopoulou, T., Stamatoyannopoulos, G., and Groudine, M. (1987). Evidence for a locus activation region: The formation of developmentally stable hypersensitive sites in globin-expressing hybrids. *Nucleic Acids Res.* **15,** 10159–10177.

Forrester, W. C., Epner, E., Driscoll, M. C., Enver, T., Brice, M., Papayannopoulou, T., and Groudine, M. (1990). A deletion of the human beta-globin locus activation region causes a major alteration in chromatin structure and replication across the entire beta-globin locus. *Genes Dev.* **4,** 1637–1649.

Forsberg, E. C., and Bresnick, E. H. (2001). Histone acetylation beyond promoters: Long-range acetylation patterns in the chromatin world. *Bioessays* **23,** 820–830.

Forsberg, E. C., Johnson, K., Zaboikina, T. N., Mosser, E. A., and Bresnick, E. H. (1999). Requirement of an E1A-sensitive coactivator for long-range transactivation by the beta-globin locus control region. *J. Biol. Chem.* **274,** 26850–26859.

Forsberg, E. C., Downs, K. M., and Bresnick, E. H. (2000a). Direct interaction of NF-E2 with hypersensitive site 2 of the beta-globin locus control region in living cells. *Blood* **96,** 334–339.

Forsberg, E. C., Downs, K. M., Christensen, H. M., Im, H., Nuzzi, P. A., and Bresnick, E. H. (2000b). Developmentally dynamic histone acetylation pattern of a tissue-specific chromatin domain. *Proc. Natl. Acad. Sci. USA* **97,** 14494–14499.

Fujiwara, Y., Chang, A. N., Williams, A. M., and Orkin, S. H. (2004). Functional overlap of GATA-1 and GATA-2 in primitive hematopoietic development. *Blood* **103,** 583–585.

Fukugawa, T., Nogami, M., Yoshikawa, M., Ikeno, M., Okazaki, T., Takami, Y., Nakayama, T., and Oshimura, M. (2004). Dicer is essential for formation of the heterochromatin structure in vertebrate cells. *Nat. Cell Biol.* **6,** 784–791.

Grass, J. A., Boyer, M. E., Pal, S., Wu, J., Weiss, M. J., and Bresnick, E. H. (2003). GATA-1-dependent transcriptional repression of GATA-2 via disruption of positive autoregulation and domain-wide chromatin remodeling. *Proc. Natl. Acad. Sci. USA* **100,** 8811–8816.

Grass, J. A., Jing, H., Kim, S.-I., Martowicz, M. L., Pal, S., Blobel, G. A., and Bresnick, E. H. (2006). Distinct functions of dispersed GATA factor complexes at an endogenous gene locus. *Mol. Cell. Biol.* **26,** 7056–7067.

Gregory, T., Yu, C., Ma, A., Orkin, S. H., Blobel, G. A., and Weiss, M. J. (1999). GATA-1 and erythropoietin cooperate to promoter erythroid cell survival by regulating bcl-xl expression. *Blood* **94,** 87–96.

Gribnau, J., Diderich, K., Pruzina, S., Calzolari, R., and Fraser, P. (2000). Intergenic transcription and developmental remodeling of chromatin subdomains in the human beta-globin locus. *Mol. Cell* **5,** 377–386.

Grosveld, F., van Assendelft, G. B., Greaves, D. R., and Kollias, G. (1987). Position-independent, high-level expression of the human beta-globin gene in transgenic mice. *Cell* **51,** 975–985.

Groudine, M., and Weintraub, H. (1981). Activation of globin genes during chicken development. *Cell* **24,** 393–401.

Groudine, M., Das, S., Neiman, P., and Weintraub, H. (1978). Regulation of expression and chromosomal subunit conformation of avian retrovirus genomes. *Cell* **14,** 865–878.

Haussecker, D., and Proudfoot, N. J. (2005). Dicer-dependent turnover of intergenic transcripts from the human beta-globin gene cluster. *Mol. Cell. Biol.* **25,** 9724–9733.

Hebbes, T. R., Thorne, A. W., and Crane-Robinson, C. (1988). A direct link between core histone acetylation and transcriptionally active chromatin. *EMBO J.* **7,** 1395–1402.

Hodge, D., Coghill, E., Keys, J., Maguire, T., Hartmann, B., McDowall, A., Weiss, M., Grimmond, S., and Perkins, A. (2006). A global role for EKLF in definitive and primitive erythropoiesis. *Blood* **107,** 3359–3370.

Hong, W., Nakazawa, M., Chen, Y. Y., Kori, R., Vakoc, C. R., Rakowski, C., and Blobel, G. A. (2005). FOG-1 recruits the NuRD repressor complex to mediate transcriptional repression by GATA-1. *EMBO J.* **24,** 67–78.

Hug, B. A., Wesselschmidt, R. L., Fiering, S., Bender, M. A., Epner, E., Groudine, M., and Ley, T. J. (1996). Analysis of mice containing a targeted deletion of beta-globin locus control region 5″ hypersensitive site 3. *Mol. Cell. Biol.* **16,** 2906–2912.

Hung, H. L., Lau, J., Kim, A. Y., Weiss, M. J., and Blobel, G. A. (1999). CREB-binding protein acetylates hematopoietic transcription factor GATA-1 at functionally important sites. *Mol. Cell. Biol.* **19,** 3496–3505.

Hung, H. L., Kim, A. Y., Hong, W., Rakowski, C., and Blobel, G. A. (2001). Stimulation of NF-E2 DNA binding by CREB-binding protein (CBP)-mediated acetylation. *J. Biol. Chem.* **276,** 10715–10721.

Igarashi, K., Hoshino, H., Muto, A., Suwabe, N., Nishikawa, S., Nakauchi, H., and Yamamoto, M. (1998). Multivalent DNA binding complex generated by small Maf and Bach1 as a possible biochemical basis for beta-globin locus control region complex. *J. Biol. Chem.* **273,** 11783–11790.

Im, H., Grass, J. A., Christensen, H. M., Perkins, A., and Bresnick, E. H. (2002). Histone deacetylase-dependent establishment and maintenance of broad low-level histone acetylation within a tissue-specific chromatin domain. *Biochemistry* **41,** 15152–15160.

Im, H., Park, C., Feng, Q., Johnson, K. D., Kiekhaefer, C. M., Choi, K., Zhang, Y., and Bresnick, E. H. (2003). Dynamic regulation of histone H3 methylated at lysine 79 within a tissue-specific chromatin domain. *J. Biol. Chem.* **278,** 18346–18352.

Im, H., Grass, J. A., Johnson, K. D., Boyer, M. E., Wu, J., and Bresnick, E. H. (2004). Measurement of protein-DNA interactions *in vivo* by chromatin immunoprecipitation. *Methods Mol. Biol.* **284,** 129–146.

Im, H., Grass, J. A., Johnson, K. D., Kim, S.-I., Boyer, M. E., Imbalzano, A. N., Bieker, J. J., and Bresnick, E. H. (2005). Chromatin domain activation via GATA-1 utilization of a small subset of dispersed GATA motifs within a broad chromosomal region. *Proc. Natl. Acad. Sci. USA* **102,** 17065–17070.

Jacobs, S. A., Taverna, S. D., Zhang, Y., Briggs, S. D., Li, J., Eissenberg, J. C., Allis, C. D., and Khorasanizadeh, S. (2001). Specificity of the HP1 chromo domain for the methylated N-terminus of histone H3. *EMBO J.* **20,** 5232–5241.

Jenuwein, T., and Allis, C. D. (2001). Translating the histone code. *Science* **293,** 1074–1080.

Jimenez, G., Gale, K. B., and Enver, T. (1992). The mouse beta-globin locus control region: Hypersensitive sites 3 and 4. *Nucleic Acids Res.* **20,** 5797–5803.

Johnsen, O., Skammelsrud, N., Luna, L., Nishizawa, M., Prydz, H., and Kolsto, A. B. (1996). Small Maf proteins interact with the human transcription factor TCF11/Nrf1/LCR-F1. *Nucleic Acids Res.* **24,** 4289–4297.

Johnson, K. D., and Bresnick, E. H. (2002). Dissecting long-range transcriptional mechanisms by chromatin immunoprecipitation. *Methods* **26,** 27–36.

Johnson, K. D., Christensen, H. M., Zhao, B., and Bresnick, E. H. (2001). Distinct mechanisms control RNA polymerase II recruitment to a tissue-specific locus control region and a downstream promoter. *Mol. Cell* **8,** 465–471.

Johnson, K. D., Grass, J. D., Boyer, M. E., Kiekhaefer, C. M., Blobel, G. A., Weiss, M. J., and Bresnick, E. H. (2002). Cooperative activities of hematopoietic regulators recruit RNA polymerase II to a tissue-specific chromatin domain. *Proc. Natl. Acad. Sci. USA* **99,** 11760–11765.

Johnson, K. D., Grass, J. A., Im, H., Park, C., Choi, K., and Bresnick, E. H. (2003). Highly restricted localization of RNA polymerase II to the hypersensitive site cores of a tissue-specific locus control region. *Mol. Cell. Biol.* **23,** 6468–6493.

Johnson, K. D., Boyer, M. E., Kim, S.-I., Kang, S. Y., Wickrema, A., Cantor, A. B., and Bresnick, E. H. (2007). Friend of GATA-1-independent transcriptional repression: A novel mode of GATA-1 function. *Blood* **109,** 5230–5233.

Kanai, F., Marignani, P. A., Sarbassova, D., Yagi, R., Hall, R. A., Donowiwtz, M., Hisaminato, A., Fujiwara, T., Ito, Y., Cantley, L. C., and Yaffe, M. B. (2000). TAZ: A novel transcriptional coactivator regulated by interactions with 14–3-3 and PDZ domain proteins. *EMBO J.* **19,** 6778–6791.

Kanellopoulou, C., Muljo, S. A., Kung, A. L., Ganesan, S., Drapkin, R., Jenuwein, T., Livingston, D. M., and Rajewsky, K. (2005). Dicer-deficient mouse embryonic stem cells are defective in differentiation and centromeric silencing. *Genes Dev.* **19,** 489–501.

Khavari, P. A., Peterson, C. L., Tamkun, J. W., Mendel, D. B., and Crabtree, G. R. (1993). BRG1 contains a conserved domain of the SWI2/SNF2 family necessary for normal mitotic growth and transcription. *Nature* **366,** 170–174.

Kiekhaefer, C. M., Grass, J. A., Johnson, K. D., Boyer, M. E., and Bresnick, E. H. (2002). Hematopoietic activators establish an overlapping pattern of histone acetylation and methylation within a tissue-specific chromatin domain. *Proc. Natl. Acad. Sci. USA* **99,** 14309–14314.

Kiekhaefer, C. M., Grass, J. D., and Bresnick, E. H. (2004). WW domain binding motif within the activation domain of the hematopoietic activator p45/NF-E2 is essential for establishment of a tissue-specific histone modification pattern. *J. Biol. Chem.* **279,** 7456–7461.

Kim, A., and Dean, A. (2004). Developmental stage differences in chromatin subdomains of the beta-globin locus. *Proc. Natl. Acad. Sci. USA* **101,** 7028–7033.

Kim, S. -I., and Bresnick, E. H. (2007). Transcriptional control of erythropoiesis: emerging mechanisms and principles. *Oncogene* **26,** 6777–6794.

Kim, A., Kiefer, C. M., and Dean, A. (2007a). Distinctive signatures of histone methylation in transcribing coding and noncoding human beta-globin sequences. *Mol. Cell. Biol.* **27,** 1271–1279.

Kim, A., Zhao, H., Ifrim, I., and Dean, A. (2007b). Beta-globin intergenic transcription and histone acetylation dependent on an enhancer. *Mol. Cell. Biol.* **27,** 2980–2986.

Kim, S.-I., Bultman, S. J., Jing, H., Blobel, G. A., and Bresnick, E. H. (2007c). Dissecting molecular steps in chromatin domain activation during hematopoietic differentiation. *Mol. Cell. Biol.* **27,** 4551–4565.

Kingsley, P. D., Malik, J., Emerson, R. L., Bushnell, T. P., McGrath, K. E., Bloedorn, L. A., Bulger, M., and Palis, J. (2006). "Maturational" globin switching in primary primitive erythroid cells *Blood* **107,** 1665–1672.

Klingmuller, U., Wu, H., Hsiao, J. G., Toker, A., Duckworth, B. C., Cantley, L. C., and Lodish, H. F. (1997). Identification of a novel pathway important for proliferation and differentiation of primary erythroid progenitors. *Proc. Natl. Acad. Sci. USA* **94,** 3016–3021.

Ko, L. J., and Engel, J. D. (1993). DNA-binding specificities of the GATA transcription factor family. *Mol. Cell. Biol.* **13,** 4011–4022.

Kong, S., Bohl, D., Li, C., and Tuan, D. (1997). Transcription of the HS2 enhancer toward a cis-linked gene is independent of the orientation, position, and distance of the enhancer relative to the gene. *Mol. Cell. Biol.* **17,** 3955–3965.

Kotkow, K. J., and Orkin, S. H. (1995). Dependence of globin gene expression in mouse erythroleukemia cells on the NF-E2 heterodimer. *Mol. Cell. Biol.* **15,** 4640–4647.

Kouzarides, T. (2007). Chromatin modifications and their function. *Cell* **128,** 693–705.

Krogan, N. J., Kim, M., Tong, A., Golshani, A., Cagney, G., Canadien, V., Richards, D. P., Beattie, B. K., Emili, A., Boone, C., Shilatifard, A., Buratowski, S., *et al.* (2003). Methylation of histone H3 by Set2 in *Saccharomyces cerevisiae* is linked to transcriptional elongation by RNA polymerase II. *Mol. Cell. Biol.* **12,** 4207–4218.

Lachner, M., O'Carroll, D., Rea, S., Mechtler, K., and Jenuwein, T. (2001). Methylation of histone H3 lysine 9 creates a binding site for HP1 proteins. *Nature* **410,** 116–120.

Lee, D. Y., Hayes, J. J., Pruss, D., and Wolffe, A. P. (1993). A positive role for histone acetylation in transcription factor access to nucleosomal DNA. *Cell* **72,** 73–84.

Letting, D. L., Rakowski, C., Weiss, M. J., and Blobel, G. A. (2003). Formation of a tissue-specific histone acetylation pattern by the hematopoietic transcription factor GATA-1. *Mol. Cell. Biol.* **23,** 1334–1340.

Letting, D. L., Chen, Y. Y., Rakowski, C., Reedy, S., and Blobel, G. A. (2004). Context-dependent regulation of GATA-1 by friend of GATA-1. *Proc. Natl. Acad. Sci. USA* **101,** 476–481.

Li, Q., and Stamatoyannopoulos, G. (1994). Hypersensitive site 5 of the human beta locus control region functions as a chromatin insulator. *Blood* **84,** 1399–1401.

Li, Q., Peterson, K. R., Fang, X. R., and Stamatoyannopoulos, G. (2002). Locus control regions. *Blood* **100,** 3077–3086.

Ling, J., Ainol, L., Zhang, L., Yu, X., Pi, W., and Tuan, D. (2004a). HS2 enhancer function is blocked by a transcriptional terminator inserted between the enhancer and the promoter. *J. Biol. Chem.* **279,** 51704–51713.

Ling, K. W., Ottersbach, K., van Hamburg, J. P., Oziemlak, A., Tsai, F. Y., Orkin, S. H., Ploemacher, R., Hendriks, R. W., and Dzierzak, E. (2004b). GATA-2 plays two functionally distinct roles during the ontogeny of hematopoietic stem cells. *J. Exp. Med.* **200,** 871–882.

Litt, M. D., Simpson, M., Gaszner, M., Allis, C. D., and Felsenfeld, G. (2001a). Correlation between histone lysine methylation and developmental changes at the chicken {beta}-globin locus. *Science* **9,** 2453–2455.

Litt, M. D., Simpson, M., Recillas-Targa, F., Prioleau, M. N., and Felsenfeld, G. (2001b). Transitions in histone acetylation reveal boundaries of three separately regulated neighboring loci. *EMBO J.* **20,** 2224–2235.

Lu, S. J., Rowan, S., Bani, M. R., and Ben-David, Y. (1994). Retroviral integration within the Fli-2 locus results in inactivation of the erythroid transcription factor NF-E2 in Friend erythroleukemias: Evidence that NF-E2 is essential for globin expression. *Proc. Natl. Acad. Sci. USA* **91,** 8398–8402.

Lugus, J. J., Chung, Y. S., Mills, J. C., Kim, S. I., Grass, J. A., Kyba, M., Doherty, J. M., Bresnick, E. H., and Choi, K. (2007). GATA2 functions at multiple steps in hemangioblast development and differentiation. *Development* **134,** 393–405.

Marks, P. A., Richon, V. M., Miller, T., and Kelly, W. K. (2004). Histone deacetylase inhibitors. *Adv. Cancer Res.* **91,** 137–168.

Martin, F., van Deursen, J. M., Shivdasani, R. A., Jackson, C. W., Troutman, A. G., and Ney, P. A. (1998). Erythroid maturation and globin gene expression in mice with combined deficiency of NF-E2 and Nrf-2. *Blood* **91,** 3459–3466.

Martowicz, M. L., Grass, J. A., Boyer, M. E., Guend, H., and Bresnick, E. H. (2005). Dynamic GATA factor interplay at a multi-component regulatory region of the GATA-2 locus. *J. Biol. Chem.* **280,** 1724–1732.

Martowicz, M. L., Grass, J. A., and Bresnick, E. H. (2006). GATA-1-mediated transcriptional repression yields persistent transcription factor IIB—chromatin complexes. *J. Biol. Chem.* **281,** 37345–37352.

Menegishi, N., Ohta, J., Suwabe, N., Nakauchi, H., Ishihara, H., Hayashi, N., and Yamamoto, M. (1998). Alternative promoters regulate transcription of the mouse GATA-2 gene. *J. Biol. Chem.* **273,** 3625–3634.

Merika, M., and Orkin, S. H. (1993). DNA-binding specificity of GATA family transcription factors. *Mol. Cell. Biol.* **13,** 3999–4010.

Mignotte, V., Eleouet, J. F., Raich, N., and Romeo, P. H. (1989). Cis- and trans-acting elements involved in the regulation of the erythroid promoter of the human porphobilinogen deaminase gene. *Proc. Natl. Acad. Sci. USA* **86,** 6548–6552.

Miller, I. J., and Bieker, J. J. (1993). A novel, erythroid cell-specific murine transcription factor that binds to the CACCC element and is related to the Kruppel family of nuclear proteins. *Mol. Cell. Biol.* **13,** 2776–2786.

Miyamoto, T., Iwasaki, H., Reizis, B., Ye, M., Graf, T., Weissman, I. L., and Akashi, K. (2002). Myeloid or lymphoid promiscuity as a critical step in hematopoietic lineage commitment. *Dev. Cell* **3,** 137–147.

Moon, A. M., and Ley, T. J. (1991). Functional properties of the beta-globin locus control region in K562 erythroleukemia cells. *Blood* **77,** 2272–2284.

Mosser, E. A., Kasanov, J. D., Forsberg, E. C., Kay, B. K., Ney, P. A., and Bresnick, E. H. (1998). Physical and functional interactions between the transactivation domain of the hematopoietic transcription factor NF-E2 and WW domains. *Biochemistry* **37,** 13686–13695.

Nakayama, J., Rice, J. C., Strahl, B. D., Allis, C. D., and Grewal, S. I. (2001). Role of histone H3 lysine 9 methylation in epigenetic control of heterochromatin assembly. *Science* **292,** 110–113.

Ney, P. A., Sorrentino, B. P., Lowrey, C. H., and Nienhuis, A. W. (1990a). Inducibility of the HS II enhancer depends on binding of an erythroid specific nuclear protein. *Nucleic Acids Res.* **18,** 6011–6017.

Ney, P. A., Sorrentino, B. P., McDonagh, K. T., and Nienhuis, A. W. (1990b). Tandem AP-1-binding sites within the human beta-globin dominant control region function as an inducible enhancer in erythroid cells. *Genes Dev.* **4,** 993–1006.

Ney, P. A., Andrews, N. C., Jane, S. M., Safer, B., Purucker, M. E., Weremowicz, S., Morton, C. C., Goff, S. C., Orkin, S. H., and Nienhuis, A. W. (1993). Purification of the human NF-E2 complex: cDNA cloning of the hematopoietic cell-specific subunit and evidence for an associated partner. *Mol. Cell. Biol.* **13,** 5604–5612.

Nielsen, S. J., Schneider, R., Bauer, U. M., Bannister, A. J., Morrison, A., O'Carroll, D., Firestein, R., Cleary, M., Jenuwein, T., Herrera, R. E., and Kouzarides, T. (2001). Rb targets histone H3 methylation and HP1 to promoters. *Nature* **412,** 561–565.

Nilson, D. G., Sabatino, D. E., Bodine, D. M., and Gallagher, P. G. (2006). Major erythrocyte membrane protein genes in EKLF-deficient mice. *Exp. Hematol.* **34,** 705–712.

Nudel, U., Salmon, J. E., Terada, M., Bank, A., Rifkind, R. A., and Marks, P. A. (1977). Differential effects of chemical inducers on expression of beta globin genes in murine erythroleukemia cells. *Proc. Natl. Acad. Sci. USA* **74,** 1100–1104.

Ogryzko, V. V., Schiltz, R. L., Russanova, V., Howard, B. H., and Nakatani, Y. (1996). The transcriptional coactivators p300 and CBP are histone acetyltransferases. *Cell* **87,** 953–959.

Ogryzko, V. V., Kotani, T., Zhang, X., Schlitz, R. L., Howard, T., Yang, X. J., Howard, B. H., Qin, J., and Nakatani, Y. (1998). Histone-like TAFs within the PCAF histone acetylase complex. *Cell* **94,** 35–44.

Okada, Y., Feng, Q., Lin, Y., Jiang, Q., Li, Y., Coffield, V. M., Su, L., Xu, G., and Zhang, Y. (2005). hDOT1L links histone methylation to leukemogenesis. *Cell* **121,** 167–178.

Olivier, E. N., Qiu, C., Velho, M., Hirsch, R. E., and Bouhassira, E. E. (2006). Large-scale production of embryonic red blood cells from human embryonic stem cells. *Exp. Hematol.* **34,** 1635–1642.

Orkin, S. H., and Zon, L. I. (1997). Genetics of erythropoiesis: Induced mutations in mice and zebrafish. *Ann. Rev. Genet.* **31,** 33–60.

Orlando, V., Strutt, H., and Paro, R. (1997). Analysis of chromatin structure by *in vivo* formaldehyde cross-linking. *Methods* **11,** 205–214.

Pal, S., Cantor, A. B., Johnson, K. D., Moran, T., Boyer, M. E., Orkin, S. H., and Bresnick, E. H. (2004). Coregulator-dependent facilitation of chromatin occupancy by GATA-1. *Proc. Natl. Acad. Sci. USA* **101,** 980–985.

Partington, G. A., and Patient, R. K. (1999). Phosphorylation of GATA-1 increases its DNA-binding affinity and is correlated with induction of human K562 erythroleukaemia cells. *Nucleic Acids Res.* **27,** 1168–1175.

Perissi, V., Aggarwal, A., Glass, C. K., Rose, D. W., and Rosenfeld, M. G. (2004). A corepressor/coactivator exchange complex required for transcriptional activation by nuclear receptors and other regulated transcription factors. *Cell* **116,** 511–526.

Perkins, A. C., Sharpe, A. H., and Orkin, S. H. (1995). Lethal beta-thalassaemia in mice lacking the erythroid CACCC- transcription factor EKLF. *Nature* **375,** 318–322.

Pevny, L., Lin, C. S., D'Agati, V., Simon, M. C., Orkin, S. H., and Costantini, F. (1995). Development of hematopoietic cells lacking transcription factor GATA-1. *Development* **121,** 163–172.

Ragoczy, T., Telling, A., Sawado, T., Groudine, M., and Kosak, S. T. (2003). A genetic analysis of chromosome territory looping: Diverse roles for distal regulatory elements. *Chromosome Res.* **11,** 513–525.

Ragoczy, T., Bender, M. A., Telling, A., Byron, R., and Groudine, M. (2006). The locus control region is required for association of the murine beta-globin locus with engaged transcription factories during erythroid maturation. *Genes Dev.* **20,** 1447–1457.

Reik, A., Telling, A., Zitnik, G., Cimbora, D., Epner, E., and Groudine, M. (1998). The locus control region is necessary for gene expression in the human beta-globin locus but not the maintenance of an open chromatin structure in erythroid cells. *Mol. Cell. Biol.* **18,** 5992–6000.

Richmond, T. D., Chohan, M., and Barber, D. L. (2005). Turning cells red: Signal transduction mediated by erythropoietin. *Trends Cell Biol.* **15,** 146–155.

Rodriquez, P., Bonte, E., Krijgsveld, J., Kolodziej, K. E., Guyot, B., Heck, A. J., Vyas, P., de Boer, E., Grosveld, F., and Strouboulis, J. (2005). GATA-1 forms distinct activating and repressive complexes in erythroid cells. *EMBO J.* **24,** 2354–2366.

Rooke, H. M., and Orkin, S. H. (2006). Phosphorylation of Gata1 at serine residues 72, 142, and 310 is not essential for hematopoiesis *in vivo*. *Blood* **107,** 3527–3530.

Saha, A., Wittmeyer, J., and Cairns, B. R. (2006). Chromatin remodeling: The industrial revolution of DNA around histones. *Nat. Rev. Mol. Cell. Biol.* **7,** 437–447.

Sawado, T., Igarashi, K., and Groudine, M. (2001). Activation of beta-major globin gene transcription is associated with recruitment of NF-E2 to the beta-globin LCR and gene promoter. *Proc. Natl. Acad. Sci. USA* **98,** 10226–10231.

Sawado, T., Halow, J., Bender, M. A., and Groudine, M. (2003). The beta-globin locus control region (LCR) functions primarily by enhancing the transition from transcription initiation to elongation. *Genes Dev.* **17,** 1009–1118.

Schubeler, D., Groudine, M., and Bender, M. A. (2001). The murine beta-globin locus control region regulates the rate of transcription but not the hyperacetylation of histones at the active genes. *Proc. Natl. Acad. Sci. USA* **98,** 11432–11437.

Shivdasani, R. A., and Orkin, S. H. (1995). Erythropoiesis and globin gene expression in mice lacking the transcription factor NF-E2. *Proc. Natl. Acad. Sci. USA* **92,** 8690–8694.

Shivdasani, R. A., Rosenblatt, M. F., Zucker-Franklin, D., Jackson, C. W., Hunt, P., Saris, C. J., and Orkin, S. H. (1995). Transcription factor NF-E2 is required for platelet formation independent of the actions of thrombopoietin/MGDF in megakaryocyte development. *Cell* **81,** 695–704.

Stamatoyannopoulos, G. (1991). Human hemoglobin switching. *Science* **252,** 383.

Strahl, B. D., and Allis, C. D. (2000). The language of covalent histone modifications. *Nature* **403,** 41–45.

Talbot, D., and Grosveld, F. (1991). The 5′HΣ2 of the globin locus control region enhances transcription through the interaction of a multimeric complex binding at two functionally distinct NF-E2 binding sites. *EMBO J.* **10,** 1391–1398.

Tewari, R., Gillemans, N., Wijgerde, M., Nuez, B., von Lindern, M., Grosveld, F., and Philipsen, S. (1998). Erythroid Kruppel-like factor (EKLF) is active in primitive and definitive erythroid cells and is required for the function of 5′HΣ3 of the beta-globin locus control region. *EMBO J.* **17,** 2334–2341.

Tolhuis, B., Palstra, R. J., Splinter, E., Grosveld, F., and de Laat, W. (2002). Looping and interaction between hypersensitive sites in the active beta-globin locus. *Mol. Cell* **10,** 1453–1475.

Towatari, M., May, G. E., Marais, R., Perkins, G. R., Marshall, C. J., Cowley, S., and Enver, T. (1995). Regulation of GATA-2 phosphorylation by mitogen-activated protein kinase and interleukin-3. *J. Biol. Chem.* **270,** 4101–4107.

Towatari, M., Ciro, M., Ottolenghi, S., Tsuzuki, S., and Enver, T. (2004). Involvement of mitogen-activated protein kinase in the cytokine-regulated phosphorylation of transcription factor GATA-1. *Hematol. J.* **5,** 262–272.

Tsai, F.-Y., and Orkin, S. H. (1997). Transcription factor GATA-2 is required for proliferation/survival of early hematopoietic cells and mast cell formation, but not for erythroid and myeloid terminal differentiation. *Blood* **89,** 3636–3643.

Tsai, F. Y., Keller, G., Kuo, F. C., Weiss, M., Chen, J., Rosenblatt, M., Alt, F. W., and Orkin, S. H. (1994). An early haematopoietic defect in mice lacking the transcription factor GATA-2. *Nature* **371,** 221–226.

Tsai, S. F., Martin, D. I., Zon, L. I., D'Andrea, A. D., Wong, G. G., and Orkin, S. H. (1989). Cloning of cDNA for the major DNA-binding protein of the erythroid lineage through expression in mammalian cells. *Nature* **339,** 446–451.

Tsang, A. P., Visvader, J. E., Turner, C. A., Fujuwara, Y., Yu, C., Weiss, M. J., Crossley, M., and Orkin, S. H. (1997). FOG, a multitype zinc finger protein as a cofactor for transcription factor GATA-1 in erythroid and megakaryocytic differentiation. *Cell* **90,** 109–119.

Tsang, A. P., Fujiwara, Y., Hom, D. B., and Orkin, S. H. (1998). Failure of megakaryopoiesis and arrested erythropoiesis in mice lacking the GATA-1 transcriptional cofactor FOG. *Genes Dev.* **12,** 1176–1188.

Tse, C., Sera, T., Wolffe, A. P., and Hansen, J. C. (1998). Disruption of higher-order folding by core histone acetylation dramatically enhances transcription of nucleosomal arrays by RNA polymerase III. *Mol. Cell. Biol.* **18,** 4629–4638.

Tuan, D., and London, I. M. (1984). Mapping of DNase I-hypersensitive sites in the upstream DNA of human embryonic epsilon-globin gene in K562 leukemia cells. *Proc. Natl. Acad. Sci. USA* **81,** 2718–2722.

Uddin, S., Ah-Kang, J., Ulaszek, J., Mahmud, D., and Wickrema, A. (2004). Differentiation stage-specific activation of p38 mitogen-activated protein kinase isoforms in primary human erythroid cells. *Proc. Natl. Acad. Sci. USA* **101,** 147–152.

Vakoc, C. R., Letting, D. L., Gheldof, N., Sawado, T., Bender, M. A., Groudine, M., Weiss, M. J., Dekker, J., and Blobel, G. A. (2005a). Proximity among distant regulatory elements at the beta globin locus requires GATA-1 and FOG-1. *Mol. Cell* **17,** 453–462.

Vakoc, C. R., Mandat, S. A., Olenchock, B. A., and Blobel, G. A. (2005b). Histone H3 lysine 9 methylation and HP1gamma are associated with transcriptional elongation through mammalian chromatin. *Mol. Cell* **19,** 381–391.

Vakoc, C. R., Sachdeva, M. M., Wang, H., and Blobel, G. A. (2006). Profile of histone lysine methylation across transcribed mammalian chromatin. *Mol. Cell. Biol.* **26,** 9185–9195.

van Holde, K. (1989). "Chromatin." Springer-Verlag, New York.

van Leeuwen, F., Gafken, P. R., and Gottschling, D. E. (2002). Dot1p modulates silencing in yeast by methylation of the nucleosome core. *Cell* **109,** 745–756.

Verdel, A., Jia, S., Gerber, S., Sugiyama, T., Gygi, S., Grewal, S. I., and Moazed, D. (2004). RNAi-mediated targeting of heterochromatin by the RITS complex. *Science* **303,** 672–676.

Vettese-Dadey, M., Grant, P. A., Hebbes, T. R., Crane- Robinson, C., Allis, C. D., and Workman, J. L. (1996). Acetylation of histone H4 plays a primary role in enhancing transcription factor binding to nucleosomal DNA *in vitro*. *EMBO J.* **15,** 2508–2518.

Wai, A. W., Gillemans, N., Raguz-Bolognesi, S., Pruzina, S., Zafarana, G., Meijer, D., Philipsen, S., and Grosveld, F. (2003). HS5 of the human beta-globin locus control region: A developmental stage-specific border in erythroid cells. *EMBO J.* **22,** 4489–4500.

Weiss, M. J., Keller, G., and Orkin, S. H. (1994). Novel insights into erythroid development revealed through *in vitro* differentiation of GATA-1 embryonic stem cells. *Genes Dev.* **8,** 1184–1197.

Weiss, M. J., Yu, C., and Orkin, S. H. (1997). Erythroid-cell-specific properties of transcription factor GATA-1 revealed by phenotypic rescue of a gene-targeted cell line. *Mol. Cell. Biol.* **17,** 1642–1651.

Welch, J. J., Watts, J. A., Vakoc, C. R., Yao, Y., Wang, H., Hardison, R. C., Blobel, G. A., Chodosh, L. A., and Weiss, M. J. (2004). Global regulation of erythroid gene expression by transcription factor GATA-1. *Blood* **104,** 3136–3147.

Whitelaw, E., Tsai, S. F., Hogben, P., and Orkin, S. H. (1990). Regulated expression of globin chains and the erythroid transcription factor GATA-1 during erythropoiesis in the developing mouse. *Mol. Cell. Biol.* **10,** 6596–6606.

Wojda, U., Noel, P., and Miller, J. L. (2002). Fetal and adult hemoglobin production during adult erythropoiesis: Coordinate expression correlates with cell proliferation. *Blood* **99,** 3005–3013.

Wolffe, A. (1998). "Chromatin: Structure and Function," 3rd Ed. Academic Press, San Diego.

Wozniak, R. J., Boyer, M. E., Grass, J. A., Lee, Y.-S., and Bresnick, E. H. (2007). Context-dependent GATA factor function: Combinatorial requirements for transcriptional control in hematopoietic and endothelial cells. *J. Biol. Chem.* **282,** 14665–14674.

Xiang, P., Fang, X., Yin, W., Barkess, G., and Li, Q. (2006). Non-coding transcripts far upstream of the epsilon-globin gene are distinctly expressed in human primary tissues and erythroleukemia cell lines. *Biochem. Biophys. Res. Commun.* **344,** 623–630.

Yagi, R., Chen, L. F., Shigesada, K., Murakami, Y., and Ito, Y. (1999). A WW domain-containing yes-associated protein (YAP) is a novel transcriptional coactivator. *EMBO J.* **18,** 2551–2562.

Yang, X. J., Ogryzko, V. V., Nishikawa, J., Howard, B. H., and Nakatani, Y. (1996). A p300/CBP-associated factor that competes with the adenoviral oncoprotein E1A. *Nature* **382,** 319–324.

Zhang, W., Kadam, S., Emerson, B. M., and Bieker, J. J. (2001). Site-specific acetylation by p300 or CREB binding protein regulates erythroid Kruppel-like factor transcriptional activity via its interaction with the SWI-SNF complex. *Mol. Cell. Biol.* **21,** 2413–2422.

Zhao, W., Kitidis, C., Fleming, M. D., Lodish, H. F., and Ghaffari, S. (2006). Erythropoietin stimulates phosphorylation and activation of GATA-1 via the PI3-kinase/AKT signaling pathway. *Blood* **107,** 907–915.

Zhou, D., Pawlik, K. M., Ren, J., Sun, C. W., and Townes, T. M. (2006). Differential binding of erythroid Krupple-like factor to embryonic/fetal globin gene promoters during development. *J. Biol. Chem.* **281,** 16052–16057.

CHAPTER FOUR

The Role of the Epigenetic Signal, DNA Methylation, in Gene Regulation During Erythroid Development

Gordon D. Ginder,* Merlin N. Gnanapragasam,* *and* Omar Y. Mian*

Contents

* Department of Internal Medicine, Massey Cancer Center, Virginia Commonwealth University, Richmond, Virginia 23298

Current Topics in Developmental Biology, Volume 82
ISSN 0070-2153, DOI: 10.1016/S0070-2153(07)00004-X

Abstract

The sequence complexity of the known vertebrate genomes alone is insufficient to account for the diversity between individuals of a species. Although our knowledge of vertebrate biology has evolved substantially with the growing compilation of sequenced genomes, understanding the temporal and spatial regulation of genes remains fundamental to fully exploiting this information. The importance of epigenetic factors in gene regulation was first hypothesized decades ago when biologists posited that methylation of DNA could heritably alter gene expression [Holliday and Pugh, 1975. *Science* **187**(4173), 226–232; Riggs, 1975. *Cytogenet. and Cell Genet.* **14**(1), 9–25; Scarano *et al.*, 1967. *Proc. Natl. Acad. Sci. USA* **57**(5), 1394–1400)]. It was subsequently shown that vertebrate DNA methylation, almost exclusively at the 5′ position of cytosine in the dinucleotide CpG, played a role in a number of processes including embryonic development, genetic imprinting, cell differentiation, and tumorigenesis. At the time of this writing, a large and growing list of genes is known to exhibit DNA methylation-dependent regulation, and we understand in some detail the mechanisms employed by cells in using methylation as a regulatory modality. In this context, we revisit one of the original systems in which the role of DNA methylation in vertebrate gene regulation during development was described and studied: erythroid cells. We briefly review the recent advances in our understanding of DNA methylation and, in particular, its regulatory role in red blood cells during differentiation and development. We also address DNA methylation as a component of erythroid chromatin architecture, and the interdependence of CpG methylation and histone modification.

1. Introduction

1.1. DNA methylation as an epigenetic signal

As the sequences of the vertebrate genomes have become available and many of the sequence-specific binding proteins directly involved in the regulation of transcription have been characterized, it has become increasingly clear that another level of control beyond DNA base sequence exists. The broad term "epigenetics" has been used to describe heritable regulatory signals not directly encoded in DNA sequence. A functional definition of an epigenetic regulatory signal is one that can be passed on during somatic or germ line cell replication, but that can also be altered without a change in DNA sequence.

DNA methylation was one of the earliest epigenetic signals to be identified. Methylation in vertebrates appears to occur only at position 5 in the cytosine ring, and almost exclusively in the dinucleotide CpG. Histone postsynthetic modification is another well-characterized epigenetic signaling system. Because the role of the latter in erythroid development is covered elsewhere in this volume (Chapter 3 by Wozniak and Bresnick), the present chapter will be restricted primarily to DNA methylation and its role in erythroid cell development in vertebrates.

1.2. The mechanism of DNA methylation-mediated control of gene expression

DNA methylation has long been posited as a potential regulator of gene expression (Holliday and Pugh, 1975; Riggs, 1975; Scarano *et al.*, 1967). A direct inverse relationship between DNA methylation and vertebrate gene expression was first described in the case of the globin genes (Kuo *et al.*, 1979; McGhee and Ginder, 1979; Razin and Riggs, 1980; Shen and Maniatis, 1980; van der Ploeg and Flavell, 1980). Despite the many examples of such a close relationship between DNA methylation and gene transcription, there are only a limited number of well-characterized examples in which a regulatory role in developmental gene expression has been firmly established. However, DNA methylation clearly has been shown to be required for normal embryonic development on a global level (Li *et al.*, 1992; Okano *et al.*, 1999). DNA methylation also has been shown to be involved in aberrant tumor suppressor gene silencing in cancer (Baylin, 2005; Jones and Baylin, 2002), and may provide a target for molecular approaches to treating cancer (Baylin, 2005; Egger *et al.*, 2004). The finding that DNA methylation plays a key role in silencing of retroposons and repetitive DNA sequences has led some to suggest that this is the major role of DNA methylation in normal cells *in vivo* (Walsh and Bestor, 1999; Warnecke and Bestor, 2000). Nonetheless, a great deal has been learned recently about the mechanisms through which DNA methylation affects normal gene expression. These topics have been extensively reviewed recently, so they will be discussed only briefly here.

DNA methylation is felt to act primarily as an inhibitor of transcription. This effect occurs through two main mechanisms. The first, and by far the less common, is through direct interference with binding of transcription factors to the promoter region of a gene (Tate and Bird, 1993). Much more commonly, the repressive effect of DNA methylation is mediated indirectly through methyl cytosine-binding proteins (MCBPs) and their associated corepressor factors and complexes, which interpret the 5′-methyl cytosine signal. These complexes, in turn, may act through enzymatic modification of adjacent histones, chromatin remodeling proteins, or direct inhibition of transcription initiation (Fig. 4.1).

The first MCBPs were identified and initially characterized over 15 years ago (Boyes and Bird, 1991; Meehan *et al.*, 1989). Two major classes of these proteins have been characterized. The first class consists of the methyl-binding domain proteins comprising a family of proteins that share a highly conserved methyl cytosine-binding domain (MBD) (Hendrich and Bird,

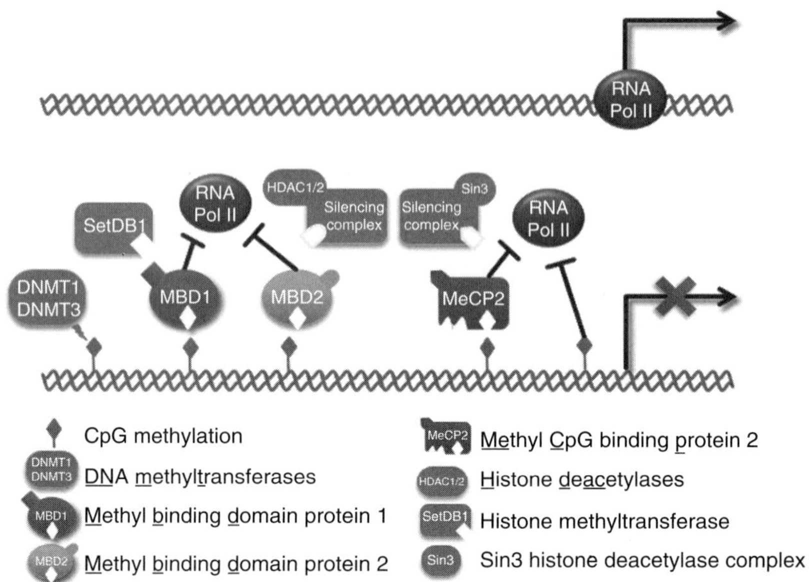

Figure 4.1 The protein mediators of DNA methylation. The top strand in the figure illustrates a transcriptionally accessible region of unmethylated DNA. The bottom strand is methylated at several CpG sites by the action of maintenance and *de novo* methyl transferases (DNMT1, DNMT3a, and DNMT3b). DNA methylation results in transcriptional repression by three principle mechanisms: (1) direct inhibition of protein binding, (2) recruitment of methyl-binding domain proteins (MBD1, MBD2, MeCP2) that block transcription, or (3) by the action of corepressor complexes (SetDB1, HDAC, Sin3) that associate with methyl-binding proteins and modify the chromatin environment (histone proteins not shown).

1998; Hendrich and Tweedie, 2003). The proteins of this family, MBD1–4 and MeCP2, have all been shown to be capable of mediating transcriptional silencing by binding to methylated CpGs. MBD4, however, appears to act predominantly as a glycosylase, which is active in DNA mismatch repair (Bellacosa *et al.*, 1999; Hendrich *et al.*, 1999; Kondo *et al.*, 2005).

Although both the mechanisms of binding of individual MBD proteins to methylated cytosine and the mechanisms of transcriptional repression vary, the common theme is that each is associated with one or more corepressor complexes. MBD2 has been shown to be associated with a complex known as MeCP1 which includes the NURD corepressor complex. It contains histone deacetylases (HDAC) 1 and 2 (Feng and Zhang, 2001; Nan *et al.*, 1998; Ng *et al.*, 1999). MBD2 binding requires multiple methylated cytosines within 150–200 bp of DNA. Although repression appears to be mediated by HDAC enzymatic activity in some cases (Feng and Zhang, 2003; Ng *et al.*, 1999), in others it is based on transcriptional repressor activity of other components of the NURD complex, independent of histone deacetylase activity (Singal *et al.*, 2002a; Wade, 2001). Moreover, NURD is known to be a somewhat heterogeneous complex (Bowen *et al.*, 2004), and in the case of the avian primary erythroid cell MeCP1-like complex, it has been shown to exhibit some degree of sequence preference compared to the MeCP1 originally isolated from cultured human tumor cells (Feng and Zhang, 2001; Singal *et al.*, 2002b).

MeCP2 was originally shown to be associated with the Sin3A corepressor complex (Nan *et al.*, 1998), and subsequently, evidence from some cell types suggested that it is associated with the BRM1 chromatin remodeling complex (Harikrishnan *et al.*, 2005), although the latter is somewhat controversial (Hu *et al.*, 2006). Although MeCP2 has been shown to be capable of binding to a sequence containing a single methylated CpG dinucleotide, all of the examples in which it has been shown to bind *in vivo* and mediate repression of transcription involve sequences containing multiple methylated CpGs.

MBD1 contains a transcriptional repressor domain (Fujita *et al.*, 2000) and is associated with a multiprotein complex containing chromatin-associated factor 1 (Reese *et al.*, 2003) and the histone lysine methylase, SETB1 (Sarraf and Stancheva, 2004). MBD3 has been shown to bind preferentially to methylated DNA in *Xenopus* but not in birds or mammals (Ballestar and Wolffe, 2001; Hendrich *et al.*, 2001). MBD3 has frequently been identified as a component of MBD2-associated NURD complexes, but is not present in the avian primary erythroid MeCP1 complex (Kransdorf *et al.*, 2006). Recent studies have suggested that the presence of MBD2 and MBD3 in NURD complexes in some cell types is mutually exclusive (LeGuezennec *et al.*, 2006). Interestingly, two isoforms of MBD3 have been described in NURD complexes. One of these, MBD3L1, appears to have repressor activity, while the other, MBD3L2, opposes MeCP1-mediated silencing (Jiang *et al.*, 2004; Jin *et al.*, 2005).

The emerging picture of transcriptional repression by MBD proteins is one of significant variations in the associated corepressor complexes, even for a specific MBD (Fig. 4.1). While some of this apparent variability may be due to the limitations of biochemical purification techniques, it is likely that much of it is real. This possibility of variability provides for a degree of tissue and gene specificity of repressor activity that far exceeds the sequence specificity of DNA binding of individual MBD proteins.

Another well-characterized group of MCBPs is the Kaiso family. These proteins bind to DNA containing methylated cytosines through a BTB/Poz domain zinc finger motif, which has independent binding sequence specificity (Daniel and Reynolds, 1999; Prokhortchouk *et al.*, 2001). Knockout of the Kaiso gene in *Xenopus* is embryonic lethal, but in mice the only detectable phenotype is resistance to intestinal cancer (Prokhortchouk *et al.*,

2006; Ruzov *et al.*, 2004). However, these proteins have been implicated in the sequence-specific repression of several genes (Filion *et al.*, 2006).

Much of the information about the mechanisms of DNA methylation-mediated repression of vertebrate gene transcription has been obtained from tumor cells and cultured cell lines, but methylation-mediated regulation in several normal tissue models also has been studied (Ehrlich, 2005; Walsh and Bestor, 1999). Of the limited number of systems in which these mechanisms have been studied in normal tissues, the hematopoietic system, including the erythroid compartment, is among the best characterized (Burns *et al.*, 1988; Ginder *et al.*, 1984; Hutchins *et al.*, 2002, 2005; Kransdorf *et al.*, 2006; Singal *et al.*, 1997, 2002b).

1.3. Acquisition and loss of DNA methylation

DNA methylation is carried out by two major classes of cytosine methyl transferases, or DNA methyl transferase (DNMT), enzymes. The first group, which includes DNMT3a and DNMT3b, carry out *de novo* methylation on completely unmethylated substrate DNA. The second group, DNMT1, copies methylation onto the newly synthesized DNA strand at replication, thus passing on the methylation pattern to daughter cells. The details of how methylation patterns are established have been recently reviewed (Klose and Bird, 2006). Briefly, there are three major possible mechanisms that have been identified. The first is through direct recognition of DNA sequence by the methyl transferase, and some evidence exist for this mechanism in the case of the PWWP domain of DNMT3b (Ge *et al.*, 2004; Qiu *et al.*, 2002). A second mechanism studied is recruitment of DNMTs by transcriptional repressors or postsynthetic modifications of chromatin proteins. Both the PML-RAR and c-MYC proteins have been shown to interact with *de novo* DNMTs (Brenner *et al.*, 2005; Di Croce *et al.*, 2002). Recently, it has been shown that histone lysine 79 methylation is a mark in embryonic stem cells that appears to recruit *de novo* methyl transferases to embryonic genes that are subsequently silenced by the polycomb protein complex in tumor cells, although the specific role that this mechanism plays in DNA methylation-mediated gene regulation during normal development has yet to be defined (Ohm *et al.*, 2007; Schlesinger *et al.*, 2007). A third mechanism, first described in plants, involves RNAi-mediated transcriptional repression-dependent DNA methylation (Matzke and Birchler, 2005). A similar relationship has been described in some mammalian cells (Kanellopoulou *et al.*, 2005; Morris *et al.*, 2004), but not in others (Murchison *et al.*, 2005; Svoboda *et al.*, 2004; Ting *et al.*, 2005). Thus, the precise role of RNAi and related microRNAs in establishing normal tissue-specific or developmentally specific DNA methylation patterns in animals in general, and in erythroid cells in particular, remains to be elucidated.

An important facet of DNA methylation in the regulation of gene expression is the process of demethylation. The mechanism by which methylation of cytosines is removed from DNA remains a topic of both active investigation and some controversy. Considerable evidence supports the role of passive loss of methylation during embryonic development at the time of replication through a competition between transcriptional machinery and DNMTs (Reik and Walter, 2001; Santos-Rosa *et al.*, 2002), but conclusive direct evidence for this mechanism is sparse. Progressive "demethylation" of the γ-globin promoter during erythroid differentiation has recently been described in baboon fetal liver and adult bone marrow (Singh *et al.*, 2007), as well as in human CD34+ hematopoetic stem cells (Mabaera *et al.*, 2007). This change in methylation correlates with a transient period of increased γ expression; however, these studies must be interpreted with caution. The nature of the proposed demethylation is unclear (i.e., primary or secondary, passive or active), and the precise cause–effect relationship between demethylation and expression remains to be established.

A number of investigators have reported that rapid demethylation can occur during development and in some tissues in the absence of DNA replication (Kafri *et al.*, 1992; Monk *et al.*, 1987). These results argue for an active demethylase enzyme, but the details of this enzymatic process remain to be fully characterized. One type of DNA demethylase that has been well characterized is a DNA glycosylase repair enzyme, first reported in chicken embryos and subsequently in mouse myoblasts (Jost, 1993; Jost *et al.*, 1995, 2001; Swisher *et al.*, 1998). A truncated form of MBD2, MBD2b, has been reported to catalyze direct removal of methyl groups from the 5-position on the cytosine ring in DNA (Bhattacharya *et al.*, 1999), but other investigators have not been able to confirm this activity (Ng *et al.*, 1999, Wolffe *et al.*, 1999). Recently, a stage-specific demethylase activity derived from definitive chicken erythroid cells was shown to target both hemimethylated and fully methylated DNA substrates *in vitro* (Ramachandran *et al.*, 2007). Another recent report describes a mediator of active DNA demethylation that operates in the absence of maintenance of DNMT (Barreto *et al.*, 2007). The nuclear protein Gadd45a, known to be involved in DNA repair and maintenance of genome stability, targets methylated regions and recruits excision repair enzymes that remove 5-methyl cytosine. Moreover, site-specific demethylation following recruitment of Gadd45a appears sufficient to relieve transcriptional repression. This unexpected observation introduces another potentially critical player in determining DNA methylation states. Taken together, the available evidence support both passive and active demethylation processes. The latter process appears to be mediated predominantly by enzymes that directly excise 5-methyl cytosine from DNA.

1.4. The interplay between DNA methylation and histone modification in controlling gene expression

It has become increasingly clear that the relationships between DNA methylation and histone modifications that affect transcription are quite complex. Studies in *Arabidopsis thaliana* and *Neurospora* demonstrated that disruption of the repressive modification, lysine K9 methylation of histone H3, caused global loss of cytosine methylation (Jackson *et al.*, 2002; Tamaru *et al.*, 2003). Conversely, several studies in mammals have shown that DNA methylation recruits methylated CpG-binding complexes that contain enzymes capable of catalyzing repressive methylation of histones. For example, the complex recruited to methyl cytosines at DNA replication by MBD1 includes SETDB1 that catalyzes H3 K9 methylation (Sarraf and Stancheva, 2004), and MeCP2 has also been shown to mediate H3 K9 methylation (Fuks *et al.*, 2003). Similarly, DNA methylation was found to control histone H3 K9 methylation in *Arabidopsis* (Fig. 4.2; Soppe *et al.*, 2002).

A number of studies have also addressed the relationships between DNA methylation and active histone epigenetic marks such as histones H3 and H4 lysine 9 acetylation and histone H3 lysine 4 methylation. Studies using

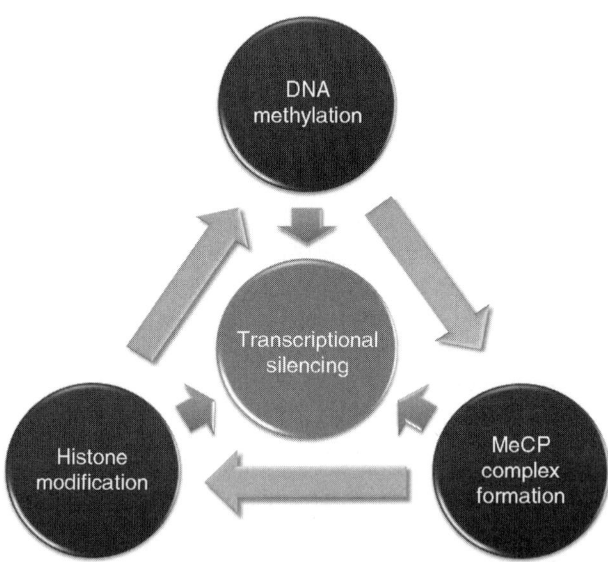

Figure 4.2 A dynamic cycle of epigenetic gene silencing. Emerging models suggest that DNA methylation and postsynthetic histone modification are dependent on one another and reinforce each other. Through recruitment of enzymes and corepressor complexes, either mark appears capable of recapitulating the other, initiating a cycle that reinforces gene silencing.

stably transfected reporter gene constructs containing the chicken β-globin locus chromatin insulator element showed that the insulator was capable of both maintaining histone acetylation and preventing extinction of transcription and subsequent DNA methylation (Mutskov *et al.*, 2002; Pikaart *et al.*, 1998). Similar conclusions about the role of histone H3/H4 acetylation in protection from DNA methylation have been presented based on studies of the erythroid-specific carbonic anhydrase II gene (Brinkman *et al.*, 2007). Conversely, recent evidence from studies of episomes in mammalian cells suggest that DNA methylation prevents the activating dimethylation modification of histone H3 K4 in nucleosomes adjacent to the DNA methylation mark, regardless of transcriptional activity in the local chromatin environment (Okitsu and Hsieh, 2007). This suggests that DNA methylation can act in a dominant fashion to extinguish activating modifications of histones, as well as to recruit enzymes that catalyze repressive modifications.

The picture that is emerging is one of a dynamic interplay between the epigenetic signals of DNA methylation and histone postsynthetic modifications (Johnson *et al.*, 2007). As depicted in Fig. 4.2, there appears to be a positive reinforcing relationship between DNA methylation and repressive histone modifications such as histone H3 K9 methylation. Conversely, activating histone modifications, such as H3/H4 K9 acetylation, oppose the process of transcriptional silencing and DNA methylation. In the case of the chicken embryonic ρ-globin gene, evidence from both transfected primary erythroid cells (Singal *et al.*, 2002b) and *in vivo* treatment of adult chickens (Ginder *et al.*, 1984) suggest that blocking histone deacetylation cannot overcome the silencing effects of DNA methylation on embryonic ρ-globin gene transcription, which occur during erythroid development.

2. DNA Methylation in Erythroid Cell Differentiation

2.1. Direct effects of DNA methylation on erythroid genes

Hematopoietic differentiation is driven by competition between lineage-specific transcription factors, as reviewed by Cantor and Orkin (2001), as well as the coordinated appearance of inhibitory factors, which prune unnecessary gene expression (Hu *et al.*, 1997; Laslo *et al.*, 2006). This is substantiated by the intriguing observation that hematopoietic stem cells, rather than displaying conservative gene expression, promiscuously express genes from a variety of lineages at low levels, which must then be silenced appropriately as the cell differentiates (Iwasaki *et al.*, 2006; Laslo *et al.*, 2006). Studies have supported the concept that tissue-specific CpG methylation can maintain a "closed" chromatin conformation and lead to transcriptional silencing, (Jaenisch and Bird, 2003). It is not entirely clear, however, whether changes in CpG methylation always precede changes in expression or whether they are a consequence of promoter activity, that is, increased occupancy of the promoter region by *trans* activator factors and transcription machinery (Enver *et al.*, 1988). Tagoh *et al.* (2004) have reported lineage-specific CpG demethylation and chromatin remodeling in common myeloid precursor (CMP) cells that precede enhancer–promoter complex formation and mRNA synthesis of cognate genes. These findings, along with similar observations made by others (Okitsu and Hsieh, 2007), support the notion that altered DNA methylation status may, at least in some instances, be an early step in regulating lineage-specific genes and may be an early determinant of cell fate in primitive erythroid precursors. As nature is parsimonious, it is likely that other genes involved in the differentiation of erythroid cells are regulated in a similar manner. For the present, the evidence in many cases supports only the coexistence of gene silencing and CpG methylation, and often only *ex vivo*. Even so, the relationship between CpG methylation and gene expression during hematopoietic differentiation is a potentially important one. This is particularly true for maintaining nonerythroid genes silent during maturation of erythroid cells and vice versa. The ETS transcription factor PU.1, for example, is expressed in a lineage-specific fashion when demethylated at key CpG dinucleotides (Amaravadi and Klemsz, 1999). Early during blood cell development, critical cytosine residues are hypomethylated, and PU.1 is expressed in CD34+ CD38− hematopoietic stem cells (Voso *et al.*, 1994). However, persistent expression of this protein prevents differentiation of erythroblasts into mature erythrocytes (Schuetze *et al.*, 1993). As such, in order for progenitor cells to undergo erythroid differentiation, PU.1 is selectively silenced and methylated. This silencing further promotes erythrocyte development by removing the antagonism of the critical erythroid transcription factor GATA1 by PU.1 (Zhang *et al.*, 2000).

2.2. Indirect effects via erythropoietin gene expression

DNA methylation also indirectly regulates erythroid differentiation and proliferation by mediating tissue-specific expression of erythropoietin (EPO) (Yin and Blanchard, 2000). Red blood cell differentiation and proliferation is largely directed by EPO, which is expressed predominantly in liver and kidney tissues in response to hypoxia. The relatively simple promoter structure of the EPO gene fails to explain the tissue specificity of its expression. A hypoxia-induced factor-1 (HIF1) response element confers upon EPO the ability to sense hypoxic stress; however, HIF1 is expressed in a large number of tissues in response to hypoxia, once again failing to account for the restricted expression of EPO. As shown by Yin and Blanchard (2000), the promoter and 5′ UTR elements of the EPO gene contain CpG-rich regions that are methylated in most tissues, leading to

MeCP binding and attendant corepressor complex formation. These sites are hypomethylated in tissues that express EPO, allowing the assembly of an HIF1-inducible active promoter complex.

3. The β-Globin Locus: A Model for the Role of DNA Methylation in Developmental Gene Regulation

The individual genes in the β-globin gene cluster have been shown to be developmentally regulated by a complex interplay between *cis* elements, *trans* factors, competition for an upstream enhancer/locus control region (LCR), and epigenetics. However, the exact mechanism of globin gene switching during development is still not completely understood. The vertebrate β-type globin genes are among a small group of genes whose normal developmental expression has been shown to depend on DNA methylation. The initial correlations between DNA methylation and gene expression arose from studies of the β-globin gene clusters of the chicken, of the rabbit, and in humans. Site-specific cytosine methylation within or adjacent to these genes was found to correlate with transcriptional repression (Ginder and McGhee, 1981; McGhee and Ginder, 1979; Razin and Riggs, 1980; Shen and Maniatis, 1980; van der Ploeg and Flavell, 1980). These β-type globin genes were also the first group of genes for which the treatment with 5-azacytidine, an irreversible inhibitor of DNMT, was shown to alter developmentally established patterns of expression by increasing levels of the fetal γ-globin gene RNA. This was first demonstrated in baboons and subsequently in patients with β-thalassemia and sickle-cell anemia (Charache *et al.*, 1983; DeSimone *et al.*, 1982; Ley *et al.*, 1982).

Although globin gene expression correlates inversely with DNA methylation, debate still exists as to whether DNA methylation is a primary mechanism involved in initiating globin gene switching (Enver *et al.*, 1988). There is, however, a general consensus that methylation can serve as a lock-off mechanism that may follow other events that initiate developmental globin gene repression. It has become clear that, once in place, DNA methylation can prevent transcription despite an optimum nuclear *trans* factor environment. Recent data from studies of human β-globin YAC transgenic mice lacking the methyl cytosine-binding protein, MBD2, revealed delayed silencing of the fetal γ-globin gene during embryonic erythroid development (Fig. 4.3), suggesting a possible role for DNA methylation in primary developmental globin gene silencing (Rupon *et al.*, 2006). Despite the absence of CpG islands in the mammalian locus, recent studies have demonstrated developmental stage-specific methylation

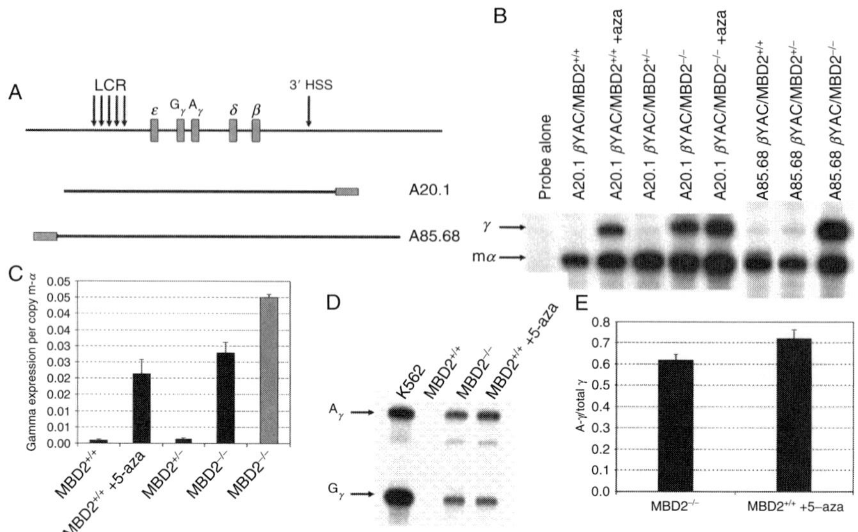

Figure 4.3 Transcription of the human γ-globin gene in wild-type and MBD2$^{-/-}$ β-YAC transgenic mice. (A) Structure of the human ρ-globin locus and construct boundaries of two independent β-YAC transgenic mouse lines, A20.1 and A85.68. (B) RNase protection assay demonstrating marked increase in γ-globin mRNA in MBD2$^{-/-}$ β-YAC compared to wild-type and MBD2$^{+/-}$ heterozygous β-YAC mouse adult erythroid cells (from Rupon *et al.*, 2006). (C) Phosphoimager quantitation of RNase protection assay of γ-globin mRNA in adult erythroid cells of control and 5-azacytidine-treated wild-type and either MBD2$^{+/-}$ or MBD2$^{-/-}$ β-YAC transgenic mice. Black bars represent the A20.1 line and gray bars represent the A85.68 line. (D) RNase protection assay showing the levels of $^{A}\gamma$- and $^{G}\gamma$-globin gene expression in untreated or 5-azacytidine-treated wild-type or MBD2$^{-/-}$ adult β-YAC transgenic mice. (E) Quantitative representation of levels of $^{A}\gamma$-globin gene relative to total γ-globin mRNA expression.

within the embryonic regions of the mouse β-globin locus in a manner that suggests a regulatory role rather than a simple reflection of primary changes in expression (Hsu *et al.*, 2007).

3.1. DNA methylation in developmental regulation of avian β-type globin genes

Among vertebrates, the role of DNA methylation in the developmental regulation of β-globin gene switching has been extensively characterized for the chicken β-globin gene cluster ($5'$ ρ-β^{H}-β^{A}-ε-$3'$) on chromosome 7. Rho (ρ) and epsilon (ε) are the embryonic globin genes; beta H (β^{H}) and beta A (β^{A}) are the adult globin genes. The ρ-globin gene is expressed abundantly in primitive erythrocytes produced in the yolk sac from 36 h of embryogenesis to day 5 and is not expressed in definitive erythrocytes, which are produced beginning at day 5 (Chan *et al.*, 1974). The transcriptional silencing of the embryonic ρ-globin gene occurs concomitantly with activation of the adult β-globin gene on day 5 of embryonic development (Groudine *et al.*, 1981). A strong inverse correlation exists between site-specific DNA methylation and expression of the chicken β-type globin genes (Ginder and McGhee, 1981; McGhee and Ginder, 1979). Moreover, early studies showed that treatment with 5-azacytidine in anemic adult chickens resulted in the activation of ρ-globin gene transcription and not that of ε-globin, suggesting a specificity of action even within the embryonic β-type globin genes (Burns *et al.*, 1988; Ginder *et al.*, 1984).

The sequences immediately upstream and downstream of the transcription initiation site of the embryonic ρ-globin gene contain CpG-rich regions that meet the criteria for CpG islands. Work in our laboratory has established that every CpG site in the 235 bp promoter and 248 bp proximally transcribed region (exon 1 and intron1) is methylated in adult erythrocytes but unmethylated in primitive embryonic erythrocytes (Ginder and McGhee, 1981; Singal *et al.*, 1997, 2002b). Singal *et al.* showed that *de novo* methylation targets the CpG-dense proximally transcribed region in the ρ-globin gene in the coding strand, followed by spreading into the 3′ region and then into the promoter region during the switch from embryonic to definitive erythropoiesis (Singal and vanWert, 2001). Methylation of the template strand lags behind that of the coding strand, and complete methylation of both strands occurs only after complete transcriptional repression has occurred. These results support the concept that methylation acts as a lock-off mechanism for ρ-globin gene transcription during erythroid development rather than the primary initiating event.

Once completely methylated, the ρ-globin promoter becomes locked off and can only be fully activated after loss of methylation. Transcription of the ρ-globin gene and a concomitant loss of CpG methylation can be induced in anemic adult chickens by treatment with the DNA methylation inhibitor 5-azacytidine (Ginder *et al.*, 1984). The level of unmethylated ρ-globin gene transcription is increased five- to tenfold further by the histone deacetylase inhibitor, butyrate, but this agent alone has no effect on expression of the methylated ρ-globin gene. Transfection of primary avian erythroid cells with ρ-globin gene constructs methylated at the same CpGs that are methylated *in vivo* and in the presence or absence of the histone deacetylase inhibitor, trichostatin (TSA), recapitulated *in vitro* the results in adult chickens, and established the dominance of DNA methylation over histone acetylation in silencing transcription of this gene (Singal *et al.*, 2002b).

3.2. Erythroid cell-methylated cytosine-binding complex

Work in our laboratory has also shown that the methyl cytosine-binding protein complex (MeCPC), an erythroid cell-derived methyl cytosine-binding protein complex containing MBD2, forms *in vitro* on the methylated ρ-globin promoter and proximal transcribed region more avidly than HeLa cell-derived MeCP1 complex (Kransdorf *et al.*, 2006; Singal *et al.*, 2002b). This result indicates that MCBPs may form tissue-restricted complexes that are targeted to specific sets of genes. Such a mechanism would result in greater specificity of the transcriptional silencing mediated by DNA methylation. We recently reported the biochemical characterization of primary adult erythroid cell-derived MeCPC. The complex is similar to MeCP1 in that it contains six components of the MeCP1/NURD complex, but differs in that it does not contain MBD3 (Kransdorf *et al.*, 2006). The chicken homologue of MBD2 is a critical component of MeCPC. Chromatin immunoprecipitation assays showed that MBD2 is bound to the silenced ρ-globin gene in adult erythroid cells *in vivo*, but not in 5-day embryonic erythroid cells (Fig. 4.4). Conversely, MBD2 is bound to the adult β^A gene in primitive embryonic cells but not in adult erythroid cells.

The MBD2-containing MeCPC is present in primary adult mouse erythroid cells and in adult phenotype murine erythroleukemia (MEL) cells, demonstrating conservation across species (Kransdorf *et al.*, 2006). Studies in a stably transfected adult erythroid MEL cell model showed that MBD2 recruits the NURD complex to the silent ρ-globin gene, and that transcriptional derepression of the methylated ρ-globin gene and loss of association of NURD components occur when MBD2 function is knocked down by siRNA (Fig. 4.5). Chromatin immunoprecipitation studies in this stably transfected cell model showed that MBD2 was bound to the methylated ρ-globin gene and that MBD2 recruits components of the NURD corepressor complex.

MeCPC appears to utilize a histone deacetylase-independent transcription inhibitory mechanism, despite the presence of HDAC1 within the complex, because repression caused by methylation of the ρ-globin gene promoter and the proximal transcribed region is not relieved by trichostatin (TSA), a histone deacetylase inhibitor, in primary erythroid cell transfection assays (Singal *et al.*, 2002b).

A proposed model for regulation of the ρ-globin gene during erythroid development is illustrated in Fig. 4.6 and is described as follows (Kransdorf *et al.*, 2006). In day 4 embryonic erythrocytes, the presence of positively acting *trans* factors such as GATA1 drives high levels of transcription from the unmethylated ρ-globin gene. Transcription of the gene is already silenced in early (embryonic day 7) definitive erythrocytes before the methylation of the ρ-globin gene is complete. In adults, where there is, by then, complete methylation of the ρ-globin gene, MBD2 binds to the methylated gene and recruits MTA2 and the other components of the MeCPC to the gene (Kransdorf *et al.*, 2006). The complex maintains transcriptional inactivity either by remodeling local chromatin into a non-permissive configuration or by direct transcriptional inhibition (Fig. 4.6).

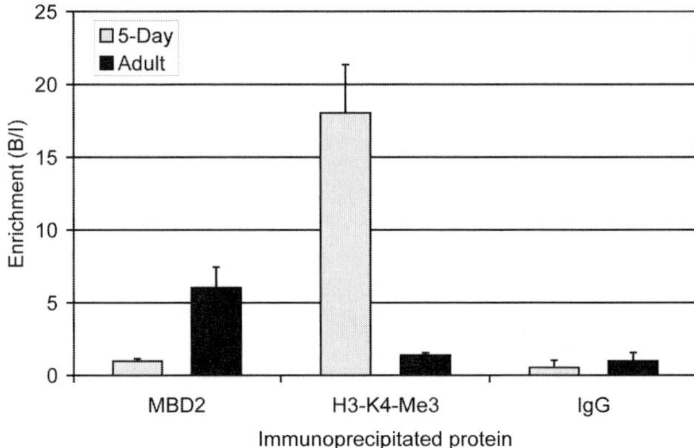

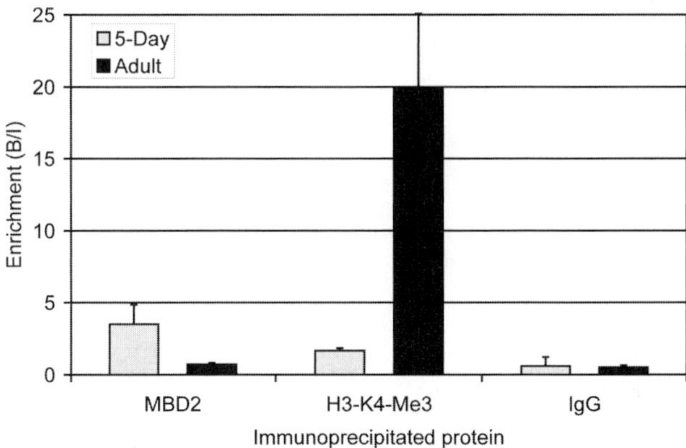

Figure 4.4 MBD2 occupancy inversely correlates with transcription and histone H3-lysine 4-trimethylation (H3-K4-Me_3) at the ρ- and β^A-globin genes. (A) Enrichment for MBD2, H3-K4-Me_3, and IgG at the ρ-globin gene in 5-day (light gray) and adult (black) erythrocytes as determined by ChIP assay. The data show that MBD2 is depleted from the transcriptionally active ρ-globin gene in 5-day erythrocytes, but enriched at the transcriptionally inactive and methylated ρ-globin gene in adult erythrocytes. In contrast, H3-K4-Me_3 is enriched at the transcriptionally active ρ-globin gene and depleted at the transcriptionally inactive gene. (B) Enrichment for MBD2, H3-K4-Me_3, and IgG at the β^A-globin gene in 5-day (light gray) and adult (black) erythrocytes as determined by ChIP assay. The data show that MBD2 is enriched at the transcriptionally inactive and methylated β^A-globin gene in 5-day erythrocytes, but depleted from the transcriptionally active β^A-globin gene in adult erythrocytes (from Kransdorf *et al.*, 2006).

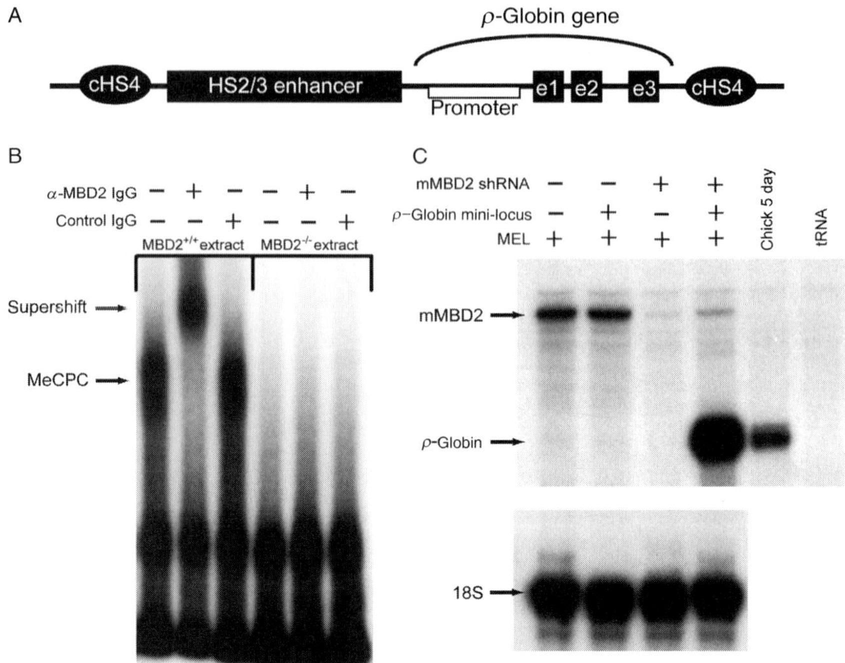

Figure 4.5 MBD2 is a critical component of the MeCPC in primary mouse splenocytes and MEL-ρ cells. (A) Graphic depiction of the ρ-globin mini-locus introduced into MEL cells. The locus contains: (1) a 4.5-kb ρ-globin genomic sequence, (2) a 4-kb chicken LCR enhancer element (HSS2 and HSS3), and (3) 5′ and 3′ cHS4 insulator elements that surround the gene and enhancer. A 2.5-kb fragment of the ρ-globin genomic sequence extending from 248 bp upstream to 2.2 kb downstream of the cap site was excised, *in vitro* methylated and religated prior to transfection into MEL cells to recapitulate methylation at the same sites as the endogenous gene in chicken adult erythroid cells. (B) EMSA performed with nuclear extract from primary mouse splenocytes. Extracts derived from spleens of $MBD2^{+/+}$ mice form the MeCPC that can be supershifted by the addition of anti-mMBD2. Corresponding extracts derived from $MBD2^{-/-}$ mice do not form any complex on the M-ρ248 probe. (C) RNase protection assay analyzing expression of ρ-globin, mMBD2, and 18S RNAs in MEL-ρ cells treated with shRNAs targeting mMBD2. Significant knockdown of mMBD2 expression is seen in MEL-ρ cells containing shRNA-expressing plasmids that target mMBD2, as compared with control cells (lanes 3 and 4 as compared with 1 and 2). No ρ-globin expression is seen in MEL-ρ cells with wild-type MBD2 expression (lane 2). In contrast, robust ρ-globin expression is seen in MEL-ρ cells in which mMBD2 expression has been knocked-down by shRNA (lane 4) (from Kransdorf *et al.*, 2006).

3.2.1. Primate β-type globin genes

A number of studies in other vertebrates support a role for DNA methylation in developmental β-type globin gene silencing. Primates, including humans, have been the most extensively studied. Baboons are considered

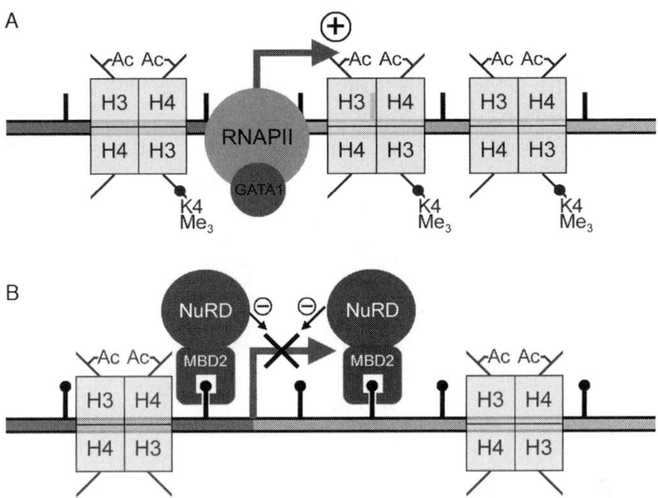

Figure 4.6 Model for the developmental regulation of ρ-globin transcription in chicken erythrocytes. (A) In embryonic day 4 primitive chicken erythrocytes, robust expression of the ρ-globin gene is seen. The presence of positively acting *trans* factors, such as GATA1, recruits RNA polymerase (RNAPII) and drives high levels of transcription from the unmethylated ρ-globin gene. Histones throughout the gene exhibit relatively high levels of active modifications. (B) No ρ-globin gene transcription is seen in embryonic day 14 definitive chicken erythrocytes in which there is dense methylation in the promoter and proximal transcribed sequences-methylated ρ-globin gene in adult cells. MBD2 recruits these components to the methylated ρ-globin gene. MBD2 binds the methylated sequences and recruits MTA2 and other components of an NURD corepressor complex. The complex maintains transcriptional inactivity by remodeling chromatin into a nonpermissive configuration and/or direct transcriptional inhibition rather than through HDAC enzymatic activity (from Kransdorf *et al.*, 2006).

excellent animal models for studies of globin gene regulation because the structure of the β-globin locus and the developmental pattern of globin gene expression are highly conserved in baboons and humans (Barrie *et al.*, 1981; DeSimone and Mueller, 1978). In baboons, an inverse correlation is observed between DNA methylation of the ε- and γ-globin gene promoters and the expression of the genes during globin gene switching (Lavelle *et al.*, 2006a). Methylation of the ε- and γ-globin gene promoters during ε-to-γ and γ-to-β switching was found to be initiated in an incomplete manner with no observed preference for any individual CpG sites. These observations support a stochastic model in which initial low levels of methylation are predicted to increase over time as a result of both *de novo* and maintenance methylation. These data again suggest that DNA methylation may not be the initiating event in globin gene silencing, but that it is involved in a lock-off mechanism.

The genes of the 70-kb human β-globin gene complex are expressed sequentially during development in the 5′ to 3′ order that they appear on chromosome 11: 5′ε-, $^{G}\gamma$-, $^{A}\gamma$-, and β-3′ (Fig. 4.6). The ε-globin gene is the major β-like globin gene expressed in primitive, yolk sac-derived red blood cells. During human development, a shift in the site of erythropoiesis from yolk sac to liver coincides with ε-globin silencing and predominant expression of γ-globin. Later in gestation, a second shift in the site of erythropoiesis from liver to bone marrow coincides with increased β-globin expression and decreased γ-globin expression. Completion of the switch from γ- to β-globin expression occurs postnatally. The regulation of the γ-globin gene is of special interest because of the ameliorating effects of the synthesis of fetal hemoglobin in patients with either β-thalassemia syndrome or sickle-cell disease (Stamatoyannopoulos, 2005).

As noted, a strong inverse correlation occurs between DNA methylation and the expression of both baboon and human β-globin genes (DeSimone and Mueller, 1978; DeSimone *et al.*, 1982; Lavelle *et al.*, 2006a,b; van der Ploeg and Flavell, 1980). In addition, it was first demonstrated in primate models and humans that inhibition of DNA methylation by 5-azacytidine partially reversed the γ- to β-globin switch (Charache *et al.*, 1983; DeSimone *et al.*, 1982; Ley *et al.*, 1982). The role of DNA methylation in silencing of the human ε-globin gene is less clear than in the case of the γ-globin genes. When adult erythroid phenotype mouse erythroleukemia (MEL) cells containing human chromosome 11 were treated with 5-azacytidine, only γ-globin and not ε-globin gene transcriptional activation was detected (Ley *et al.*, 1984). However, unpublished data from our studies of the human β-globin locus in transgenic mice support a role for DNA methylation in maintaining the extremely tight silencing of the ε-globin gene.

3.3. A special role for methyl-binding domain protein 2 (MBD2)?

The data on the composition of the MeCPC and the role of MBD2 in regulating developmental expression of the chicken β-type globin genes appear to be potentially very relevant to the regulation of globin gene switching in the human β-globin locus. It has recently been determined that the fetal γ-globin gene is expressed at high levels in adult erythroid cells from $MBD2^{-/-}$ mice that contain a single copy of a human β-globin locus YAC (β-YAC) at a level commensurate with β-YAC mice treated with 5-azacytidine (see Fig. 4.3; Pace *et al.*, 1994; Rupon *et al.*, 2006) Moreover, in the absence of MBD2, the developmental silencing of the γ-globin gene is delayed (Fig. 4.7).

Taken together, these results indicate that DNA methylation may contribute to γ-to-β-globin gene switching through MBD2. However, MBD2 does not bind near the γ-globin promoter (Rupon *et al.*, 2006), which is not surprising because there are no true CpG islands within 6 kb of the γ-globin

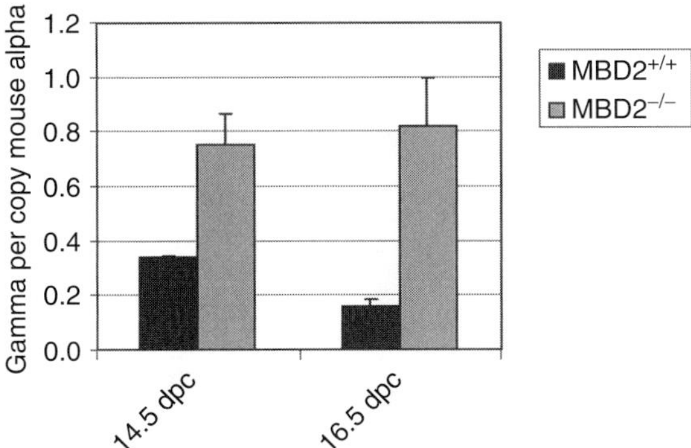

Figure 4.7 Loss of methyl cytosine-binding protein 2 delays developmental γ-to-β switching in β-YAC transgenic mice. Quantitative representation of the level of γ-globin gene mRNA in wild-type or MBD2$^{-/-}$ β-YAC mouse fetal liver erythroid cells from day 14.5 and day 16.5 dpc embryos as determined by RNase protection assay (from Rupon *et al.*, 2006).

genes, and MBD2 has been shown to bind *in vivo* only to methylated CpG island sequences. In contrast, the chicken ρ-globin gene promoter and its proximal transcribed region contain two CpG islands that are methylated in adult erythroid cells. Based on these results, it seems likely that loss of MBD2 in the human βYAC transgenic mouse model results in transcriptional activation of a CpG-rich gene or genes that are normally silent in adult erythroid cells. The product of this gene(s) would, in turn, result in transcriptional activation of the γ-globin gene. Alternatively, MBD2 might either bind in *cis* to the γ-globin locus at a site very distal to the promoter, or might participate in a methyl cytosine-associated repressor complex in which it does not bind directly to DNA. The finding that 5-azacytidine causes only a minor increase in γ-globin gene expression in MBD2 knockout βYAC mice (Fig. 4.8) supports the notion that much of the effect of 5-azacytidine on stimulating γ-globin gene expression in adult stage erythroid cells is through its inhibitory effects on DNA methylation. This contrasts with the alternative hypothesis that the myelotoxic effects are the predominant contributor to the high level of induction *in vivo* of HbF in patients and baboons treated with 5-azacytidine (Charache *et al.*, 1983; DeSimone *et al.*, 1982; Ley *et al.*, 1982; Stamatoyannopoulos, 2005).

Many compounds currently available to treat patients with hemoglobinopathies carry short- or long-term potential risks of toxicity and, in addition, the responses to the agents are variable. Although 5-azacytidine is successful in inducing fetal hemoglobin (HbF) in patients with sickle-cell anemia and β-thalassemia, it has been used primarily in older patients with severe disease because of the concerns associated with its risks of carcinogenicity. Because MBD2 is found to increase HbF to levels similar to those

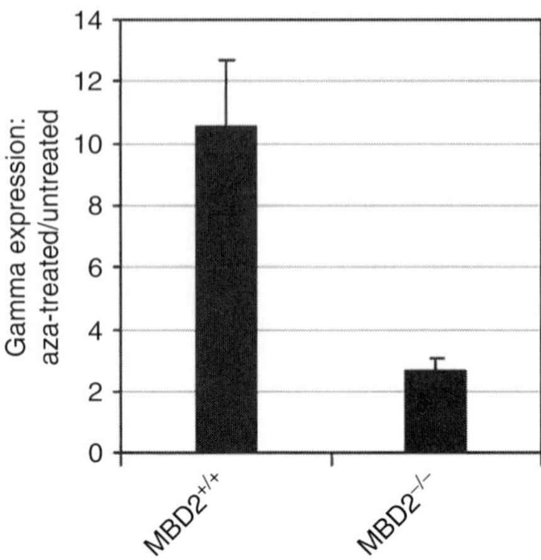

Figure 4.8 5-Azacytidine induction of γ-globin expression in adult stage erythrocytes mediated by methyl cytosine-binding protein 2. Quantitative real-time PCR was used to determine γ-globin mRNA induction in erythroid cells by 5-azacytidine treatment of wild-type versus MBD2$^{-/-}$ β-YAC adult mice (from Rupon *et al.*, 2006).

attained with 5-azacytidine treatment in β-YAC transgenic mice, and because MBD2 is not required for normal mammalian embryonic development and confers a minimally abnormal phenotype in null mice (Feng and Zhang, 2001), MBD2 may be an excellent target for therapeutic modulation aimed at altering developmental gene expression patterns such as γ-globin gene reactivation in adults.

4. DNA Methylation of the Human α-Globin Locus

4.1. Comparison to β-locus methylation

Examinations of CpG methylation at the human α-globin cluster on chromosome 16 followed shortly after the initial characterization of the association of methylation with β-globin gene regulation. When originally studied in chickens (Haigh *et al.*, 1982), it appeared that CpG methylation might account for a measure of the tissue specificity and developmental switching of α-globin gene expression in all vertebrates in a manner analogous to that described for the β-globin cluster (Ginder and McGhee, 1981; McGhee and Ginder, 1979; Shen and Maniatis, 1980; van der Ploeg and Flavell, 1980). However, closer inspection of the human α-globin locus revealed CpG clusters on either side of the transcription start site that remained unmethylated in nonerythroid tissues (Bird *et al.*, 1987). This is in contrast to the human β-type globin genes which are relatively CpG-depleted and remain heavily methylated in nonerythroid tissues (van der Ploeg and Flavell, 1980). As observed by Bird *et al.*, the divergent DNA methylation pattern of these two loci may explain in part their very different behavior: definitive α-globin gene expression occurs very early in development, while β-globin gene expression occurs much later (Peschle *et al.*, 1985). In addition, optimum expression of the β-globin genes requires an upstream *cis* LCR. The analogous enhancer 40 kb upstream from the α-globin locus (HS-40) does not appear necessary for high levels of α-globin gene expression in transfected cells (Mellon *et al.*, 1981), and it also differs in its ability to confer copy number-dependent expression in transgenic mice (Sharpe *et al.*, 1993). Interestingly, an α-globin pseudo-gene, Ψ-α-1, though bearing high homology to the native α-1 and α-2 genes, is deficient in the CpG-rich regions present in the active gene loci. This suggests the possibility that this gene was inactivated, and coincident heavy methylation of the CpG-rich regions in the germ line led to eventual depletion of these dinucleotides by deamination of 5-methyl-cytosine (Bird, 1980).

4.2. Differences in avian α-globin locus methylation

The situation in chickens appears to differ from that in mammals. An upstream CpG region (−3.5 to 4 kb) was found to be methylated in nonerythroid cells and unmethylated in erythroid cells in chicken (Razin *et al.*, 2000). Although the significance of this site is not known, given that the α-globin genes lie in a region of relatively open/active chromatin in a variety of tissues, this region may be important for silencing expression in nonerythroid tissues. It has been demonstrated that recapitulation of developmental stage-specific methylation of the chicken embryonic type alpha (α^{π}) promoter region is sufficient to suppress transcription in transfected adult primary chicken erythroid cells (Singal *et al.*, 2002a). An MeCP complex containing MBD2 and HDAC1, similar to the ρ-globin gene-binding complex (Kransdorf *et al.*, 2006), was shown to associate with methylated, but not with unmethylated, embryonic π'-globin gene promoter proximal sequences. A similar mechanism appears improbable for silencing human embryonic α-type globin zeta(ζ)-gene expression, given the absence of promoter CpG methylation in both erythroid and nonerythroid tissues (Bird *et al.*, 1987).

4.3. Indirect effects of DNA methylation on α-globin genes

One gene of particular interest in the human α-globin cluster region is ggPRX, a housekeeping gene that overlaps the upstream regulatory region of the α-globin gene locus and is transcribed in the opposite direction (Sjakste *et al.*, 2000). Klochkov *et al.* (2006) recently proposed a CpG methylation-mediated CCCTC-binding factor (CTCF)-dependent silencing mechanism for normalizing expression of ggPRX in erythroid cells. A putative CTCF silencer-binding site exists in the CpG-rich upstream regulatory region, and Klochkov *et al.* have shown that CTCF binds only in the demethylated state, which occurs in erythroid cells, but not in lymphoid, and presumably other tissues, where α-globin gene expression is silenced. CTCF binding suppresses the ggPRX gene, effectively normalizing expression in erythroid cells where α-globin gene enhancer elements are active and may otherwise drive overexpression of the overlapping ggPRX gene.

Mutations in human ATRX, classified as a member of the SWI/SNF family of chromatin remodeling proteins, produce α-thalassemia in association with two syndromes: X-linked mental retardation (ATRX) (Gibbons *et al.*, 1995) and myelodysplastic syndrome (ATMDS) (Gibbons *et al.*, 2003). The exact nature of ATRX's effect on α-globin gene expression, however, remains undetermined. Interestingly, ATRX contains a putative N-terminal PHD domain similar to the DNMT3 family of *de novo* methyl transferases, and mutations in this protein are associated with changes in the pattern of DNA methylation (Gibbons *et al.*, 2000). As noted by Higgs *et al.* (2005), ATRX's selective effects on α-globin levels, but not β-globin levels, may hold the key to understanding its mechanism, particularly given that these two loci have very different methylation patterns.

Studies of human α-thalassemia have focused on inherited and acquired mutations in promoter/enhancer elements or the α-globin genes themselves. Barbour *et al.* (2000) have described an 18 kb deletion 3′ to the α-cluster that abolishes expression without disturbing any known positive regulator sequences or the globin genes themselves. This deletion exhibits a so-called chromosomal position effect by juxtaposing the α-globin genes to a heavily methylated downstream alu-rich region. These observations, taken as a whole, highlight the importance of the chromosomal environment for normal expression of α-globin genes in erythroid and nonerythroid cells.

5. DNA Methylation of Other Erythroid-Specific Genes

As mentioned above, the role that DNA methylation plays in regulating gene expression is often uncertain, even amidst evidence that it correlates well with transcriptional repression. Such a correlation has been established for the ABO blood group genes encoding A and B glycosyltransferases (Kominato *et al.*, 1999). The functional role of DNA methylation at this locus and its relevance in establishing normal ABO expression patterns remains to be determined. The ABO proximal promoter lies in a CpG island that is unmethylated in tissues that express these genes, notably erythrocytes and some epithelial cells. Conversely, this region is often hypermethylated in cultured nonerythroid cells and in human epithelial cancers where ABO expression is absent (Chihara *et al.*, 2005; Gao *et al.*, 2004). Although aberrant CpG island hypermethylation of tumor suppressor genes is typical in neoplastic tissues (Baylin, 2005; Jones, 2002; Jones and Baylin, 2002), its role is less broadly defined in establishing tissue-specific and developmental expression patterns in normal tissues. Nevertheless, it is tempting to postulate a role for DNA methylation in limiting ABO gene expression, particularly given the unusual positioning of a CpG island near this highly tissue-restricted gene. Furthermore, a distal promoter containing an alternate transcription start site upstream from the CpG-rich region has been described for the ABO locus that also displays a correlation between hypermethylation and transcriptional repression (Kominato *et al.*, 2002). As is the case for many differentially methylated genes, more investigation is necessary before a conclusive functional relationship can be substantiated.

6. Conclusions and Future Directions

There has emerged an increasing appreciation of the role of DNA methylation in normal developmental gene regulation, primarily as a lock-off mechanism for maintaining transcriptional silencing. Among the best characterized examples of the relatively small group of genes whose normal developmental expression is controlled, at least in part, by DNA methylation are the vertebrate globin genes. Another well-characterized example is the regulation of the IL-4 gene during murine T-lymphocyte development (Hutchins *et al.*, 2002, 2005). Interestingly, the methylated cytosine-binding protein, MBD2, plays a key role in mediating the repressive effect of DNA methylation in both of these intensively investigated cases. This relationship highlights the general importance of the MCBPs and their associated corepressor complexes in interpreting the epigenetic mark of DNA methylation to impart developmental tissue specificity of gene regulation.

Recently, genomic approaches have identified a large number of tissue-specific and differentially methylated regions, including promoter-associated CpG islands, in primary murine tissues (Song *et al.*, 2005). Analysis of CpG methylation patterns of human chromosomes 6, 20, and 22 for comparisons in 12 different tissues from human and mouse led to an estimate that up to 70% of 5′UTR or promoter loci that are orthologous between human and mouse may have conserved tissue-specific DNA methylation profiles (Eckhardt *et al.*, 2006). Studies such as these suggest that many other genes may be regulated by methylation during cell differentiation and development in erythroid and other cell types. Future studies are needed to validate the cause–effect relationship between DNA methylation and transcriptional silencing in the many specific genes involved, as well as to characterize the specific mechanism(s) of such silencing, and to define the complex interplay between DNA methylation and other epigenetic signals. While the level of complexity involved may appear daunting, these studies should reveal the diversity of epigenetic interactions that contribute exquisite specificity to the regulation of gene expression patterns during development in erythroid cells and other tissues.

ACKNOWLEDGMENTS

The authors thank Catharine W. Tucker for assistance in the preparation of this manuscript. They also apologize to those whose work has not been cited due to space constraints or oversight. This work was supported by the National Institutes of Health (R01 DK29902) and the Massey Cancer Center at Virginia Commonwealth University.

Figures 4.4, 4.5, and 4.6 were originally published in BLOOD. Kransdorf *et al.*, MBD2 is a critical component of a methyl cytosine binding protein complex isolated from primary erythroid cells. Blood 2004; 108(8) 2836–2845 © The American Society of Hematology.

REFERENCES

Amaravadi, L., and Klemsz, M. J. (1999). DNA methylation and chromatin structure regulate PU.1 expression. *DNA Cell Biol.* **18**(12), 875–884.

Ballestar, E., and Wolffe, A. (2001). Methyl-CpG-binding proteins: Targeting specific gene repression. *Eur. J. Biochem.* **268,** 1–6.

Barbour, V. M., Tufarelli, C., Sharpe, J. A., Smith, Z. E., Ayyub, H., Heinlein, C. A., Sloane-Stanley, J., Indrak, K., Wood, W. G., and Higgs, D. R. (2000). Alpha-thalassemia resulting from a negative chromosomal position effect. *Blood* **96**(3), 800–807.

Barreto, G., Schafer, A., Marhold, J., Stach, D., Swaminathan, S. K., Handa, V., Doderlein, G., Maltry, N., Wu, W., Lyko, F., and Niehrs, C. (2007). Gadd45a promotes epigenetic gene activation by repair-mediated DNA demethylation. *Nature* **445**(7128), 671–675.

Barrie, P. A., Jeffreys, A. J., and Scott, A. F. (1981). Evolution of the beta-globin gene cluster in man and the primates. *J. Mol. Biol.* **149**(3), 319–336.

Baylin, S. B. (2005). DNA methylation and gene silencing in cancer. *Nat. Clin. Pract. Oncol.* **2**(Suppl. 1), S4–S11.

Bellacosa, A., Cicchillitti, L., Schepis, F., Riccio, A., Yeung, A. T., Matsumoto, Y., Golemis, E. A., Genuardi, M., and Neri, G. (1999). MED1, a novel human methyl-CpG-binding endonuclease, interacts with DNA mismatch repair protein MLH1. *Proc. Natl. Acad. Sci. USA* **96**(7), 3969–3974.

Bhattacharya, S. K., Ramchandani, S., Cervoni, N., and Szyf, M. (1999). A mammalian protein with specific demethylase activity for mCpG DNA. *Nature* **397**(6720), 579–583.

Bird, A. P. (1980). DNA methylation and the frequency of CpG in animal DNA. *Nucleic Acids Res.* **8**(7), 1499–1504.

Bird, A. P., Taggart, M. H., Nicholls, R. D., and Higgs, D. R. (1987). Non-methylated CpG-rich islands at the human alpha-globin locus: Implications for evolution of the alpha-globin pseudogene. *EMBO J.* **6**(4), 999–1004.

Bowen, N., Fujita, N., Kajita, M., and Wade, P. (2004). Mi-2/NuRD: Multiple complexes for many purposes. *Biochemica et Biophysica Acta* **1677,** 52–57.

Boyes, J., and Bird, A. (1991). DNA methylation inhibits transcription indirectly via a methyl-CpG binding protein. *Cell* **64**(6), 1123–1134.

Brenner, C., Deplus, R., Didelot, C., Loriot, A., Vire, E., De Smet, C., Gutierrez, A., Danovi, D., Bernard, D., Boon, T., Pelicci, P. G., Amati, B., *et al.* (2005). Myc represses transcription through recruitment of DNA methyltransferase corepressor. *EMBO J.* **24**(2), 336–346.

Brinkman, A. B., Pennings, S. W., Braliou, G. G., Rietveld, L. E., and Stunnenberg, H. G. (2007). DNA methylation immediately adjacent to active histone marking does not silence transcription. *Nucleic Acids Res.* **35**(3), 801–811.

Burns, L. J., Glauber, J. G., and Ginder, G. D. (1988). Butyrate induces selective transcriptional activation of a hypomethylated embryonic globin gene in adult erythroid cells. *Blood* **72**(5), 1536–1542.

Cantor, A. B., and Orkin, S. H. (2001). Hematopoietic development: A balancing act. *Curr. Opin. Genet. Dev.* **11**(5), 513–519.

Chan, L., Weidmann, M., and Ingram, V. (1974). Regulation of specific gene expression during embryonic development: Synthesis of globin messenger RNA during red cell formation in chick embryos. *Dev. Biol.* **40**(1), 174–185.

Charache, S., Dover, G. J., Smith, K., Talbot, C. C., Jr., Moyer, M., and Boyer, S. (1983). Treatment of sickle cell anemia with 5-azacytidine results in increased fetal hemoglobin production and is associated with nonrandom hypomethylation of DNA around the gamma-delta-beta-globin gene complex. *Proc. Natl. Acad. Sci. USA* **80**(15), 4842–4846.

Chihara, Y., Sugano, K., Kobayashi, A., Kanai, Y., Yamamoto, H., Nakazono, M., Fujimoto, H., Kakizoe, T., Fujimoto, K., Hirohashi, S., and Hirao, Y. (2005). Loss of blood group A antigen expression in bladder cancer caused by allelic loss and/or methylation of the ABO gene. *Lab. Invest.* **85**(7), 895–907.

Daniel, J. M., and Reynolds, A. B. (1999). The catenin p120(ctn) interacts with Kaiso, a novel BTB/POZ domain zinc finger transcription factor. *Mol. Cell. Biol.* **19**(5), 3614–3623.

DeSimone, J., and Mueller, A. L. (1978). Fetal hemoglobin synthesis in baboons (Papio cynocephalus). *J. Lab. Clin. Med.* **91**(6), 862–871.

DeSimone, J., Heller, P., Hall, L., and Zwiers, D. (1982). 5-Azacytidine stimulates fetal hemoglobin synthesis in anemic baboons. *Proc. Natl. Acad. Sci. USA* **79**(14), 4428–4431.

Di Croce, L., Raker, V. A., Corsaro, M., Fazi, F., Fanelli, M., Faretta, M., Fuks, F., Lo, C. F., Kouzarides, T., Nervi, C., Minucci, S., and Pelicci, P. G. (2002). Methyltransferase recruitment and DNA hypermethylation of target promoters by an oncogenic transcription factor. *Science* **295**(5557), 1079–1082.

Eckhardt, F., Lewin, J., Cortese, R., Rakyan, V. K., Attwood, J., Burger, M., Burton, J., Cox, T. V., Davies, R., Down, T. A., Haefliger, C., Horton, R., *et al.* (2006). DNA methylation profiling of human chromosomes 6, 20 and 22. *Nat. Genet.* **38**(12), 1378–1385.

Egger, G., Liang, G., Aparicio, A., and Jones, P. (2004). Epigenetics in human disease and prospects for epigenetic therapy. *Nature* **429**(6990), 457–463.

Ehrlich, M. (2005). The controversial denouement of vertebrate DNA methylation research. *Biochemistry (Mosc.)* **70**(5), 568–575.

Enver, T., Zhang, J. W., Papayannopoulou, T., and Stamatoyannopoulos, G. (1988). DNA methylation: A secondary event in globin gene switching?. *Genes Dev.* **2**(6), 698–706.

Feng, Q., and Zhang, Y. (2001). The MeCP1 complex represses transcription through preferential binding, remodeling, and deacetylating methylated nucleosomes. *Genes Dev.* **15**(7), 827–832.

Feng, Q., and Zhang, Y. (2003). The NuRD complex: Linking histone modification to nucleosome remodeling. *Curr. Top. Microbiol. Immunol.* **274,** 269–290.

Filion, G. J., Zhenilo, S., Salozhin, S., Yamada, D., Prokhortchouk, E., and Defossez, P. A. (2006). A family of human zinc finger proteins that bind methylated DNA and repress transcription. *Mol. Cell. Biol.* **26**(1), 169–181.

Fujita, N., Shimotake, N., Ohki, I., Chiba, T., Saya, H., Shirakawa, M., and Nakao, M. (2000). Mechanism of transcriptional regulation by methyl-CpG binding protein MBD1. *Mol. Cell. Biol.* **20**(14), 5107–5118.

Fuks, F., Hurd, P. J., Wolf, D., Nan, X., Bird, A. P., and Kouzarides, T. (2003). The methyl-CpG-binding protein MeCP2 links DNA methylation to histone methylation. *J. Biol. Chem.* **278**(6), 4035–4040.

Gao, S., Worm, J., Guldberg, P., Eiberg, H., Krogdahl, A., Liu, C. J., Reibel, J., and Dabelsteen, E. (2004). Genetic and epigenetic alterations of the blood group ABO gene in oral squamous cell carcinoma. *Int. J. Cancer* **109**(2), 230–237.

Ge, Y. Z., Pu, M. T., Gowher, H., Wu, H. P., Ding, J. P., Jeltsch, A., and Xu, G. L. (2004). Chromatin targeting of *de novo* DNA methyltransferases by the PWWP domain. *J. Biol. Chem.* **279**(24), 25447–25454.

Gibbons, R. J., Picketts, D. J., Villard, L., and Higgs, D. R. (1995). Mutations in a putative global transcriptional regulator cause X-linked mental retardation with alpha-thalassemia (ATR-X syndrome). *Cell* **80**(6), 837–845.

Gibbons, R. J., McDowell, T. L., Raman, S., O'Rourke, D. M., Garrick, D., Ayyub, H., and Higgs, D. R. (2000). Mutations in ATRX, encoding a SWI/SNF-like protein, cause diverse changes in the pattern of DNA methylation. *Nat. Genet.* **24**(4), 368–371.

Gibbons, R. J., Pellagatti, A., Garrick, D., Wood, W. G., Malik, N., Ayyub, H., Langford, C., Boultwood, J., Wainscoat, J. S., and Higgs, D. R. (2003). Identification of acquired somatic mutations in the gene encoding chromatin-remodeling factor ATRX in the alpha-thalassemia myelodysplasia syndrome (ATMDS). *Nat. Genet.* **34**(4), 446–449.

Ginder, G. D., and McGhee, J. D. (1981). DNA methylation in the chicken adult beta-globin gene: A relationship with gene expression. *In "Organization and Expression of Globin Genes,"* p. 191. Liss, New York.

Ginder, G. D., Whitters, M. J., and Pohlman, J. K. (1984). Activation of a chicken embryonic globin gene in adult erythroid cells by 5-azacytidine and sodium butyrate. *Proc. Natl. Acad. Sci. USA* **81**(13), 3954–3958.

Groudine, M., Peretz, M., and Weintraub, H. (1981). Transcriptional regulation of hemoglobin switching in chicken embryos. *Mol. Cell. Biol.* **1**(3), 281–288.

Haigh, L. S., Owens, B. B., Hellewell, O. S., and Ingram, V. M. (1982). DNA methylation in chicken alpha-globin gene expression. *Proc. Natl. Acad. Sci. USA* **79**(17), 5332–5336.

Harikrishnan, K., Chow, M., Baker, E. P. S., Bassal, S., Brasacchio, D., Wang, L., Craig, J., Jones, P., Sif, S., and El-Osta, A. (2005). Brahma links the SWI/SNF chromatin-remodeling complex with MeCP2-dependent transcriptional silencing. *Nat. Genet.* **37**(3), 254–264.

Hendrich, B., and Bird, A. (1998). Identification and characterization of a family of mammalian methyl-CpG binding proteins. *Mol. Cell. Biol.* **18**(11), 6538–6547.

Hendrich, B., and Tweedie, S. (2003). The methyl-CpG binding domain and the evolving role of DNA methylation in animals. *Trends Genet.* **19**(5), 269–277.

Hendrich, B., Hardeland, U., Ng, H. H., Jiricny, J., and Bird, A. (1999). The thymine glycosylase MBD4 can bind to the product of deamination at methylated CpG sites. *Nature* **401**(6750), 301–304.

Hendrich, B., Guy, J., Ramsahoye, B., Wilson, V. A., and Bird, A. (2001). Closely related proteins MBD2 and MBD3 play distinctive but interacting roles in mouse development. *Genes Dev.* **15**(6), 710–723.

Higgs, D. R., Garrick, D., Anguita, E., De Gobbi, M., Hughes, J., Muers, M., Vernimmen, D., Lower, K., Law, M., Argentaro, A., Deville, M. A., and Gibbons, R.

(2005). Understanding alpha-globin gene regulation: Aiming to improve the management of thalassemia. *Ann. NY Acad. Sci.* **1054,** 92–102.

Holliday, R., and Pugh, J. E. (1975). DNA modification mechanisms and gene activity during development. *Science* **187**(4173), 226–232.

Hsu, M., Mabaera, R., Lowrey, C. H., Martin, D. I., and Fiering, S. (2007). CpG hypomethylation in a large domain encompassing the embryonic {beta}-like globin genes in primitive erythrocytes. *Mol. Cell. Biol.* **27**(13), 5047–5054.

Hu, K., Nan, X., Bird, A., and Wang, W. (2006). Testing for association between MeCP2 and the brahma-associated SWI/SNF chromatin-remodeling complex. *Nat. Genet.* **38**(9), 962–964.

Hu, M., Krause, D., Greaves, M., Sharkis, S., Dexter, M., Heyworth, C., and Enver, T. (1997). Multilineage gene expression precedes commitment in the hemopoietic system. *Genes Dev.* **11**(6), 774–785.

Hutchins, A. S., Mullen, A. C., Lee, H. W., Sykes, K. J., High, F. A., Hendrich, B. D., Bird, A. P., and Reiner, S. L. (2002). Gene silencing quantitatively controls the function of a developmental trans-activator. *Mol. Cell* **10**(1), 81–91.

Hutchins, A. S., Artis, D., Hendrich, B. D., Bird, A. P., Scott, P., and Reiner, S. L. (2005). Cutting edge: A critical role for gene silencing in preventing excessive type 1 immunity. *J. Immunol.* **175**(9), 5606–5610.

Iwasaki, H., Mizuno, S., Arinobu, Y., Ozawa, H., Mori, Y., Shigematsu, H., Takatsu, K., Tenen, D. G., and Akashi, K. (2006). The order of expression of transcription factors directs hierarchical specification of hematopoietic lineages. *Genes Dev.* **20**(21), 3010–3021.

Jackson, J. P., Lindroth, A. M., Cao, X., and Jacobsen, S. E. (2002). Control of CpNpG DNA methylation by the KRYPTONITE histone H3 methyltransferase. *Nature* **416** (6880), 556–560.

Jaenisch, R., and Bird, A. (2003). Epigenetic regulation of gene expression: How the genome integrates intrinsic and environmental signals. *Nat. Genet.* **33**(Suppl), 245–254.

Jiang, C.-L., Jin, S.-G., and Pfeifer, G. (2004). MBD3L1 is a transcriptional repressor that interacts with methyl-CpG-binding protein 2 (MBD2) and components of the NuRD complex. *J. Biol. Chem.* **279**(80), 51456–52464.

Jin, S. G., Jiang, C. L., Rauch, T., Li, H., and Pfeifer, G. P. (2005). MBD3L2 interacts with MBD3 and components of the NuRD complex and can oppose MBD2-MeCP1-mediated methylation silencing. *J. Biol. Chem.* **280**(13), 12700–12709.

Johnson, L. M., Bostick, M., Zhang, X., Kraft, E., Henderson, I., Callis, J., and Jacobsen, S. E. (2007). The SRA methyl cytosine binding domain links DNA and histone methylation. *Curr. Biol.* **17**(4), 379–384.

Jones, P. A. (2002). DNA methylation and cancer. *Oncogene* **21**(35), 5358–5360.

Jones, P. A., and Baylin, S. B. (2002). The fundamental role of epigenetic events in cancer. *Nat. Rev. Genet.* **3**(6), 415–428.

Jost, J. P. (1993). Nuclear extracts of chicken embryos promote an active demethylation of DNA by excision repair of 5-methyldeoxycytidine. *Proc. Natl. Acad. Sci. USA* **90**(10), 4684–4688.

Jost, J. P., Siegmann, M., Sun, L., and Leung, R. (1995). Mechanisms of DNA demethylation in chicken embryos. Purification and properties of a 5-methylcytosine-DNA glycosylase. *J. Biol. Chem.* **270**(17), 9734–9739.

Jost, J. P., Oakeley, E. J., Zhu, B., Benjamin, D., Thiry, S., Siegmann, M., and Jost, Y. C. (2001). 5-Methylcytosine DNA glycosylase participates in the genome-wide loss of DNA methylation occurring during mouse myoblast differentiation. *Nucleic Acids Res.* **29**(21), 4452–4461.

Kafri, T., Ariel, M., Brandeis, M., Shemer, R., Urven, L., McCarrey, J., Cedar, H., and Razin, A. (1992). Developmental pattern of gene-specific DNA methylation in the mouse embryo and germ line. *Genes Dev.* **6**(5), 705–714.

Kanellopoulou, C., Muljo, S. A., Kung, A. L., Ganesan, S., Drapkin, R., Jenuwein, T., Livingston, D. M., and Rajewsky, K. (2005). Dicer-deficient mouse embryonic stem cells are defective in differentiation and centromeric silencing. *Genes Dev.* **19**(4), 489–501.

Klochkov, D., Rincon-Arano, H., Ioudinkova, E. S., Valadez-Graham, V., Gavrilov, A., Recillas-Targa, F., and Razin, S. V. (2006). A CTCF-dependent silencer located in the differentially methylated area may regulate expression of a housekeeping gene overlapping a tissue-specific gene domain. *Mol. Cell. Biol.* **26**(5), 1589–1597.

Klose, R. J., and Bird, A. P. (2006). Genomic DNA methylation: The mark and its mediators. *Trends Biochem. Sci.* **31**(2), 89–97.

Kominato, Y., Hata, Y., Takizawa, H., Tsuchiya, T., Tsukada, J., and Yamamoto, F. (1999). Expression of human histo-blood group ABO genes is dependent upon DNA methylation of the promoter region. *J. Biol. Chem.* **274**(52), 37240–37250.

Kominato, Y., Hata, Y., Takizawa, H., Matsumoto, K., Yasui, K., Tsukada, J., and Yamamoto, F. (2002). Alternative promoter identified between a hypermethylated upstream region of repetitive elements and a CpG island in human ABO histo-blood group genes. *J. Biol. Chem.* **277**(40), 37936–37948.

Kondo, E., Gu, Z., Horii, A., and Fukushige, S. (2005). The thymine DNA glycosylase MBD4 represses transcription and is associated with methylated p16(INK4a) and hMLH1 genes. *Mol. Cell. Biol.* **25**(11), 4388–4396.

Kransdorf, E. P., Wang, S. Z., Zhu, S. Z., Langston, T. B., Rupon, J. W., and Ginder, G. D. (2006). MBD2 is a critical component of a methyl cytosine binding protein complex isolated from primary erythroid cells. *Blood* **108**(8), 2836–2845.

Kuo, M. T., Mandel, J. L., and Chambon, P. (1979). DNA methylation: Correlation with DNase I sensitivity of chicken ovalbumin and conalbumin chromatin. *Nucleic Acids Res.* **7**(8), 2105–2113.

Laslo, P., Spooner, C. J., Warmflash, A., Lancki, D. W., Lee, H. J., Sciammas, R., Gantner, B. N., Dinner, A. R., and Singh, H. (2006). Multilineage transcriptional priming and determination of alternate hematopoietic cell fates. *Cell* **126**(4), 755–766.

Lavelle, D., Vaitkus, K., Hankewych, M., Singh, M., and De Simone, J. (2006a). Developmental changes in DNA methylation and covalent histone modifications of chromatin associated with the epsilon-, gamma-, and beta-globin gene promoters in Papio anubis. *Blood Cells Mol. Dis.* **36**(2), 269–278.

Lavelle, D., Vaitkus, K., Hankewych, M., Singh, M., and DeSimone, J. (2006b). Effect of 5-aza-2′-deoxycytidine (Dacogen) on covalent histone modifications of chromatin associated with the epsilon-, gamma-, and beta-globin promoters in Papio anubis. *Exp. Hematol.* **34**(3), 339–347.

LeGuezennec, X., Vermeulen, M., Brinkman, A. B., Hoeijmakers, W. A., Cohen, A., Lasonder, E., and Stunnenberg, H. G. (2006). MBD2/NuRD and MBD3/NuRD, two distinct complexes with different biochemical and functional properties. *Mol. Cell. Biol.* **26**(3), 843–851.

Ley, T. J., DeSimone, J., Anagnou, N. P., Keller, G. H., Humphries, R. K., Turner, P. H., Young, N. S., Keller, P., and Nienhuis, A. W. (1982). 5-Azacytidine selectively increases gamma-globin synthesis in a patient with beta$^+$ thalassemia. *N. Engl. J. Med.* **307**(24), 1469–1475.

Ley, T. J., Chiang, Y. L., Haidaris, D., Anagnou, N. P., Wilson, V. L., and Anderson, W. F. (1984). DNA methylation and regulation of the human beta-globin-like genes in mouse erythroleukemia cells containing human chromosome 11. *Proc. Natl. Acad. Sci. USA* **81**(21), 6618–6622.

Li, E., Bestor, T. H., and Jaenisch, R. (1992). Targeted mutation of the DNA methyltransferase gene results in embryonic lethality. *Cell* **69**(6), 915–926.

Mabaera, R. H., Richardson, C. A., Johnson, K. J., Hsu, M., Fiering, S. H., and Lowrey, C. H. (2007). Developmental and differentiation-specific patterns of human {gamma} and {beta}-globin promoter DNA methylation. *Blood* **110,** 1343–1352.

Matzke, M. A., and Birchler, J. A. (2005). RNAi-mediated pathways in the nucleus. *Nat. Rev. Genet.* **6**(1), 24–35.

McGhee, J. D., and Ginder, G. D. (1979). Specific DNA methylation sites in the vicinity of the chicken beta-globin genes. *Nature* **280**(5721), 419–420.

Meehan, R. R., Lewis, J. D., McKay, S., Kleiner, E. L., and Bird, A. P. (1989). Identification of a mammalian protein that binds specifically to DNA containing methylated CpGs. *Cell* **58**(3), 499–507.

Mellon, P., Parker, V., Gluzman, Y., and Maniatis, T. (1981). Identification of DNA sequences required for transcription of the human alpha 1-globin gene in a new SV40 host–vector system. *Cell* **27**(2 Pt 1), 279–288.

Monk, M., Boubelik, M., and Lehnert, S. (1987). Temporal and regional changes in DNA methylation in the embryonic, extraembryonic and germ cell lineages during mouse embryo development. *Development* **99**(3), 371–382.

Morris, K. V., Chan, S. W., Jacobsen, S. E., and Looney, D. J. (2004). Small interfering RNA-induced transcriptional gene silencing in human cells. *Science* **305**(5688), 1289–1292.

Murchison, E. P., Partridge, J. F., Tam, O. H., Cheloufi, S., and Hannon, G. J. (2005). Characterization of Dicer-deficient murine embryonic stem cells. *Proc. Natl. Acad. Sci. USA* **102**(34), 12135–12140.

Mutskov, V. J., Farrell, C. M., Wade, P. A., Wolffe, A. P., and Felsenfeld, G. (2002). The barrier function of an insulator couples high histone acetylation levels with specific protection of promoter DNA from methylation. *Genes Dev.* **16**(12), 1540–1554.

Nan, X., Ng, H. H., Johnson, C. A., Laherty, C. D., Turner, B. M., Eisenman, R. N., and Bird, A. (1998). Transcriptional repression by the methyl-CpG-binding protein MeCP2 involves a histone deacetylase complex. *Nature* **393**(6683), 386–389.

Ng, H. H., Zhang, Y., Hendrich, B., Johnson, C. A., Turner, B. M., Erdjument-Bromage, H., Tempst, P., Reinberg, D., and Bird, A. (1999). MBD2 is a transcriptional repressor belonging to the MeCP1 histone deacetylase complex. *Nat. Genet.* **23**(1), 58–61.

Ohm, J. E., McGarvey, K. M., Yu, X., Cheng, L., Schuebel, K. E., Cope, L., Mohammad, H. P., Chen, W., Daniel, V. C., Yu, W., Berman, D. M., Jenuwein, T., *et al.* (2007). A stem cell-like chromatin pattern may predispose tumor suppressor genes to DNA hypermethylation and heritable silencing. *Nat. Genet.* **39**(2), 237–242.

Okano, M., Bell, D. W., Haber, D. A., and Li, E. (1999). DNA methyltransferases Dnmt3a and Dnmt3b are essential for *de novo* methylation and mammalian development. *Cell* **99**(3), 247–257.

Okitsu, C. Y., and Hsieh, C.-L. (2007). DNA methylation dictates histone H3K4 methylation. *Mol. Cell. Biol.* **27**(7), 2746–2757.

Pace, B., Li, Q., Peterson, K., and Stamatoyannopoulos, G. (1994). Alpha-amino butyric acid cannot reactivate the silenced gamma gene of the beta locus YAC transgenic mouse. *Blood* **84**(12), 4344–4353.

Peschle, C., Mavilio, F., Care, A., Migliaccio, G., Migliaccio, A. R., Salvo, G., Samoggia, P., Petti, S., Guerriero, R., Marinucci, M., Lazzaro, D., and Russo, G. (1985). Haemoglobin switching in human embryos: Asynchrony of zeta–alpha and epsilon–gamma-globin switches in primitive and definite erythropoietic lineage. *Nature* **313**(5999), 235–238.

Pikaart, M. J., Recillas-Targa, F., and Felsenfeld, G. (1998). Loss of transcriptional activity of a transgene is accompanied by DNA methylation and histone deacetylation and is prevented by insulators. *Genes Dev.* **12**(18), 2852–2862.

Prokhortchouk, A., Hendrich, B., Jorgensen, H., Ruzov, A., Wilm, M., Georgiev, G., Bird, A., and Prokhortchouk, E. (2001). The p120 catenin partner Kaiso is a DNA methylation-dependent transcriptional repressor. *Genes Dev.* **15**(13), 1613–1618.

Prokhortchouk, A., Sansom, O., Selfridge, J., Caballero, I. M., Salozhin, S., Aithozhina, D., Cerchietti, L., Meng, F. G., Augenlicht, L. H., Mariadason, J. M., Hendrich, B., Melnick, A., *et al.* (2006). Kaiso-deficient mice show resistance to intestinal cancer. *Mol. Cell. Biol.* **26**(1), 199–208.

Qiu, C., Sawada, K., Zhang, X., and Cheng, X. (2002). The PWWP domain of mammalian DNA methyltransferase Dnmt3b defines a new family of DNA-binding folds. *Nat. Struct. Biol.* **9**(3), 217–224.

Ramachandran, K., Wert, J. V., Gopisetty, G., and Singal, R. (2007). Developmentally regulated demethylase activity targeting the beta(A)-globin gene in primary avian erythroid cells. *Biochemistry* **46**(11), 3416–3422.

Razin, A., and Riggs, A. D. (1980). DNA methylation and gene function. *Science* **210**(4470), 604–610.

Razin, S. V., Ioudinkova, E. S., and Scherrer, K. (2000). Extensive methylation of a part of the CpG island located 3.0–4.5 kbp upstream to the chicken alpha-globin gene cluster may contribute to silencing the globin genes in non-erythroid cells. *J. Mol. Biol.* **299**(4), 845–852.

Reese, B. E., Bachman, K. E., Baylin, S. B., and Rountree, M. R. (2003). The methyl-CpG binding protein MBD1 interacts with the p150 subunit of chromatin assembly factor 1. *Mol. Cell. Biol.* **23**(9), 3226–3236.

Reik, W., and Walter, J. (2001). Genomic imprinting: Parental influence on the genome. *Nat. Rev. Genet.* **2**(1), 21–32.

Riggs, A. D. (1975). X inactivation, differentiation, and DNA methylation. *Cytogenet. Cell Genet.* **14**(1), 9–25.

Rupon, J. W., Wang, S. Z., Gaensler, K., Lloyd, J., and Ginder, G. D. (2006). Methyl binding domain protein 2 mediates {gamma}-globin gene silencing in adult human betaYAC transgenic mice. *Proc. Natl. Acad. Sci. USA* **103**(17), 6617–6622.

Ruzov, A., Dunican, D. S., Prokhortchouk, A., Pennings, S., Stancheva, I., Prokhortchouk, E., and Meehan, R. R. (2004). Kaiso is a genome-wide repressor of transcription that is essential for amphibian development. *Development* **131**(24), 6185–6194.

Santos-Rosa, H., Schneider, R., Bannister, A. J., Sherriff, J., Bernstein, B. E., Emre, N. C., Schreiber, S. L., Mellor, J., and Kouzarides, T. (2002). Active genes are tri-methylated at K4 of histone H3. *Nature* **419**(6905), 407–411.

Sarraf, S. A., and Stancheva, I. (2004). Methyl-CpG binding protein MBD1 couples histone H3 methylation at lysine 9 by SETDB1 to DNA replication and chromatin assembly. *Mol. Cell* **15**(4), 595–605.

Scarano, E., Iaccarino, M., Grippo, P., and Parisi, E. (1967). The heterogeneity of thymine methyl group origin in DNA pyrimidine isostichs of developing sea urchin embryos. *Proc. Natl. Acad. Sci. USA* **57**(5), 1394–1400.

Schlesinger, Y., Straussman, R., Keshet, I., Farkash, S., Hecht, M., Zimmerman, J., Eden, E., Yakhini, Z., Ben Shushan, E., Reubinoff, B. E., Bergman, Y., Simon, I., *et al.* (2007). Polycomb-mediated methylation on Lys27 of histone H3 pre-marks genes for *de novo* methylation in cancer. *Nat. Genet.* **39**(2), 232–236.

Schuetze, S., Stenberg, P. E., and Kabat, D. (1993). The Ets-related transcription factor PU.1 immortalizes erythroblasts. *Mol. Cell. Biol.* **13**(9), 5670–5678.

Sharpe, J. A., Wells, D. J., Whitelaw, E., Vyas, P., Higgs, D. R., and Wood, W. G. (1993). Analysis of the human alpha-globin gene cluster in transgenic mice. *Proc. Natl. Acad. Sci. USA* **90**(23), 11262–11266.

Shen, C. K., and Maniatis, T. (1980). Tissue-specific DNA methylation in a cluster of rabbit beta-like globin genes. *Proc. Natl. Acad. Sci. USA* **77**(11), 6634–6638.

Singal, R., and vanWert, J. M. (2001). *De novo* methylation of an embryonic globin gene during normal development is strand specific and spreads from the proximal transcribed region. *Blood* **98**(12), 3441–3446.

Singal, R., Ferris, R., Little, J. A., Wang, S. Z., and Ginder, G. D. (1997). Methylation of the minimal promoter of an embryonic globin gene silences transcription in primary erythroid cells. *Proc. Natl. Acad. Sci. USA* **94**(25), 13724–13729.

Singal, R., vanWert, J., and Ferdinand, L., Jr. (2002a). Methylation of alpha-type embryonic globin gene alpha-pi represses transcription in primary erythroid cells. *Blood* **100**(12), 4217–4222.

Singal, R., Wang, S. Z., Sargent, T., Zhu, S. Z., and Ginder, G. D. (2002b). Methylation of promoter proximal-transcribed sequences of an embryonic globin gene inhibits transcription in primary erythroid cells and promotes formation of a cell type-specific methyl cytosine binding complex. *J. Biol. Chem.* **277**(3), 1897–1905.

Singh, M., Lavelle, D., Vaitkus, K., Mahmud, N., Hankewych, M., and DeSimone, J. (2007). The gamma-globin gene promoter progressively demethylates as the hematopoietic stem progenitor cells differentiate along the erythroid lineage in baboon fetal liver and adult bone marrow. *Exp. Hematol.* **35**(1), 48–55.

Sjakste, N., Iarovaia, O. V., Razin, S. V., Linares-Cruz, G., Sjakste, T., Le Gac, V., Zhao, Z., and Scherrer, K. (2000). A novel gene is transcribed in the chicken alpha-globin gene domain in the direction opposite to the globin genes. *Mol. Gen. Genet.* **262**(6), 1012–1021.

Song, F., Smith, J. F., Kimura, M. T., Morrow, A. D., Matsuyama, T., Nagase, H., and Held, W. A. (2005). Association of tissue-specific differentially methylated regions (TDMs) with differential gene expression. *Proc. Natl. Acad. Sci. USA* **102**(9), 3336–3341.

Soppe, W. J., Jasencakova, Z., Houben, A., Kakutani, T., Meister, A., Huang, M. S., Jacobsen, S. E., Schubert, I., and Fransz, P. F. (2002). DNA methylation controls histone H3 lysine 9 methylation and heterochromatin assembly in Arabidopsis. *EMBO J.* **21**(23), 6549–6559.

Stamatoyannopoulos, G. (2005). Control of globin gene expression during development and erythroid differentiation. *Exp. Hematol.* **33**(3), 259–271.

Svoboda, P., Stein, P., Filipowicz, W., and Schultz, R. M. (2004). Lack of homologous sequence-specific DNA methylation in response to stable dsRNA expression in mouse oocytes. *Nucleic Acids Res.* **32**(12), 3601–3606.

Swisher, J. F., Rand, E., Cedar, H., and Marie, P. A. (1998). Analysis of putative RNase sensitivity and protease insensitivity of demethylation activity in extracts from rat myoblasts. *Nucleic Acids Res.* **26**(24), 5573–5580.

Tagoh, H., Melnik, S., Lefevre, P., Chong, S., Riggs, A. D., and Bonifer, C. (2004). Dynamic reorganization of chromatin structure and selective DNA demethylation prior to stable enhancer complex formation during differentiation of primary hematopoietic cells *in vitro*. *Blood* **103**(8), 2950–2955.

Tamaru, H., Zhang, X., McMillen, D., Singh, P. B., Nakayama, J., Grewal, S. I., Allis, C. D., Cheng, X., and Selker, E. U. (2003). Trimethylated lysine 9 of histone H3 is a mark for DNA methylation in Neurospora crassa. *Nat. Genet.* **34**(1), 75–79.

Tate, P. H., and Bird, A. P. (1993). Effects of DNA methylation on DNA-binding proteins and gene expression. *Curr. Opin. Genet. Dev.* **3**(2), 226–231.

Ting, A. H., Schuebel, K. E., Herman, J. G., and Baylin, S. B. (2005). Short double-stranded RNA induces transcriptional gene silencing in human cancer cells in the absence of DNA methylation. *Nat. Genet.* **37**(8), 906–910.

van der Ploeg, L. H., and Flavell, R. A. (1980). DNA methylation in the human gamma delta beta-globin locus in erythroid and nonerythroid tissues. *Cell* **19**(4), 947–958.

Voso, M. T., Burn, T. C., Wulf, G., Lim, B., Leone, G., and Tenen, D. G. (1994). Inhibition of hematopoiesis by competitive binding of transcription factor PU.1. *Proc. Natl. Acad. Sci. USA* **91**(17), 7932–7936.

Wade, P. (2001). Methyl CpG-binding proteins and transcriptional repression. *BioEssays* **23** (12), 1131–1137.

Walsh, C. P., and Bestor, T. H. (1999). Cytosine methylation and mammalian development. *Genes Dev.* **13**(1), 26–34.

Warnecke, P. M., and Bestor, T. H. (2000). Cytosine methylation and human cancer. *Curr. Opin. Oncol.* **12**(1), 68–73.

Wolffe, A. P., Jones, P. L., and Wade, P. A. (1999). DNA demethylation. *Proc. Natl. Acad. Sci. USA* **96**(11), 5894–5896.

Yin, H., and Blanchard, K. L. (2000). DNA methylation represses the expression of the human erythropoietin gene by two different mechanisms. *Blood* **95**(1), 111–119.

Zhang, P., Zhang, X., Iwama, A., Yu, C., Smith, K. A., Mueller, B. U., Narravula, S., Torbett, B. E., Orkin, S. H., and Tenen, D. G. (2000). PU.1 inhibits GATA-1 function and erythroid differentiation by blocking GATA-1 DNA binding. *Blood* **96**(8), 2641–2648.

CHAPTER FIVE

Three-Dimensional Organization of Gene Expression in Erythroid Cells

Wouter de Laat,* Petra Klous,* Jurgen Kooren,*
Daan Noordermeer,* Robert-Jan Palstra,* Marieke Simonis,*
Erik Splinter,* *and* Frank Grosveld*

Contents

Abstract

The history of globin research is marked by a series of contributions seminal to our understanding of the genome, its function, and its relation to disease. For example, based on studies on hemoglobinopathies, it was understood that gene expression can be under the control of DNA elements that locate away from the genes on the linear chromosome template. Recent technological developments have allowed the demonstration that these regulatory DNA elements communicate with the genes through physical interaction, which loops out the intervening chromatin fiber. Subsequent studies showed that the spatial organization of the β-globin locus dynamically changes in relation

* Department of Cell Biology and Genetics, Erasmus MC, PO Box 2040, 3000 CA Rotterdam, The Netherlands

Current Topics in Developmental Biology, Volume 82
ISSN 0070-2153, DOI: 10.1016/S0070-2153(07)00005-1

to differences in gene expression. Moreover, it was shown that the β-globin locus adopts a different position in the nucleus during development and erythroid maturation. Here, we discuss the most recent insight into the three-dimensional organization of gene expression.

1. Introduction: Three-Dimensional Studies in a Historical Context

The history of studying globin genes goes back to the earliest days of molecular biology. This is not surprising because globins are the most abundant molecules in blood, and blood is easily obtained from and regenerated by organisms. Importantly its mRNA is abundant that allowed nucleic acid-based studies even before the advent of recombinant DNA technology. Moreover, hemoglobinopathies, diseases caused by a defect in globin production, form the most common single-gene disorder worldwide and have therefore traditionally been subject of intense research. The history of globin research is marked by a series of contributions seminal to our understanding of the genome, its function, and its relation to disease. More than half a century ago, Pauling *et al.* (1949) found that globin molecules from patients with sickle-cell disease were different than those from healthy persons, thereby establishing for the first time a molecular basis for disease (Pauling *et al.*, 1949). Ingram (1957), 8 years later, demonstrated that sickle-cell anemia is caused by globin gene mutations, thus providing the first example of a genetic disorder, while Max Perutz elucidated the three-dimensional (3D) structure of hemoglobin in the 1950s and 1960s, a study he started in the late 1930s (Bernal *et al.*, 1938). In the early 1970s, the globin cDNAs were among the first eukaryotic genes to be sequenced (Marotta *et al.*, 1974; Poon *et al.*, 1974; Proudfoot and Brownlee, 1974) and cloned into bacterial vectors (Rougeon *et al.*, 1975). The existence of intervening sequences, or introns, in eukaryotic genes was first demonstrated in the globin genes (Jeffreys and Flavell, 1977; Tilghman *et al.*, 1978), and the rabbit β-globin gene was the first gene that transcribed under the control of its own *cis*-regulatory sequences in transfection experiments (Mantei *et al.*, 1979). In the 1980s, transgenic mouse studies using human β-globin gene constructs led to the discovery of the locus control region (LCR), a dominant regulatory region that confers position-independent and copy number-dependent expression to linked transgenes (Grosveld *et al.*, 1987). Finally, intense research in the 1990s focusing on transgenic and knockout β-globin mouse models provided fundamental insight into the developmental regulation of gene expression, gene competition, transcriptional pulsing, and the function of distal transcription regulatory DNA elements.

Recent technical developments have enabled studies of the intricate folding of DNA in the cell nucleus. Nuclear architecture is an emerging key contributor to genome function and again, pioneering DNA-folding studies were first done on the β-globin gene locus.

This chapter will summarize our current knowledge on the 3D organization of DNA in the cell nucleus. We will explain the novel technologies that have recently boosted progress in this field, discuss the relevance of DNA topology for genome function, and mention the protein factors that have been implicated in DNA folding. First, we will summarize some original observations that led to the idea that the 3D organization of DNA may be important for gene regulation.

2. Long-Range Gene Activation and Gene Competition: DNA-Folding Matters?

It has long been recognized that in higher eukaryotes, transcription regulatory DNA elements, such as enhancers, are located away from the genes that they control (Banerji *et al.*, 1981; Wasylyk *et al.*, 1983), in some cases up to hundreds of kilobases. For example, the key regulatory element of the β-globin locus, the LCR, is positioned 50 kb away from the most distal β-globin gene. One of the most spectacular examples of transcription control over distance is the regulator of *SHH* expression in the developing limb, which lies 1 Mb away from the *SHH* gene (Lettice *et al.*, 2003), but many other regulatory sequences have been described that locate at a large distance from their target gene (reviewed in Kleinjan and van Heyningen, 2005).

Various models have been proposed in the past to explain how remote control elements communicate with their target genes. Most of these models assume that enhancers emit some signal that travels along the intervening chromatin fiber toward the gene that gets activated. The looping model, first put forward by Ptashne, states that enhancers and promoters communicate through direct interactions between proteins bound to these DNA elements, with the intervening DNA looping out (Ptashne, 1986). This model assumes flexibility of the chromatin fiber and as a consequence dispersed DNA fragments will randomly collide. Collision frequency will depend on the physical constraints of the chromatin fiber and is inversely correlated to the genomic site separation, that is the distance between two DNA segments on the linear chromosome template (measured in kilobases). Productive loop formation depends on affinities between proteins bound to colliding DNA segments. The model was originally based on work on bacterial and phage repressor proteins, like the Gal-, AraC, and l repressor proteins, which were found to function only when homomultimerized and bound to two separate operator sites. Electron microscopy visually demonstrated the DNA in between to loop out (reviewed in Ptashne, 1986). This looping model was the first to consider DNA folding to be important for gene regulation.

Eukaryotes have more complex gene clusters with regulatory elements functioning over much greater distances, and for a long time, direct evidence for chromatin looping to occur between enhancers and their genes was absent. However, with respect to transcription, a number of observations could only be explained satisfactory by the "looping model."

The first involved studies on *trans*-activation, that is, the ability of an enhancer to activate a promoter present on a physically separate DNA molecule. Most important in this respect is the naturally occurring phenomenon of transvection in *Drosophila,* where an enhancer can activate a promoter on the, paired, homologous allele (Bickel and Pirrotta, 1990). While pairing of homologues frequently occurs in *Drosophila*, this phenomenon is not observed in higher eukaryotes like mice and man, and until recently evidence for *trans*-communication *in vivo* was lacking in these species (see below). However, Schaffner and coworkers demonstrated *in vitro* that enhancers can stimulate transcription *in trans*, by coupling an enhancer- to a promoter-containing plasmid via a biotin–streptavidin bridge (Mueller-Storm *et al.*, 1989). *Trans*-activation of transcription was also observed when enhancer-containing and promoter-containing plasmids were injected as intertwined catenates into frog oocytes (Dunaway and Droge, 1989). Similarly, but still artificially, more recent transient transfection assays with reporter plasmids and GAGA as a DNA-bridging factor also demonstrated transcriptional activation *in trans* in mammalian cells (Mahmoudi *et al.*, 2002). All these studies on *trans*-activation demonstrate that a *cis* configuration of enhancer and promoter is not an absolute prerequisite for interaction, as is only predicted by the "looping model."

Other observations on gene regulation that supported the looping model are related to gene competition. Multiple genes can compete for a single enhancer for their activation, as was demonstrated originally by transfection assays using plasmids containing different numbers of genes and enhancers (de Villiers *et al.*, 1983; Wasylyk *et al.*, 1983). For β-globin, this was first demonstrated by experiments done by Choi and Engel, showing that two developmentally regulated chicken β-globin genes compete for a shared enhancer that is located in between the genes. The authors proposed that transcriptional activation of the genes depends on their ability to physically contact the enhancer, with a looped conformation of the DNA being the outcome of such interaction (Choi and Engel, 1988).

This mechanism was further investigated in transgenic mice carrying modified human β-globin locus constructs. Unlike the chicken locus, the mouse and human β-globin locus have the key transcription regulatory DNA element, the LCR, located upstream of a series of developmentally regulated β-globin genes (Fig. 5.1). The human locus contains five β-like

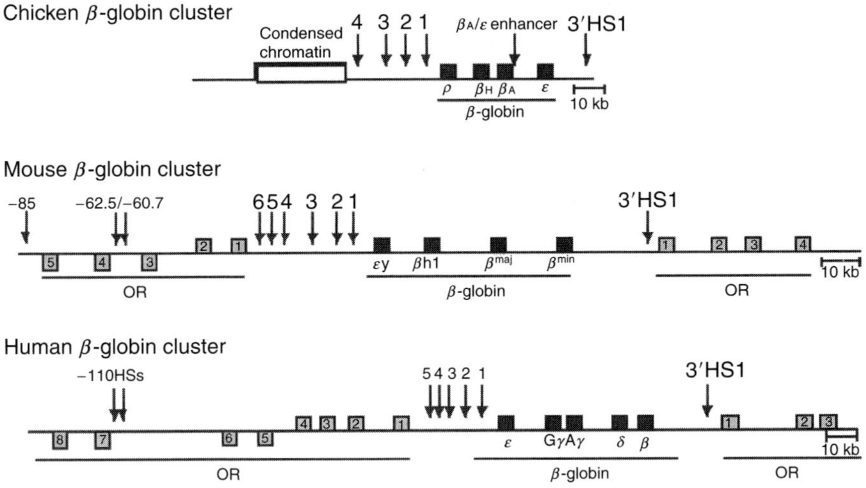

Figure 5.1 The β-globin loci. Indicated are the hypersensitive sites (arrows), the β-globin genes (black rectangles), and olfactory receptor genes (gray rectangles).

globin genes, named ε-, ${}^{G}\gamma$-, ${}^{A}\gamma$-, δ-, and β-globin that are expressed in the embryo (ε), in the fetus (${}^{G}\gamma$, ${}^{A}\gamma$) and at adult stages (δ, β), respectively. The genes are arranged on the linear chromosome template in the order of their expression during development, with ε-globin located most proximal to the LCR. Transgenic experiments that changed the order of fetal γ-globin and adult β-globin caused premature activation of the adult β-globin gene and earlier silencing of the fetal γ-globin gene in mice (Hanscombe *et al.*, 1991). This demonstrated that the adult β-globin gene can express in the fetus, but the presence of a competing γ-globin gene more proximal to the LCR prevents it from doing so. Thus, correct developmental expression depends on the gene order and the relative distance of genes to the LCR (Hanscombe *et al.*, 1991). The authors proposed that the β-globin genes compete for contacting the LCR for their activation, with proximal genes having a competitive advantage over more distal genes (Fig. 5.2). On gradual silencing of the proximal genes, the LCR can contact the more distal genes more often.

The phenomenon of gene competition was further investigated by means of fluorescence *in situ* hybridization (FISH) studies analyzing ongoing transcription of two β-globin genes in single cells under the microscope. Cells were taken at a developmental stage that allowed expression of both fetal and adult β-globin genes and the question was asked whether two genes present on the same allele can both be actively expressed at the same time or not. Strikingly, it was shown that either the one, or the other, but not both genes were active at a given time, with the key experiment showing that the fetal γ-globin gene was transcribed in cells that already

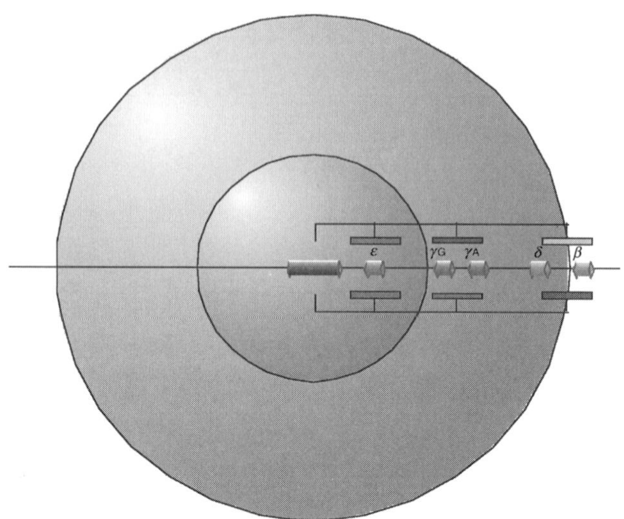

Figure 5.2 Spatial distance matters. Original model by Hanscombe *et al.* (1991), stating that correct developmental expression depends on the gene order and the relative spatial distance of genes to the LCR, which would function as a holocomplex.

contained adult β-globin gene transcripts (Wijgerde *et al.*, 1995). This observation was interpreted to suggest that at the time of β-globin gene switching, the LCR alternates between the fetal and adult genes for their activation and most importantly that the process of transcription is intermittent. This implied that the multiple regulatory elements that together build up the LCR spatially cluster to form a holocomplex (Ellis *et al.*, 1993) that can only contact a single gene at a given time.

The competitive advantage of an enhancer proximal gene is lost when genes are closely spaced at further distance from the regulator, as predicted by the looping model, but not by any other model. Again, this was first suggested using plasmid-based assays (Heuchel *et al.*, 1989) and later shown *in vivo* using large modified β-globin locus constructs in transgenic mice. A marked adult β-globin gene inserted at the position of the embryonic ε-globin gene had a clear preference to be transcribed over the endogenous adult β-globin gene located away from the LCR. However, this competitive advantage was lost when the marked β-globin gene was moved away from the LCR and positioned just in front of the endogenous β-globin gene (Dillon *et al.*, 1997).

Although all these experiments supported the idea that regulatory DNA elements physically contact genes to regulate their expression and that DNA topology therefore is important for genomic function, none of them directly showed *in vivo* that two distal elements linked in *cis* come in close spatial proximity, with intervening DNA looping out. Attempts failed to directly demonstrate DNA looping *in vivo* by visualization of enhancer, its target gene, and a looped out intervening sequence. The 3D resolution of microscopes was simply too limited while the resolution of the electron microscope was too high too analyze a sufficiently large number of cells. The development, 5 years ago, of two completely unrelated technologies unexpectedly opened the possibility to uncover the intricate folding of such relatively small DNA loci.

3. Novel Biochemical Approaches to Study DNA Topology: Insight into the Spatial Organization of the β-Globin Locus

In 2002, two very different biochemical approaches were described that would boost progress in our understanding of nuclear architecture: RNA-TRAP (Carter *et al.*, 2002) and 3C technology (Dekker *et al.*, 2002). Both techniques provided unprecedented insight into the intricate folding of gene loci and confirmed *in vivo* that regulatory DNA elements communicate with genes through looping, although it has yet to be shown directly that the gene is actively transcribed only when it is looped.

RNA TRAP involves the targeting of horseradish peroxidase (HRP)-labeled probes to nascent RNA transcripts, followed by quantification of HRP-catalyzed biotin deposition on chromatin nearby. It was originally applied to an actively transcribed mouse β-globin gene, and strikingly, a peak of biotin deposition was observed 50 kb away at hypersensitive site 2 (HS2) of the LCR. This implied that HS2 is in close spatial proximity to the actively transcribed β-globin gene (Carter *et al.*, 2002), which was in agreement with genetic data suggesting that HS2 is the most prominent enhancer element in the β-globin LCR (Fiering *et al.*, 1995; Fraser *et al.*, 1993).

3C (chromosome conformation capture) technology involves the cross-linking of DNA fragments that are together in the nuclear space of living cells, followed by restriction enzyme digestion and ligation between cross-linked DNA fragments. Quantitative polymerase chain reaction across ligation sites with primers diagnostic for selected DNA fragments then gives a measure for proximity frequencies between these fragments (Dekker *et al.*, 2002). The technique was originally applied to study the conformation of a yeast chromosome (Dekker *et al.*, 2002), and adapted to study the intricate folding of a mammalian gene locus, the β-globin locus (Tolhuis *et al.*, 2002). Currently, 3C technology is the preferred technique to study DNA interactions, possibly because RNA-TRAP technology is limited to genes such as β-globin that express at very high levels (so that sufficient primary transcripts are at the locus for efficient biotin deposition). An additional advantage of 3C technology is that this technique can analyze interactions of any (nontranscribed) sequence, while RNA-TRAP can only be applied to transcribed sequences.

When 3C was applied to the endogenous mouse β-globin locus, it was found that the actively transcribed adult β-globin genes contacted the HSs of the LCR, with intervening DNA containing the embryonic β-globin genes looping out. No such interactions were found in control nonexpressing tissue (Tolhuis *et al.*, 2002). Two independent observations, 3C and RNA-TRAP, therefore provided the first direct evidence that regulatory DNA elements in an endogenous mammalian gene locus loop toward the target genes that they activate in cells where the gene is actively transcribed.

In addition to the LCR-gene contacts, 3C technology also revealed long-range interactions with HSs upstream (HS-85 and HS-62/60) and downstream (3′HS1) of the β-globin locus. The function of these HSs was, and still is, unknown, but the data suggested that these outer HSs spatially cluster with the LCR and the active β-globin genes to form what is called an active chromatin hub (ACH) (Fig. 5.3). Intervening DNA containing inactive genes would be looped out from this configuration.

Significantly, in cells that do not express the β-globin genes, no such spatial conformation is observed. The inactive locus appears to adopt a seemingly linear conformation, without sites in the locus showing long-range interactions.

It is important to realize that 3C technology gives a steady-state average structure as present in the population of cells in the analysis. Thus, the fact that 3C technology identifies long-range interactions between all HSs and the active genes of the β-globin locus in red blood cells does not automatically mean that each locus in every cell adopts such a structure, nor does it show how many cells in the population have such interactions.

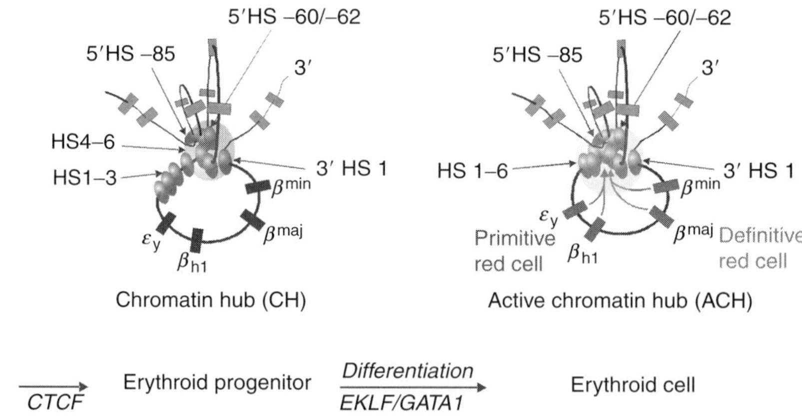

Figure 5.3 Formation of the β-globin active chromatin hub during erythroid differentiation. Indicated are the transcription factors that are required for each conformational transition. (See Color Insert.)

In fact, it is thought that these interactions are dynamic and their stability will depend on affinities between the proteins bound to these sites. Therefore, many different conformations are expected to be present in individual cells at any given time.

3C studies on the β-globin locus at earlier stages in development, when the embryonic β-globin genes are expressed in primitive erythroid cells, demonstrated that the β-globin genes switch their interaction with the ACH in relation to their expression status. Thus, in these primitive red blood cells, the embryonic (active), and not the adult (inactive), β-globin genes spatially clustered with the LCR and the outer HSs. This developmental change in the organization of DNA topology appeared conserved between the murine and human β-globin locus (Palstra *et al.*, 2003).

At each stage of development, erythroid progenitor cells mature into fully differentiated erythroid cells, being enucleated cells in the case of definitive erythropoiesis. Erythroid differentiation is accompanied by a dramatic increase in transcription rates of the β-globin genes. In erythroid progenitor cells, the β-globin genes express at levels that are comparable to those seen for most housekeeping genes. Later during differentiation, β-globin transcription efficiency is increased up to a 100-fold, reaching expression levels hardly ever observed with other genes. The β-globin genes need the LCR to reach these exceptionally high transcription rates. Thus, when the LCR is deleted from the endogenous β-globin locus in mice, transcription of all β-globin genes drops to 1–4% of the levels observed in wild-type animals (Epner *et al.*, 1998; Schubeler *et al.*, 2001). Interestingly, when 3C was applied to erythroid progenitor cells that express the β-globin genes at basal levels, a prestructure was found consisting of interactions exclusively between the outer HSs and the 5′ side of the LCR, while the β-globin genes, as well as HS1–3 of the LCR, appeared to loop out from this chromatin hub. The structural transition that occurs during erythroid differentiation and that establishes contacts between the LCR and the active genes thus coincides with an increase in β-globin gene expression levels, going from basal levels, that are LCR-independent, to extremely high levels that are LCR-dependent (Palstra *et al.*, 2003).

4. The Functional Significance of Long-Range Interactions Between Regulatory Sequences

All these data strongly suggest that physical contacts between the LCR and the genes are required for the LCR to regulate gene expression rates. It is therefore interesting to further think about the functional significance of long-range contacts between regulatory DNA elements and the formation of an ACH.

Enhancers, promoters, and other transcription regulatory DNA elements have in common that they bind often partially overlapping sets of transcription factors that will locally disrupt the nucleosome fiber, rendering these sites hypersensitive to nuclease digestion. In case of the β-globin locus, well-known erythroid-specific transcription factors that bind to the β-globin LCR and gene promoters and that are required for or have been implicated in β-globin gene expression are GATA-1 (Pevny *et al.*, 1991), EKLF (Nuez *et al.*, 1995; Perkins *et al.*, 1995), and NF-E2 (Andrews *et al.*, 1993). Spatial clustering of transcription regulatory DNA elements results in a high local concentration of binding sites for cognate transcription factors, which consequently accumulate at the site. Efficiency of transcription is proportional to the concentration of transcription factors involved (Droge and Muller-Hill, 2001), and the DNA contacts formed in the context of the β-globin ACH may therefore be necessary to drive efficient transcription of β-globin genes. Possibly, this may be analogous to the microscopically visible nucleolus, where nucleolar-organizing regions from different chromosomes (rDNA) physically meet to form a nuclear entity dedicated to the efficient transcription of the ribosomal DNA genes by RNA polymerase I. The β-globin ACH would be a first example of a structural entity dedicated to the efficient gene transcription by RNA polymerase II (de Laat and Grosveld, 2003).

Interestingly, structural analysis of the β-globin locus in cells lacking the transcription factors EKLF or GATA-1 showed that both are required to establish LCR-gene contacts, which may explain why β-globin transcription efficiency is severely reduced in their absence (Drissen *et al.*, 2004). The contacts between the outer HSs and the 5′ side of the LCR remain established at least when EKLF is absent, suggesting that EKLF, and possibly also GATA-1, is required for the β-globin locus to proceed from a chromatin hub to a fully functional ACH during erythroid differentiation. Both EKLF and GATA-1 are transcription factors that directly bind to DNA, but they seem to lack the capacity to form homodimers and therefore unlikely serve as bona fide bridging factors that bring together distant DNA elements. More likely, the ACH is a complex entity harboring multiple DNA-binding sites and transcription factors; the removal of one of these factors may, but not necessarily need to, cause destabilization and the adoption of a structure that can be formed independent of this factor. In two independent studies using transgenic mouse models containing modified human β-globin constructs, it was shown that deletion of a larger region containing HS3 is not sufficient for the ACH to collapse. One study showed that deletion of the β-globin gene promoter, in combination with HS3 but not alone, disrupted ACH formation (Patrinos *et al.*, 2004), while the other study suggested that removal of only core-HS3 destabilized the ACH (Fang *et al.*, 2007). The implication of these studies is that the promoter plays an important role in stabilizing the interactions.

While the functional significance of LCR-gene contacts seems clear, the relevance of the outer HSs participating in the ACH is less well understood. HS-85, HS-62/60, and 3′HS1 do not seem to be targets for binding of EKLF, GATA-1, or NF-E2, the factors that have been implicated in β-globin gene regulation. Instead, all these sites, as well as HS5 in the LCR, bind the transcription factor CTCF (Bulger *et al.*, 2003). CTCF (CCCTC-binding factor) is the prototype vertebrate protein exhibiting insulator activity that can act as an enhancer blocker *in vivo* and was suggested to also function as a barrier against repressive forces from nearby heterochromatin *in vitro* [Defossez and Gilson, 2002; Recillas-Targa *et al.*, 2002; reviewed by West *et al.* (2002)]. The fact that it binds to either end of the locus has led to the idea that CTCF may serve as an enhancer blocker that prevents the inappropriate activation of surrounding mouse olfactory receptor genes by the β-globin LCR in erythroid cells (Farrell *et al.*, 2002) or may be required to prevent spreading of heterochromatin into the β-globin locus. It is easy to envision how CTCF-mediated chromatin loops may accomplish such insulation. CTCF was indeed shown to be required for the long-range interactions between cognate binding sites in the β-globin locus, as was demonstrated by the depletion of the protein in conditional knockout mouse models as well as by the site-specific disruption of a CTCF-binding site in the β-globin locus (Simonis *et al.*, 2006). Disruption of CTCF-binding to the β-globin locus, however, had no effect on the expression of the β-globin genes nor did it cause activation of surrounding mouse olfactory receptor genes. The balance between active and repressive histone modifications changed but only locally at the binding sites and not elsewhere in the locus, demonstrating that CTCF does not serve to insulate the β-globin locus in erythroid cells (Simonis *et al.*, 2006). Expression of the β-globin genes was also unaltered in mouse models carrying combinations of deletions of the outer HSs (Bender *et al.*, 2006), leaving open the question why these evolutionary conserved sites participate in the tissue-specific interactions with the LCR and the active β-globin genes. Possibly, β-globin gene expression benefits from the CTCF-mediated loops to an extent that is sufficient for evolutionary selection but too limited to be detected by current technologies. Alternatively, it is a structural factor that binds to many loci merely for the folding of their DNA and that would only have a specialized function in "imprinted" regions, where it has been shown to act as an enhancer blocker (Bell and Felsenfeld, 2000; Hark *et al.*, 2000).

5. Long-Range Contacts in Other Gene Loci

After the first studies on the β-globin locus, 3C technology has been applied to many other gene loci in many different cell types. A few examples are sufficient to illustrate the conclusion that DNA folding and long-range

DNA contacts are important for the regulation of gene expression by remote control elements.

The interleukin locus has been analyzed extensively with respect to its spatial organization. The locus contains the cytokine genes 4, 5, and 13 that are expressed in T helper type 2 (TH2) cells under the control of an LCR that is located in between the genes. A poised structure was observed in various cell types, formed by contacts between the promoters of these genes. The LCR participated in these interactions in a more lineage-restricted manner and the transcription factors GATA-3 and STAT6 were shown to be required for these interactions (Spilianakis and Flavell, 2004). In an independent series of experiments on the same cytokine locus, special AT-rich sequence-binding protein 1 (SATB1) was demonstrated to be important for the formation of a unique transcriptionally active chromatin structure composed of numerous small loops, all anchored to SATB1 at their base. SATB1-mediated loop formation was suggested to be important for expression of the interleukin genes because SATB1-knockdown cells lacked these small loops and did not express the cytokine genes properly (Cai *et al.*, 2006). It is interesting to mention that SATB1 shows a very peculiar, cage-like, nuclear distribution (Cai *et al.*, 2003) and, like CTCF, has been suggested to be a nuclear matrix protein. In yet another study, deletion of a HS in the LCR (HS7) was shown to reduce contacts of the LCR with the TH2 cytokine genes and cause decreased expression of these genes (Lee *et al.*, 2005), showing again that regulatory DNA element functions over distance by contacting the target genes.

The same conclusion, but with a different twist, can be drawn from a study that combined 3C technology with a GAL4 knockin approach to analyze interactions in the imprinted *H19-Igf2* locus. It was demonstrated that the imprinting control region interacts with other differentially methylated regions in this locus in an allele-specific manner, such that the *H19* gene on the maternal allele and the *Igf2* gene on the paternal allele are brought in proximity with the enhancers that activate these genes on these respective alleles (Murrell *et al.*, 2004). CTCF, which binds exclusively to the maternal imprinting control region (Bell and Felsenfeld, 2000; Hark *et al.*, 2000), was shown to play a role in setting up a structure on the maternal allele that prevents the enhancers from interacting with the *Igf2* allele (Kurukuti *et al.*, 2006).

Evidence for DNA loops playing a role in gene silencing was previously found at the imprinted *Dlx5-Dlx6* locus. Here, the formation of a silent-chromatin loop was shown to depend on the methyl DNA-binding protein MeCP2; in the absence of MeCP2, the loop disappeared and gene contacts were made with distant activating sequences, resulting in the upregulation of *Dlx5* and *Dlx6* expression (Horike *et al.*, 2005).

Unlike the β-globin locus that resides in a large cluster of olfactory receptor genes that are inactive in erythroid cells, the α-globin gene locus is located in a gene-dense region full of housekeeping genes. When active in erythroid cells, the α-globin genes were found to contact key regulatory DNA elements as well as the promoters of surrounding housekeeping genes, while in nonexpressing cells the housekeeping gene, promoters still contact each other but not the α-globin genes (Zhou *et al.*, 2006).

The most spectacular example of long-range contacts between an enhancer and its target genes was recently provided by a report that suggested that olfactory receptor genes, present in clusters on different chromosomes, compete with each other to contact a single enhancer element for their activation, thereby ensuring that each olfactory neuron only expresses 1 allele of 1 of 1300 olfactory receptor genes (Lomvardas *et al.*, 2006).

A final observation based on 3C analysis that is relevant to mention here is that in yeast the promoter and terminator regions of active genes and genes poised to be transcribed were found to form a DNA loop (O'Sullivan *et al.*, 2004), suggesting perhaps that DNA topology facilitates transcription reinitiation by the recycling of polymerases.

6. Gene Positioning in the Nucleus: Spatial Coordination of Functionally Related Genes?

Probably the most important contribution of 3C technology so far has been its demonstration that DNA folds into configurations that brings together regulatory DNA elements to control gene expression, establishing the functional significance of chromatin architecture at least at the level of single gene loci. A major challenge for the future now seems to understand the significance of even more distant contacts, not between regulatory DNA elements that collectively control the expression of target genes, but between complete gene loci that appear to be in proximity in the nuclear space. In other words, how functionally important is a gene's exact position in the nuclear space and are the contacts it may have with other genes elsewhere on the same chromosome or even on other chromosomes?

Microscopy studies established that genomes are nonrandomly arranged in the nuclear space. This originally became apparent from studies showing that densely packed heterochromatin separates from more open euchromatin. More recently, FISH techniques showed that chromosomes occupy distinct territories in the nuclear space (Cremer and Cremer, 2001). Although transcription occurs throughout the nuclear interior (Cmarko *et al.*, 1999; Kimura *et al.*, 2002; Verschure *et al.*, 2003), active genes that cluster on chromosomes preferentially locate at the periphery or outside of their chromosome territory (Mahy *et al.*, 2002). Individual genes may migrate on changes in their transcription status, as measured against relatively large nuclear landmarks such as chromosome territories, centromeres, or the nuclear periphery (Brown *et al.*, 1997; Chambeyron and Bickmore, 2004; Grogan *et al.*, 2001; Merkenschlager *et al.*, 2004; Volpi *et al.*, 2000). Besides transcription, genomic organization is associated with the coordination of replication (Li *et al.*, 2003), recombination, the probability of loci to translocate (Roix *et al.*, 2003) (which can lead to malignancies), and the setting and resetting of epigenetic programs (Lemaitre *et al.*, 2005). Based on such observations, large-scale chromatin architecture is thought to be a key contributor to genomic function.

The differentiation of an erythroid cell is accompanied by heterochromatinization of the nuclear DNA content, shrinking of its cell nucleus and, finally, eviction of the nucleus. Interestingly, while most gene loci shut down in this increasingly repressive nuclear environment, the globin gene loci stay extremely active. It is tempting to assume that their nuclear position enables their efficient transcription also at later stages of differentiation.

Several observations suggest that the β-globin LCR can reposition physically linked DNA loci in the nuclear space. The functionally active HS2 of the β-globin LCR was shown to reposition a transgene away from centromeres, while inactivating mutations in HS2 abrogated its ability to do so, suggesting that a functional enhancer antagonizes silencing of gene loci by preventing their localization near repressive centromeric heterochromatin (Francastel *et al.*, 1999). The full LCR was suggested to position the β-globin locus outside its chromosome territory that reportedly would only occur in proerythroblasts that do not yet fully express the β-globin genes (Ragoczy *et al.*, 2003). An independent study, however, suggested that the β-globin locus always remains inside its own chromosome territory at all stages of erythroid differentiation (Brown *et al.*, 2006), making the relevance of the first observation uncertain. Based mostly on yeast studies, the nuclear periphery is thought to constitute an environment repressive for transcription, and possibly in agreement, the LCR has also been implicated in repositioning of the endogenous β-globin locus toward the nuclear interior in differentiating erythroid cells (Ragoczy *et al.*, 2006). Repositioning was not a prerequisite for transcription though, as the authors showed, the locus to be transcribed also at the nuclear rim. Colocalization with transcription factories, being nuclear sites of increased RNA polymerase II concentrations, also increased in an LCR-dependent manner (Ragoczy *et al.*, 2006), but it is not clear yet whether this reflects a more efficient *de novo* assembly of the transcription machinery onto the β-globin locus or a more efficient association with preassembled transcription factories (Chakalova *et al.*, 2005).

The idea that the nucleus contains preassembled transcription factories that may be important for the folding of the DNA in the nucleus has regained interest mostly based on studies that performed immunostaining on fixed cells with antibodies against RNA polymerase II. In one such study, a number of selected loci were shown to colocalize with the active β-globin locus in erythroid cells, despite their chromosomal location being tens of megabases apart. Colocalization occurred when the loci were actively transcribed and these interactions took place at transcription factories, as visualized by an antibody recognizing not only the transcriptionally active, but also the inactive form of RNA polymerase II. Moreover, two of the four genes that colocalized with β-globin were other erythroid-specific genes (Osborne *et al.*, 2004). Physical association between functionally related genes located at different positions in the genome has also been observed in other cell systems. For example, a study in naive T cells showed the association of, and possibly also cross talk between, two lineage-specific loci located on different chromosomes (Spilianakis *et al.*, 2005). Other recent reports suggested the interchromosomal association between two imprinted loci that would be dependent on the transcription factor CTCF (Ling *et al.*, 2006) and the interaction between the two X-inactivation centers early during female embryonic stem cell differentiation (Xu *et al.*, 2006).

Based on these reports, the idea seems to emerge that chromatin in the nucleus is shaped to a large extent by networks of functionally related genes that form long-range contacts at transcription factories to fine-tune each other's expression. However, the generality of this concept and the functional relevance of such long-range gene associations remain to be established. For example, live cell studies on the dynamics of RNAP II do not particularly support the idea of preassembled transcription factories (Kimura *et al.*, 2002) and it is important to realize that the nuclear distribution of RNA polymerase II observed after immunostaining differs depending on the fixation procedure and antibody used (Martin and Pombo, 2003). The issue whether the foci observed by immunostaining truly represent preassembled transcription sites reflect the transcription machinery assembled *de novo* on the chromatin template (Dundr *et al.*, 2002) or simply are a fixation artifact is still open. Moreover, it is rarely checked if specific gene associations are conserved in other species, which would be expected if they were to be critical for gene regulation. In the exceptional case where such controls were performed, the data argued against functional relevance of such interactions. For example, the active α- and β-globin genes, located on different chromosomes, often come together in human erythroid cells, but not in mouse cells, showing that spatial proximity is not important for α- and β-globin gene regulation (Brown *et al.*, 2006).

Clearly, a disadvantage of FISH studies is that they are biased toward the loci selected for that study and that they only allow for the analysis of a limited number of loci simultaneously. Results obtained by this technique are therefore largely anecdotal. In order to get a comprehensive understanding of nuclear architecture, high throughput methods, like the novel 4C technology, were recently developed to screen for interactions in the genome in an unbiased manner.

7. Future Directions for Studies on Nuclear Architecture: 4C Technology

4C technology (Fig. 5.4) combines 3C technology with microarray or sequencing approaches to allow for the unbiased identification of DNA elements that interact with a target sequence in the nuclear space (Ling *et al.*, 2006; Lomvardas *et al.*, 2006; Simonis *et al.*, 2006; Wurtele and Chartrand, 2006; Zhao *et al.*, 2006). The technique is comparable to chromatin immunoprecipitation-on-chip, but interrogates cross-linked DNA–DNA, instead of protein–DNA, interactions. So far, 4C has only been applied to a few target sequences (Ling *et al.*, 2006; Lomvardas *et al.*, 2006; Wurtele and Chartrand, 2006; Zhao *et al.*, 2006). Mostly, this involved the sequencing of a limited amount of interacting DNA elements, ranging from three to several hundreds of sequences analyzed, and probably therefore results were not necessarily compatible, even when the same target sequence was analyzed (Ling *et al.*, 2006; Zhao *et al.*, 2006). High-throughput sequencing approaches that collect thousands of sequences are expected to provide datasets that can be analyzed in a statistically meaningful way that is required to provide robust information about interacting DNA sequences.

In a 4C study focusing on the β-globin locus, tailored microarrays were used that simultaneously analyzed 400,000 possible interactions across 7 complete mouse chromosomes (Simonis *et al.*, 2006). Here, mathematic algorithms were applied to define the interacting DNA sequences. Subsequent analysis of these interactions by a novel high-resolution microscopy approach, called cryo-FISH (Branco and Pombo, 2006), confirmed that this strategy accurately identified long-range DNA interactions, even when such regions were together only 5% of the time.

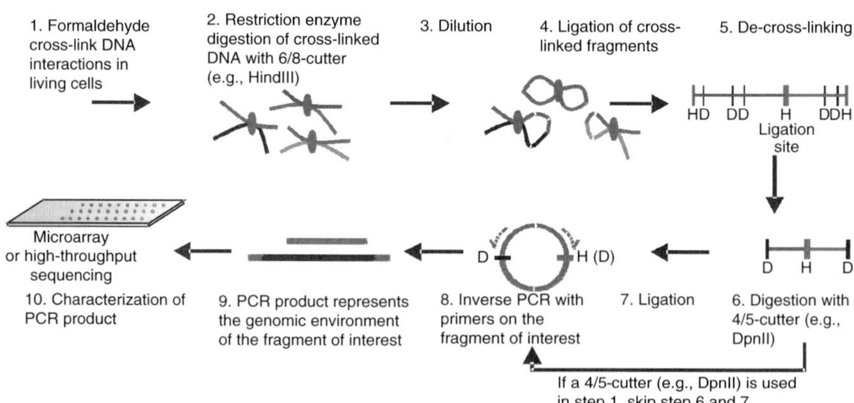

Figure 5.4 4C technology. Outline of the 4C procedure; see text for details. (See Color Insert.)

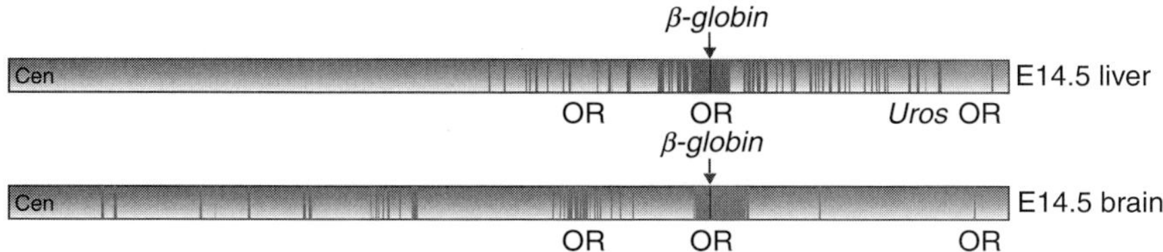

Figure 5.5 Long-range interactions within the β-globin locus. Schematic representation of regions of interaction with active (fetal liver, top) and inactive (fetal brain, bottom) *β-globin* on chromosome 7, as identified by 4C technology. Note that interacting regions are on average 150–200 kb and not drawn to scale. Taken from Simonis *et al.* (2006).

Strikingly, based on this analysis, it was found that the active β-globin locus in erythroid cells contacted a completely different set of loci than the inactive locus in brain cells (Fig. 5.5). When active, contacts were preferentially made with other actively transcribed loci located elsewhere on the chromosome, often tens of megabases away from β-globin and mostly located toward the telomere of the acrocentric mouse chromosome. When inactive, interactions occurred over similarly large distances, but now predominantly with loci that were not transcribed and that mostly located toward the centromeric side of the chromosome. In both tissues, tens of regions were identified to interact with β-globin, each spanning ~200 kb, meaning that interactions were not confined to single genes or gene promoters. The erythroid-specific genes that were previously identified by FISH to interact with the active β-globin locus (Osborne *et al.*, 2004) were also identified by 4C technology, but 4C also showed that these genes lie in larger interacting regions that additionally contain other, unrelated, genes. Overall, it was found that regions interacting with the active β-globin locus did not preferentially contain erythroid-specific genes. When 4C technology was applied to a housekeeping gene located in a gene-dense cluster of other active genes, this gene was similarly found to contact many regions in *cis* (on its own chromosome), but also in *trans* (on other chromosomes), and for this, gene contacts were essentially the same between the two tissues. Thus, the switch in genomic environments observed for β-globin between erythroid and brain cells was not due to a peculiarity of the tissues but most likely is directly related to the altered transcriptional status of the locus in the two tissues (Simonis *et al.*, 2006).

Clearly, this unbiased high-throughput technique provides new insight into DNA folding. 4C technology is expected to contribute importantly to a comprehensive understanding of nuclear architecture, picking up interactions not previously anticipated and putting the relative frequency of interactions in perspective. FISH is not very well suited for screening purposes, but will be required for validation and, most importantly, for single-cell analysis. A current problem of FISH approaches, however, is that resolution, even of confocal microscopes, is still poor in molecular terms, meaning that perfect colocalization of two spots measured by FISH does not need to imply direct, or functionally significant, contact; in fact, they could still be more than 200 nm apart. Thus, there is an important need for high-resolution microscopy approaches that better identify truly interacting sequences. In this respect, 4 pi microscopy is worth mentioning that provides an almost 10 times better resolution compared to confocal microscopes and should also allow for life cell-imaging studies (Egner and Hell, 2005). The combination of novel biochemical methodologies, such as 4C technology, and higher resolution microscopy techniques, such as 4 pi microscopy, should uncover the significance of long-range gene contacts and reveal whether DNA–DNA interactions beyond those observed between regulatory DNA elements that collectively regulate the expression of a target gene are functionally meaningful.

REFERENCES

Andrews, N. C., Erdjument-Bromage, H., Davidson, M. B., Tempst, P., and Orkin, S. H. (1993). Erythroid transcription factor NF-E2 is a haematopoietic-specific basic-leucine zipper protein. *Nature* **362,** 722–728.

Banerji, J., Rusconi, S., and Schaffner, W. (1981). Expression of a beta-globin gene is enhanced by remote SV40 DNA sequences. *Cell* **27,** 299–308.

Bell, A. C., and Felsenfeld, G. (2000). Methylation of a CTCF-dependent boundary controls imprinted expression of the Igf2 gene. *Nature* **405,** 482–485.

Bender, M. A., Byron, R., Ragoczy, T., Telling, A., Bulger, M., and Groudine, M. (2006). Flanking HS-62.5 and 3′ HS1, and regions upstream of the LCR, are not required for beta-globin transcription. *Blood* **108,** 1395–1401.

Bernal, J. D., Fankuchen, I., and Perutz, M. F. (1938). X-ray study of chymotrypsin and hemoglobin. *Nature* **141,** 523–524.

Bickel, S., and Pirrotta, V. (1990). Self-association of the *Drosophila* zeste protein is responsible for transvection effects. *EMBO J.* **9,** 2959–2967.

Branco, M. R., and Pombo, A. (2006). Intermingling of chromosome territories in interphase suggests role in translocations and transcription-dependent associations. *PLoS Biol.* **4,** e138.

Brown, J. M., Leach, J., Reittie, J. E., Atzberger, A., Lee-Prudhoe, J., Wood, W. G., Higgs, D. R., Iborra, F. J., and Buckle, V. J. (2006). Coregulated human globin genes are frequently in spatial proximity when active. *J. Cell Biol.* **172,** 177–187.

Brown, K. E., Guest, S. S., Smale, S. T., Hahm, K., Merkenschlager, M., and Fisher, A. G. (1997). Association of transcriptionally silent genes with Ikaros complexes at centromeric heterochromatin. *Cell* **91,** 845–854.

Bulger, M., Schubeler, D., Bender, M. A., Hamilton, J., Farrell, C. M., Hardison, R. C., and Groudine, M. (2003). A complex chromatin landscape revealed by patterns of nuclease sensitivity and histone modification within the mouse beta-globin locus. *Mol. Cell. Biol.* **23,** 5234–5244.

Cai, S., Han, H. J., and Kohwi-Shigematsu, T. (2003). Tissue-specific nuclear architecture and gene expression regulated by SATB1. *Nat. Genet.* **34,** 42–51.

Cai, S., Lee, C. C., and Kohwi-Shigematsu, T. (2006). SATB1 packages densely looped, transcriptionally active chromatin for coordinated expression of cytokine genes. *Nat. Genet.* **38,** 1278–1288.

Carter, D., Chakalova, L., Osborne, C. S., Dai, Y. F., and Fraser, P. (2002). Long-range chromatin regulatory interactions *in vivo. Nat. Genet.* **32,** 623–626.

Chakalova, L., Carter, D., Debrand, E., Goyenechea, B., Horton, A., Miles, J., Osborne, C., and Fraser, P. (2005). Developmental regulation of the beta-globin gene locus. *Prog. Mol. Subcell. Biol.* **38,** 183–206.

Chambeyron, S., and Bickmore, W. A. (2004). Chromatin decondensation and nuclear reorganization of the HoxB locus upon induction of transcription. *Genes Dev.* **18,** 1119–1130.

Choi, O. R., and Engel, J. D. (1988). Developmental regulation of beta-globin gene switching. *Cell* **55,** 17–26.

Cmarko, D., Verschure, P. J., Martin, T. E., Dahmus, M. E., Krause, S., Fu, X. D., van Driel, R., and Fakan, S. (1999). Ultrastructural analysis of transcription and splicing in the cell nucleus after bromo-UTP microinjection. *Mol. Biol. Cell* **10,** 211–223.

Cremer, T., and Cremer, C. (2001). Chromosome territories, nuclear architecture and gene regulation in mammalian cells. *Nat. Rev. Genet.* **2,** 292–301.

de Laat, W., and Grosveld, F. (2003). Spatial organization of gene expression: The active chromatin hub. *Chromosome. Res.* **11,** 447–459.

de Villiers, J., Olson, L., Banerji, J., and Schaffner, W. (1983). Analysis of the transcriptional enhancer effect. *Cold Spring Harb. Symp. Quant. Biol.* **47**(Pt 2), 911–919.

Defossez, P. A., and Gilson, E. (2002). The vertebrate protein CTCF functions as an insulator in Saccharomyces cerevisiae. *Nucleic Acids Res.* **30,** 5136–5141.

Dekker, J., Rippe, K., Dekker, M., and Kleckner, N. (2002). Capturing chromosome conformation. *Science* **295,** 1306–1311.

Dillon, N., Trimborn, T., Strouboulis, J., Fraser, P., and Grosveld, F. (1997). The effect of distance on long-range chromatin interactions. *Mol. Cell* **1,** 131–139.

Drissen, R., Palstra, R. J., Gillemans, N., Splinter, E., Grosveld, F., Philipsen, S., and de Laat, W. (2004). The active spatial organization of the beta-globin locus requires the transcription factor EKLF. *Genes Dev.* **18,** 2485–2490.

Droge, P., and Muller-Hill, B. (2001). High local protein concentrations at promoters: Strategies in prokaryotic and eukaryotic cells. *Bioessays* **23,** 179–183.

Dunaway, M., and Droge, P. (1989). Transactivation of the Xenopus rRNA gene promoter by its enhancer. *Nature* **341,** 657–659.

Dundr, M., Hoffmann-Rohrer, U., Hu, Q., Grummt, I., Rothblum, L. I., Phair, R. D., and Misteli, T. (2002). A kinetic framework for a mammalian RNA polymerase *in vivo. Science* **298,** 1623–1626.

Egner, A., and Hell, S. W. (2005). Fluorescence microscopy with super-resolved optical sections. *Trends Cell Biol.* **15,** 207–215.

Ellis, J., Talbot, D., Dillon, N., and Grosveld, F. (1993). Synthetic human beta-globin 5′HS2 constructs function as locus control regions only in multicopy transgene concatamers. *EMBO J.* **12,** 127–134.

Epner, E., Reik, A., Cimbora, D., Telling, A., Bender, M. A., Fiering, S., Enver, T., Martin, D. I., Kennedy, M., Keller, G., and Groudine, M. (1998). The beta-globin LCR is not necessary for an open chromatin structure or developmentally regulated transcription of the native mouse beta-globin locus. *Mol. Cell* **2,** 447–455.

Fang, X., Xiang, P., Yin, W., Stamatoyannopoulos, G., and Li, Q. (2007). Cooperativeness of the higher chromatin structure of the beta-globin locus revealed by the deletion mutations of DNase I hypersensitive site 3 of the LCR. *J. Mol. Biol.* **365,** 31–37.

Farrell, C. M., West, A. G., and Felsenfeld, G. (2002). Conserved CTCF insulator elements flank the mouse and human beta-globin loci. *Mol. Cell. Biol.* **22,** 3820–3831.

Fiering, S., Epner, E., Robinson, K., Zhuang, Y., Telling, A., Hu, M., Martin, D. I., Enver, T., Ley, T. J., and Groudine, M. (1995). Targeted deletion of 5′HS2 of the murine beta-globin LCR reveals that it is not essential for proper regulation of the beta-globin locus. *Genes Dev.* **9,** 2203–2213.

Francastel, C., Walters, M. C., Groudine, M., and Martin, D. I. (1999). A functional enhancer suppresses silencing of a transgene and prevents its localization close to centrometric heterochromatin. *Cell* **99,** 259–269.

Fraser, P., Pruzina, S., Antoniou, M., and Grosveld, F. (1993). Each hypersensitive site of the human beta-globin locus control region confers a different developmental pattern of expression on the globin genes. *Genes Dev.* **7,** 106–113.

Grogan, J. L., Mohrs, M., Harmon, B., Lacy, D. A., Sedat, J. W., and Locksley, R. M. (2001). Early transcription and silencing of cytokine genes underlie polarization of T helper cell subsets. *Immunity* **14,** 205–215.

Grosveld, F., van Assendelft, G. B., Greaves, D. R., and Kollias, G. (1987). Position-independent, high-level expression of the human beta-globin gene in transgenic mice. *Cell* **51,** 975–985.

Hanscombe, O., Whyatt, D., Fraser, P., Yannoutsos, N., Greaves, D., Dillon, N., and Grosveld, F. (1991). Importance of globin gene order for correct developmental expression. *Genes Dev.* **5,** 1387–1394.

Hark, A. T., Schoenherr, C. J., Katz, D. J., Ingram, R. S., Levorse, J. M., and Tilghman, S. M. (2000). CTCF mediates methylation-sensitive enhancer-blocking activity at the H19/Igf2 locus. *Nature* **405,** 486–489.

Heuchel, R., Matthias, P., and Schaffner, W. (1989). Two closely spaced promoters are equally activated by a remote enhancer: Evidence against a scanning model for enhancer action. *Nucleic Acids Res.* **17,** 8931–8947.

Horike, S., Cai, S., Miyano, M., Cheng, J. F., and Kohwi-Shigematsu, T. (2005). Loss of silent-chromatin looping and impaired imprinting of DLX5 in Rett syndrome. *Nat. Genet.* **37,** 31–40.

Ingram, V. M. (1957). Gene mutations in human haemoglobin: The chemical difference between normal and sickle cell haemoglobin. *Nature* **180,** 326–328.

Jeffreys, A. J., and Flavell, R. A. (1977). The rabbit beta-globin gene contains a large insert in the coding sequence. *Cell* **12,** 1097–1108.

Kimura, H., Sugaya, K., and Cook, P. R. (2002). The transcription cycle of RNA polymerase II in living cells. *J. Cell Biol.* **159,** 777–782.

Kleinjan, D. A., and van Heyningen, V. (2005). Long-range control of gene expression: Emerging mechanisms and disruption in disease. *Am. J. Hum. Genet.* **76,** 8–32.

Kurukuti, S., Tiwari, V. K., Tavoosidana, G., Pugacheva, E., Murrell, A., Zhao, Z., Lobanenkov, V., Reik, W., and Ohlsson, R. (2006). CTCF binding at the H19 imprinting control region mediates maternally inherited higher-order chromatin conformation to restrict enhancer access to Igf2. *Proc. Natl. Acad. Sci. USA* **103,** 10684–10689.

Lee, G. R., Spilianakis, C. G., and Flavell, R. A. (2005). Hypersensitive site 7 of the TH2 locus control region is essential for expressing TH2 cytokine genes and for long-range intrachromosomal interactions. *Nat. Immunol.* **6,** 42–48.

Lemaitre, J. M., Danis, E., Pasero, P., Vassetzky, Y., and Mechali, M. (2005). Mitotic remodeling of the replicon and chromosome structure. *Cell* **123,** 787–801.

Lettice, L. A., Heaney, S. J., Purdie, L. A., Li, L., de Beer, P., Oostra, B. A., Goode, D., Elgar, G., Hill, R. E., and de Graaff, E. (2003). A long-range Shh enhancer regulates expression in the developing limb and fin and is associated with preaxial polydactyly. *Hum. Mol. Genet.* **12,** 1725–1735.

Li, F., Chen, J., Solessio, E., and Gilbert, D. M. (2003). Spatial distribution and specification of mammalian replication origins during G1 phase. *J. Cell Biol.* **161,** 257–266.

Ling, J. Q., Li, T., Hu, J. F., Vu, T. H., Chen, H. L., Qiu, X. W., Cherry, A. M., and Hoffman, A. R. (2006). CTCF mediates interchromosomal colocalization between Igf2/H19 and Wsb1/Nf1. *Science* **312,** 269–272.

Lomvardas, S., Barnea, G., Pisapia, D. J., Mendelsohn, M., Kirkland, J., and Axel, R. (2006). Interchromosomal interactions and olfactory receptor choice. *Cell* **126,** 403–413.

Mahmoudi, T., Katsani, K. R., and Verrijzer, C. P. (2002). GAGA can mediate enhancer function in trans by linking two separate DNA molecules. *EMBO J.* **21,** 1775–1781.

Mahy, N. L., Perry, P. E., and Bickmore, W. A. (2002). Gene density and transcription influence the localization of chromatin outside of chromosome territories detectable by FISH. *J. Cell Biol.* **159,** 753–763.

Mantei, N., Boll, W., and Weissmann, C. (1979). Rabbit beta-globin mRNA production in mouse L cells transformed with cloned rabbit beta-globin chromosomal DNA. *Nature* **281,** 40–46.

Marotta, C. A., Forget, B. G., Weissman, S. M., Verma, I. M., McCaffrey, R. P., and Baltimore, D. (1974). Nucleotide sequences of human globin messenger RNA. *Proc. Natl. Acad. Sci. USA* **71,** 2300–2304.

Martin, S., and Pombo, A. (2003). Transcription factories: Quantitative studies of nanostructures in the mammalian nucleus. *Chromosome. Res.* **11,** 461–470.

Merkenschlager, M., Amoils, S., Roldan, E., Rahemtulla, A., O'Connor, E., Fisher, A. G., and Brown, K. E. (2004). Centromeric repositioning of coreceptor loci predicts their stable silencing and the CD4/CD8 lineage choice. *J. Exp. Med.* **200,** 1437–1444.

Mueller-Storm, H. P., Sogo, J. M., and Schaffner, W. (1989). An enhancer stimulates transcription in trans when attached to the promoter via a protein bridge. *Cell* **58,** 767–777.

Murrell, A., Heeson, S., and Reik, W. (2004). Interaction between differentially methylated regions partitions the imprinted genes Igf2 and H19 into parent-specific chromatin loops. *Nat. Genet.* **36,** 889–893.

Nuez, B., Michalovich, D., Bygrave, A., Ploemacher, R., and Grosveld, F. (1995). Defective haematopoiesis in fetal liver resulting from inactivation of the EKLF gene. *Nature* **375,** 316–318.

O'Sullivan, J. M., Tan-Wong, S. M., Morillon, A., Lee, B., Coles, J., Mellor, J., and Proudfoot, N. J. (2004). Gene loops juxtapose promoters and terminators in yeast. *Nat. Genet.* **36,** 1014–1018.

Osborne, C. S., Chakalova, L., Brown, K. E., Carter, D., Horton, A., Debrand, E., Goyenechea, B., Mitchell, J. A., Lopes, S., Reik, W., and Fraser, P. (2004). Active genes dynamically colocalize to shared sites of ongoing transcription. *Nat. Genet.* **36,** 1065–1071.

Palstra, R. J., Tolhuis, B., Splinter, E., Nijmeijer, R., Grosveld, F., and de Laat, W. (2003). The beta-globin nuclear compartment in development and erythroid differentiation. *Nat. Genet.* **35,** 190–194.

Patrinos, G. P., de Krom, M., de Boer, E., Langeveld, A., Imam, A. M., Strouboulis, J., de Laat, W., and Grosveld, F. G. (2004). Multiple interactions between regulatory regions are required to stabilize an active chromatin hub. *Genes Dev.* **18,** 1495–1509.

Pauling, L., Itano, H. A., Singer, S. J., and Wells, I. C. (1949). Sickle cell anemia a molecular disease. *Science* **110,** 543–548.

Perkins, A. C., Sharpe, A. H., and Orkin, S. H. (1995). Lethal beta-thalassaemia in mice lacking the erythroid CACCC-transcription factor EKLF. *Nature* **375,** 318–322.

Pevny, L., Simon, M. C., Robertson, E., Klein, W. H., Tsai, S. F., D'Agati, V., Orkin, S. H., and Costantini, F. (1991). Erythroid differentiation in chimaeric mice blocked by a targeted mutation in the gene for transcription factor GATA-1. *Nature* **349,** 257–260.

Poon, R., Paddock, G. V., Heindell, H., Whitcome, P., Salser, W., Kacian, D., Bank, A., Gambino, R., and Ramirez, F. (1974). Nucleotide sequence analysis of RNA synthesized from rabbit globin complementary DNA. *Proc. Natl. Acad. Sci. USA* **71,** 3502–3506.

Proudfoot, N. J., and Brownlee, G. G. (1974). Sequence at the 3′ end of globin mRNA shows homology with immunoglobulin light chain mRNA. *Nature* **252,** 359–362.

Ptashne, M. (1986). Gene regulation by proteins acting nearby and at a distance. *Nature* **322,** 697–701.

Ragoczy, T., Telling, A., Sawado, T., Groudine, M., and Kosak, S. T. (2003). A genetic analysis of chromosome territory looping: Diverse roles for distal regulatory elements. *Chromosome. Res.* **11,** 513–525.

Ragoczy, T., Bender, M. A., Telling, A., Byron, R., and Groudine, M. (2006). The locus control region is required for association of the murine beta-globin locus with engaged transcription factories during erythroid maturation. *Genes Dev.* **20,** 1447–1457.

Recillas-Targa, F., Pikaart, M. J., Burgess-Beusse, B., Bell, A. C., Litt, M. D., West, A. G., Gaszner, M., and Felsenfeld, G. (2002). Position-effect protection and enhancer blocking by the chicken beta-globin insulator are separable activities. *Proc. Natl. Acad. Sci. USA* **99,** 6883–6888.

Roix, J. J., McQueen, P. G., Munson, P. J., Parada, L. A., and Misteli, T. (2003). Spatial proximity of translocation-prone gene loci in human lymphomas. *Nat. Genet.* **34,** 287–291.

Rougeon, F., Kourilsky, P., and Mach, B. (1975). Insertion of a rabbit beta-globin gene sequence into an E. coli plasmid. *Nucleic Acids Res.* **2,** 2365–2378.

Schubeler, D., Groudine, M., and Bender, M. A. (2001). The murine beta-globin locus control region regulates the rate of transcription but not the hyperacetylation of histones at the active genes. *Proc. Natl. Acad. Sci. USA* **98,** 11432–11437.

Simonis, M., Klous, P., Splinter, E., Moshkin, Y., Willemsen, R., de Wit, E., van Steensel, B., and de Laat, W. (2006). Nuclear organization of active and inactive chromatin domains uncovered by chromosome conformation capture-on-chip (4C). *Nat. Genet.* **38,** 1348–1354.

Spilianakis, C. G., and Flavell, R. A. (2004). Long-range intrachromosomal interactions in the T helper type 2 cytokine locus. *Nat. Immunol.* **5,** 1017–1027.

Spilianakis, C. G., Lalioti, M. D., Town, T., Lee, G. R., and Flavell, R. A. (2005). Interchromosomal associations between alternatively expressed loci. *Nature* **435,** 637–645.

Tilghman, S. M., Tiemeier, D. C., Seidman, J. G., Peterlin, B. M., Sullivan, M., Maizel, J. V., and Leder, P. (1978). Intervening sequence of DNA identified in the structural portion of a mouse beta-globin gene. *Proc. Natl. Acad. Sci. USA* **75,** 725–729.

Tolhuis, B., Palstra, R. J., Splinter, E., Grosveld, F., and de Laat, W. (2002). Looping and interaction between hypersensitive sites in the active beta-globin locus. *Mol. Cell* **10,** 1453–1465.

Verschure, P. J., van der Kraan, I., Manders, E. M., Hoogstraten, D., Houtsmuller, A. B., and van Driel, R. (2003). Condensed chromatin domains in the mammalian nucleus are accessible to large macromolecules. *EMBO Rep.* **4,** 861–866.

Volpi, E. V., Chevret, E., Jones, T., Vatcheva, R., Williamson, J., Beck, S., Campbell, R. D., Goldsworthy, M., Powis, S. H., Ragoussis, J., Trowsdale, J., and Sheer, D. (2000). Large-scale chromatin organization of the major histocompatibility complex and other regions of human chromosome 6 and its response to interferon in interphase nuclei. *J. Cell Sci.* **113**(Pt 9), 1565–1576.

Wasylyk, B., Wasylyk, C., Augereau, P., and Chambon, P. (1983). The SV40 72 bp repeat preferentially potentiates transcription starting from proximal natural or substitute promoter elements. *Cell* **32,** 503–514.

West, A. G., Gaszner, M., and Felsenfeld, G. (2002). Insulators: Many functions, many mechanisms. *Genes Dev.* **16,** 271–288.

Wijgerde, M., Grosveld, F., and Fraser, P. (1995). Transcription complex stability and chromatin dynamics *in vivo*. *Nature* **377,** 209–213.

Wurtele, H., and Chartrand, P. (2006). Genome-wide scanning of HoxB1-associated loci in mouse ES cells using an open-ended chromosome conformation capture methodology. *Chromosome Res.* **14,** 477–495.

Xu, N., Tsai, C. L., and Lee, J. T. (2006). Transient homologous chromosome pairing marks the onset of X inactivation. *Science* **311,** 1149–1152.

Zhao, Z., Tavoosidana, G., Sjolinder, M., Gondor, A., Mariano, P., Wang, S., Kanduri, C., Lezcano, M., Sandhu, K. S., Singh, U., Pant, V., Tiwari, V., *et al.* (2006). Circular chromosome conformation capture (4C) uncovers extensive networks of epigenetically regulated intra- and interchromosomal interactions. *Nat. Genet.* **38,** 1341–1347.

Zhou, G. L., Xin, L., Song, W., Di, L. J., Liu, G., Wu, X. S., Liu, D. P., and Liang, C. C. (2006). Active chromatin hub of the mouse alpha-globin locus forms in a transcription factory of clustered housekeeping genes. *Mol. Cell. Biol.* **26,** 5096–5105.

CHAPTER SIX

Iron Homeostasis and Erythropoiesis

Diedra M. Wrighting* *and* Nancy C. Andrews*

Contents

Abstract

Erythrocytes require iron to perform their duty as oxygen carriers. Mammals have evolved a mechanism to maintain systemic iron within an optimal range that fosters erythroid development and function while satisfying other body iron needs. This chapter reviews erythroid iron uptake and utilization as well as systemic factors that influence iron availability. One of these factors is hepcidin, a circulating peptide hormone that maintains iron homeostasis. Elevated levels of hepcidin in the bloodstream effectively shut off iron absorption by disabling the iron exporter ferroportin. Conversely, low levels of circulating hepcidin allow ferroportin to export iron into the bloodstream. Aberrations in hepcidin expression or responsiveness to hepcidin result in disorders of iron deficiency and iron overload. It is clear that erythroid precursors communicate their iron needs to the liver to influence the production of hepcidin and thus the amount of iron

* Children's Hospital Boston and Harvard Medical School, Boston, Massachusetts 02115

Current Topics in Developmental Biology, Volume 82
ISSN 0070-2153, DOI: 10.1016/S0070-2153(07)00006-3

available for use. However, the mechanism by which erythroid cells accomplish this remains unclear and is an area of active investigation.

1. Introduction

Iron is an essential nutrient that is required for the oxygen-carrying capacity of hemoglobin. Failure to incorporate adequate iron into heme results in impaired erythrocyte maturation, leading to microcytic, hypochromic anemia. Therefore, circulating factors that modulate iron availability are of major importance in erythropoiesis.

Normally, the total body iron endowment is maintained within a tight range between 3 and 5 g (Bothwell *et al.*, 1979). Systemic iron is distributed among erythrocyte precursors in the bone marrow, tissue macrophages, liver, and all other tissues, with the largest amount found in circulating erythrocytes. Homeostasis is maintained by regulating the levels of plasma iron. Hepcidin, a circulating peptide hormone, has recently emerged as a key modulator of plasma iron concentration, and, thus, a central regulator of iron homeostasis. This chapter will review erythroid iron uptake and utilization, and the systemic factors that affect it.

2. Iron Metabolism and Homeostasis

2.1. Iron metabolism

Nearly all circulating iron is bound by the abundant serum glycoprotein transferrin. Transferrin can carry one (monoferric transferrin) or two (holotransferrin) atoms of iron per protein molecule. Erythroid precursors are the primary consumers of circulating transferrin-bound iron. During differentiation, changing rates of hemoglobin production correlate with variations in the cell surface complement of transferrin receptor-1 (TFR1) (Chan and Gerhardt, 1992). Fluctuations in TFR1 expression control the amount of transferrin-bound iron entering into erythroid cells.

The process by which transferrin delivers iron to these cells is called the transferrin cycle (Klausner *et al.*, 1983). Upon binding to TFR1 at the cell surface, transferrin and its iron cargo are endocytosed. These endosomal compartments are actively acidified by proton pumps. Acidification facilitates iron release from transferrin because low pH decreases the affinity of the protein for iron (Dautry-Varsat *et al.*, 1983). Iron then leaves the endosome through divalent metal ion transporter 1 (DMT1, also known as Nramp2 and SLC11A2) to become available for heme biosynthesis (Canonne-Hergaux *et al.*, 2001; Fleming *et al.*, 1998; Su *et al.*, 1998).

The insertion of iron into protoporphyrin IX, the final step in heme biosynthesis, occurs in the mitochondrion. Mitoferrin, a protein mutated in anemic zebra fish, was recently shown to act as a mitochondrial iron importer necessary for heme biosynthesis (Shaw *et al.*, 2006). However, another group has postulated that iron is transferred from endosomes directly into mitochondria through direct membrane contacts between the organelles (Sheftel *et al.*, 2007). The discrepancies between these two models of mitochondrial iron uptake have not yet been resolved. The fates of TFR1 and apotransferrin are more certain—they are recycled to the cell surface and circulation, respectively, where they repeat the cycle.

The amount of circulating, transferrin-bound iron is determined by three coordinated processes: macrophage iron recycling, duodenal iron absorption, and hepatic iron storage. When iron is administered therapeutically, it is assimilated by one or more of these three tissues, which play critical roles in iron metabolism and the maintenance of iron homeostasis. As discussed later, human iron disorders may result from perturbation of overall body iron amounts, or tissue iron distribution, or both.

2.2. Macrophage iron recycling

Normal human erythrocytes have a finite life span of approximately 4 months. Tissue macrophages remove senescent and damaged erythrocytes from circulation and breakdown hemoglobin to recycle iron, supplying most of the requirement for new erythropoiesis (Knutson and Wessling-Resnick, 2003). The process by which macrophages distinguish aged and damaged erythrocytes is not fully understood, but it is likely that morphological changes in the erythrocyte membrane facilitate recognition and uptake by macrophages (Bratosin *et al.*, 1998). Binding of erythrocytes to the macrophage cell surface initiates phagocytosis and lysosome-mediated degradation of the erythrocyte membrane. Heme oxygenases catalyze the oxidation of heme to biliverdin, free iron, and carbon monoxide. Because iron accumulates in macrophages of heme oxygenase-1 (HO-1) knockout mice and HO-1-deficient humans, it appears that HO-1 is required for recycling of the metal from heme (Poss and Tonegawa, 1997; Yachie *et al.*, 1999). Similar to the transferrin cycle, liberated iron may be pumped from the phagosome into the cytoplasm by DMT1, though this has not been definitively established (Jabado *et al.*, 2002). Heme-derived iron can be utilized by the cell, sequestered within the multimeric iron storage protein ferritin or exported into the plasma.

The transmembrane transporter ferroportin (also known as IREG1, SLC40A1, MTP1) is activated in macrophages after erythrophagocytosis (Canonne-Hergaux *et al.*, 2006). Ferroportin is the only cellular iron exporter that has been identified in vertebrates (Abboud and Haile, 2000; Donovan *et al.*, 2000; McKie *et al.*, 2000). It is expressed in cells of the extraembryonic visceral endoderm that provide nourishment to the developing embryo, in the intestinal epithelium and in spleen and liver macrophages that recycle iron (Donovan *et al.*, 2005). Global disruption of the murine ferroportin gene results in early embryonic lethality due to an inability of the developing embryo to acquire iron. However, when ferroportin is deleted in the embryo proper, sparing the extraembryonic visceral endoderm, mice survive until birth, but rapidly become anemic when they must rely on intestinal absorption for iron accumulation (Donovan *et al.*, 2005). Under these circumstances, tissue macrophages contain large amounts of iron, apparently because they are unable to export it to the circulation. These results support the idea that most iron liberated from heme in macrophages is mobilized through ferroportin-mediated iron export to be utilized for erythropoiesis. Because ferroportin transports ferrous iron, iron must be oxidized to its ferric form in order to bind circulating transferrin. A circulating ferroxidase, ceruloplasmin, is thought to carry out the oxidation of iron exported from macrophages (Harris *et al.*, 1999). Recently, however, an additional role for ceruloplasmin was discovered. Ceruloplasmin appears to be necessary to keep ferroportin on the cell surface (De Domenico *et al.*, 2007a). For both of these reasons, it is not surprising that ceruloplasmin deficiency (aceruloplasminemia) leads to tissue iron loading with low transferrin saturation and, often, mild anemia (Harris *et al.*, 1995).

2.3. Duodenal iron absorption

In contrast to some other metals, there is no regulated mechanism for iron excretion through the liver or kidneys. Nonetheless, macrophage-mediated iron recycling alone cannot sustain erythropoiesis over the long term. Early in life, the overall iron endowment must be increased to support growth. Later, small obligatory iron losses from bleeding and exfoliation of skin and mucosal cells would lead to negative iron balance if not offset by continuous iron intake. Thus, iron balance is achieved through regulated dietary iron absorption (Andrews, 2000). Dietary iron absorption occurs in the most proximal part of the duodenum, the first section of the small intestine (Muir and Hopfer, 1985). There, acidity from stomach acid aids in the absorption of both the heme iron, primarily derived from hemoglobin and myoglobin in meats and the inorganic iron, from other food sources. Recently, a molecule postulated to be an intestinal heme transporter was identified (Shayeghi *et al.*, 2005), but subsequent studies showed that its primary role is in high affinity folate transport (Qiu *et al.*, 2007). On the other hand, inorganic iron absorption is now well understood.

Inorganic iron in the intestinal lumen is primarily in its ferric, oxidized form. In order for iron to be absorbed, it must be reduced to the ferrous form. Reduction of iron can be carried out by an enterocyte apical membrane protein duodenal cytochrome *b* (Dcytb) (McKie *et al.*, 2001). However, genetic disruption of Dcytb in mice revealed that the protein was not required for efficient iron absorption, suggesting that other mechanisms of ferrireduction must exist, at least in mice (Gunshin *et al.*, 2005b). Once in its reduced form, the iron transporter DMT1 transfers iron across the apical membrane into the enterocyte (Fleming *et al.*, 1997; Gunshin *et al.*, 1997, 2005a). The fact that the same iron transporter is used for cellular iron uptake both in transferrin cycle endosomes and at the apical surface of the intestinal epithelium is somewhat surprising. These two membrane locales are quite different and must be reached through distinct targeting signals. However, both sites are within a low-pH milieu, important because DMT1 uses cotransport of protons to move iron across the membrane (Gunshin *et al.*, 1997; Sacher *et al.*, 2001; Xu *et al.*, 2004).

Depending on body iron needs, intracellular iron can be stored within the enterocyte or exported into the plasma. Net absorption requires transfer across the basolateral surface of the epithelium—iron retained within enterocytes is lost from the body when those cells senesce and are shed into the gut lumen. It appears that the transmembrane transporter ferroportin is responsible for most if not all basolateral iron transfer, because intestine-specific inactivation of the ferroportin gene in mice results in the rapid onset of anemia with abundant stainable iron in enterocytes (Donovan *et al.*, 2005). A ferroxidase activity acts in concert with ferroportin-mediated export. This can be supplied by the enterocyte-associated multicopper oxidase, hephaestin (Vulpe *et al.*, 1999), or by its circulating homologue, ceruloplasmin.

2.4. Hepatocyte iron storage

Hepatocytes serve many important functions including detoxification of the blood production of proteins that aid in host defense and storage of essential nutrients such as glucose and iron. Excess circulating iron in both transferrin-bound and nontransferrin-bound forms can be taken up by hepatocytes. TFR1 is expressed by hepatocytes. However, the fact that mice lacking TFR1, or transferrin, accumulate iron in their livers suggests that hepatocytes do not require the transferrin cycle to import iron from the plasma (Levy *et al.*, 1999; Trenor *et al.*, 2000). Similarly, DMT1 cannot account for all hepatocyte iron uptake because DMT1 knockout mice and human patients carrying loss-of-function mutations in DMT1 accumulate iron in the liver (Gunshin *et al.*, 2005a; Iolascon *et al.*, 2006; Mims *et al.*, 2004). Other hepatic iron importers must exist to participate in efficient iron storage. Hepatocyte membrane proteins such as megalin, TFR2, CD163, and L-type calcium channels are candidate iron importers that employ a variety of molecular mechanisms to bring iron into cells (Borregaard and Cowland, 2006; Kawabata *et al.*, 1999; Kozyraki *et al.*, 2001; Kristiansen *et al.*, 2001;

Oudit *et al.*, 2003; Yang *et al.*, 2002). Although hepatocyte iron uptake is not fully understood, it is clear that once inside the cell iron can be utilized in cellular processes or sequestered in ferritin. When iron loss or demand is too great to rely solely on recycling and absorption, iron is mobilized from hepatic storage to sustain erythropoiesis. Ferroportin and ceruloplasmin are believed to be involved in the process of iron export from hepatocytes, but this has not been shown directly.

2.5. Systemic iron homeostasis

Systemic iron homeostasis maintains body iron content within tolerable limits and dictates iron distribution. As the largest consumer of iron, the erythron is particularly sensitive to iron insufficiency. When body iron stores are depleted, iron deficiency anemia ensues. Rarely, iron deficiency anemia can be caused by genetic lesions that interfere with intestinal iron absorption, erythroid iron assimilation, or both. The best characterized of these are mutations preventing the production of transferrin (Beutler *et al.*, 2000; Hamill *et al.*, 1991) or inactivating DMT1 (Iolascon *et al.*, 2006; Mims *et al.*, 2004). Far more commonly, deficiency results from an imbalance between increased iron requirements associated with growth and blood loss and iron acquisition from the diet. In this case, iron deficiency anemia is characterized by low plasma iron, decreased iron stores, and the accumulation of iron-free protoporphyrins. Iron-deficient erythrocytes are small (microcytic), pale (hypochromic), and relatively fragile. Decreased oxygen-carrying capability caused by iron deficiency has measurable effects on quality of life, associated with symptoms including fatigue and tachycardia. Iron replacement therapy, through dietary supplementation, intravenous iron, or transfusion of iron-rich erythrocytes, is the only treatment for iron deficiency anemia.

Although essential for several important cellular functions, iron's capacity to donate and accept electrons makes the metal toxic. Several proteins such as transferrin, ferritin, lactoferrin, lipocalin, and myoglobin exist to bind iron in a variety of contexts. Binding to these proteins renders iron less able to react with its environment. Thus, a safe upper limit of body iron for each individual is set by the capacity of iron-binding proteins to sequester iron. Once this capacity is breached free iron accumulates within the plasma and cells. The accumulation of excess iron is undesirable because ferrous iron reacts with hydrogen peroxide via the Fenton reaction to produce hydroxyl radicals (Gutteridge *et al.*, 1981). These radicals damage macromolecules resulting in cellular and tissue dysfunction and organ failure associated with iron overload disorders (Gutteridge *et al.*, 1985). Complications associated with iron overload include liver fibrosis and cirrhosis, cardiomyopathy, diabetes, hypogonadotropic hypogonadism, arthritis, and hyperpigmentation. Iron overload can be caused by transfusion therapy or by genetic mutations that result in abnormally increased intestinal iron absorption. Iron overload can be managed with phlebotomy in patients with normal erythropoiesis or by chelation in patients with anemia disorders.

In order to regulate iron homeostasis, there must be sensing mechanisms that gauge the amount of iron needed to support erythropoiesis and other functions relative to iron availability (Finch, 1994). For many years, it was believed that enterocytes of the duodenum-sensed circulating iron and maintained systemic iron homeostasis accordingly (Britton *et al.*, 2002; Townsend and Drakesmith, 2002). Within the last few years, however, there has been a shift from the intestine to the liver as the hypothesized site of iron homeostatic control. Maintaining iron stores and sampling both transferrin- and nontransferrin-bound iron in the circulation, the liver is in a unique position to sense systemic iron availability. The mechanisms of plasma iron sensing and homeostatic response are not fully understood, but are under active investigation. The most compelling argument for the liver as the "ferrostat" organ is the fact that hepcidin, a key homeostatic regulator, is primarily expressed by hepatocytes.

3. Hepcidin

3.1. Hepcidin and iron homeostasis

Hepcidin was first described as a cysteine-rich peptide with modest antimicrobial activity isolated from human plasma and urine (Krause *et al.*, 2000; Park *et al.*, 2001). The circulating form of hepcidin, cleaved from an 84-amino acid precursor that is expressed primarily in the liver, contains 25-amino acids and forms four disulfide bonds (Hunter *et al.*, 2002; Nemeth *et al.*, 2006). Over the last few years, it has become clear that hepcidin functions predominantly as a regulator of iron homeostasis.

The first link between hepcidin and iron metabolism came from a screen for genes upregulated by iron excess. Pigeon *et al.* (2001) loaded mice with carbonyl iron and compared gene expression with untreated controls using suppressive subtractive hybridization. They found that hepcidin was robustly overexpressed under these conditions. They also showed that lipopolysaccharide, an inducer of inflammation, stimulated hepcidin expression in primary hepatocytes. This latter finding is consistent with the described antimicrobial activity of hepcidin and its induction by inflammation in other species (Krause *et al.*, 2000; Park *et al.*, 2001; Shike *et al.*, 2002).

A compelling rationale for why hepcidin, an antimicrobial peptide, would be regulated by both iron and inflammation is that nearly all microorganisms require iron to proliferate. Individuals with iron overload are more susceptible to bacterial infection than normal individuals (Weinberg, 1978). Thus, depriving pathogens of iron during an acute inflammatory response would prove evolutionarily advantageous to the host. However, sustained iron sequestration, as seen in chronic inflammation, has the adverse effect of host iron deprivation (Weiss and Goodnough, 2005).

The first direct evidence of a role for hepcidin in regulating iron homeostasis came from a fortuitous observation in mice carrying a targeted disruption in the transcription factor upstream stimulatory factor 2 (USF2) (Nicolas *et al.*, 2001). Inadvertently, the gene-targeting event inactivated the gene encoding hepcidin, immediately adjacent to the USF2 gene. Mutant animals developed severe iron overload with a pattern of tissue distribution resembling hemochromatosis, a form of genetic iron overload prevalent in humans. Diminished hepcidin expression was associated with increased iron absorption and recycling, hyperferremia, and severe tissue iron overload. The iron overload phenotype in these mice was characterized by a lack of iron in splenic macrophages and duodenal enterocytes despite increased intestinal iron absorption. Any doubt that the iron phenotype resulted from loss of hepcidin dissipated after the same authors showed that targeted inactivation of the hepcidin gene itself produced the same phenotype (Viatte *et al.*, 2005). Moreover, human mutations in the hepcidin gene that disrupt protein function lead to an early onset, more severe form of genetic hemochromatosis, juvenile hemochromatosis (Roetto *et al.*, 2003).

Conversely, transgenic mice overexpressing hepcidin have marked iron deficiency and anemia because of an inability to respond appropriately to body iron status (Nicolas *et al.*, 2002a; Rivera *et al.*, 2005a; Roy *et al.*, 2007). In combination with other studies, these observations suggested that increased hepcidin expression effectively shuts off iron absorption and macrophage iron recycling, trapping iron within enterocytes and macrophages. This inhibition of iron entry into circulation stalls erythropoiesis leading to a microcytic, hypochromic anemia. Taken together, these studies suggested that hepcidin is a negative regulator of cellular iron egress.

3.2. Molecular mechanism of hepcidin action

The lack of iron in enterocytes and macrophages in hepcidin deficiency, and the accumulation of iron in these cell types when hepcidin is in excess, led to the hypothesis that hepcidin functions to block ferroportin activity. Nemeth *et al.* developed a polarized human embryonic kidney cell line that expressed ferroportin tagged with green fluorescent protein and treated them with purified, active, hepcidin peptide. Initially, ferroportin was expressed on the cell membrane. However, within hours of hepcidin treatment, ferroportin was internalized into the cytoplasm where it was targeted for lysosome-mediated degradation. Binding studies indicated that hepcidin attached directly to ferroportin to trigger these events (Nemeth *et al.*, 2004b).

Other investigators have corroborated these results by showing that hepcidin binds to the cell surface of iron-loaded macrophage cell lines, resulting in cellular iron retention (Delaby *et al.*, 2005). A detailed model for hepcidin-mediated degradation of ferroportin has been proposed, based on *in vitro* studies (De Domenico *et al.*, 2007b). Hepcidin binding leads to tyrosine phosphorylation. Phosphorylated ferroportin is internalized, dephosphorylated, and then decorated with ubiquitin. Ubiquitinated ferroportin is trafficked through multivesicular bodies to the lysosome. While this series of events has not yet been observed in animals, the accumulation of synthetic hepcidin in ferroportin-expressing organs when administered to mice suggests that the model is legitimate *in vivo* (Rivera *et al.*, 2005b).

Additional understanding about ferroportin regulation can be gleaned from spontaneous mutations observed in human patients with abnormal iron homeostasis. Genetic lesions in ferroportin lead to two distinct, autosomal dominant disorders. A minority of patients have a clinical picture characterized by hepatic iron loading and elevated plasma iron, very similar to genetic hemochromatosis (discussed later). However, most patients have a different presentation, with iron loading in tissue macrophages ("ferroportin disease"). With few exceptions, the pathophysiology of these diseases can be explained by the effects of the mutations on ferroportin activity with respect to subcellular localization, iron export capability, and ability to interact with hepcidin.

Class I ferroportin mutations are associated with the hemochromatosis-like phenotype. They include Y64D, N144H, N144D, Q182H, C326S, and S338R (De Domenico *et al.*, 2005, 2006; Drakesmith *et al.*, 2005; Liu *et al.*, 2005; Schimanski *et al.*, 2005; Sham *et al.*, 2005; Wallace *et al.*, 2007). These have been characterized as gain of function mutations because they render the ferroportin protein resistant to hepcidin regulation in spite of normal or increased hepcidin levels, resulting in uncontrolled iron export. Class II ferroportin mutations are associated with ferroportin disease (De Domenico *et al.*, 2005, 2006; Drakesmith *et al.*, 2005; Liu *et al.*, 2005; Schimanski *et al.*, 2005; Sham *et al.*, 2005; Zoller *et al.*, 2005). They include A77D, D157G, V162del, G323V, and G490D. These mutations result in a loss of protein function and/or aberrant subcellular localization. A recently described mouse mutant appears to have this type of mutation, in a protein with dominant negative function (Zohn *et al.*, 2007). Several known ferroportin mutations cannot easily be placed into either of these categories.

Understanding that hepcidin maintains iron homeostasis through a physical interaction with ferroportin has led to a plausible model for the normal maintenance of iron homeostasis (Fig. 6.1) and the disruption of homeostasis in human disease. Increased plasma iron may be sensed by liver hepatocytes leading to the induction of hepcidin expression and to increased hepcidin secretion into the plasma. Circulating hepcidin then binds to ferroportin on the surface of iron-exporting cells, triggering

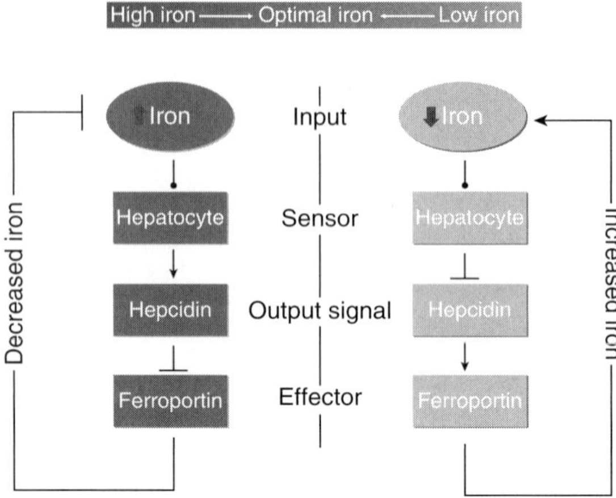

Figure 6.1 Normal iron homeostasis mediated by an iron-sensing feedback loop. When systemic iron (sensed from iron stores and or circulating iron-bound transferrin) is elevated above a preset threshold, hepatocytes sense the elevation and upregulate hepcidin accordingly (left side, top to bottom). Hepcidin, in turn, travels in the circulation to tissues expressing ferroportin. Hepcidin binds ferroportin and targets it for degradation. Decreased cell surface ferroportin leads to decreased circulating iron. When systemic iron falls below a set threshold, the liver senses this change and downregulates hepcidin (right side, top to bottom). Less circulating hepcidin allows newly synthesized ferroportin molecules to repopulate the cell membrane and export iron, increasing both macrophage iron recycling and intestinal absorption. Each branch of this regulatory loop drives systemic iron levels toward optimal amounts to sustain erythropoiesis and other body iron needs. An inability to sense high iron and/or to downregulate ferroportin leads to hemochromatosis. Conversely, inappropriately high expression of hepcidin in the setting of decreased iron availability results in the anemia of chronic disease.

internalization and degradation, rendering it unable to export iron. At the other extreme, increased iron utilization, particularly by developing erythrocytes, would deplete plasma iron, signaling for decreased hepcidin production. As the residual circulating hepcidin is cleared from the plasma through binding to ferroportin and by urinary excretion, ferroportin is replaced on the membrane. Iron export into the plasma returns plasma iron levels to normal. The anemia of inflammation and genetic hemochromatosis, discussed in the following sections, are thought to result from perturbations in this homeostatic regulatory mechanism.

3.3. Hepcidin and the pathogenesis of the anemia of inflammation

The anemia of inflammation, also known as the anemia of chronic disease, is an acquired condition that affects patients with a variety of inflammatory disorders including infection, arthritis, inflammatory bowel disease, trauma, organ failure, and cancer. It is characterized by mild-to-moderate anemia, low serum iron, macrophage iron retention, and impaired intestinal iron absorption. This disorder was first linked to aberrant hepcidin expression through the study of an unusual group of patients with a congenital metabolic disorder complicated by anemia and hepatic adenomas (Weinstein and Wolfsdorf, 2002; Wolfsdorf and Crigler, 1999). The patients had clinical features of severe anemia of inflammation, attributable to a constitutive high level of hepcidin expression by hepatocytes comprising the adenomas (Weinstein *et al.*, 2002). It was unclear why the adenoma tissue failed to regulate hepcidin expression appropriately, but the inappropriately high level of hepcidin readily accounted for the anemia. Furthermore, iron homeostasis returned to normal upon removal of adenomas (Roy *et al.*, 2003; Weinstein *et al.*, 2002).

Subsequent studies further supported the link between increased hepcidin expression and the anemia of inflammation in more routine clinical settings (Nemeth *et al.*, 2003). Induction of an inflammatory response using a variety of methods increased hepcidin expression in both mice and humans (Kemna *et al.*, 2005; Lee *et al.*, 2004; Nicolas *et al.*, 2002b; Roy *et al.*, 2004). Furthermore, a mouse model of induced hepcidin overexpression recapitulates hallmark signs of the anemia of inflammation (Roy *et al.*, 2007).

A mechanistic link between inflammation and hepcidin was forged with the discovery that hepcidin is an acute phase protein upregulated by interleukin-6 (IL-6) in isolated hepatocytes and in hepatocyte-like cell lines (Nemeth *et al.*, 2003). Administration of IL-6 also caused increased hepcidin production and consequent hypoferremia in both mice and humans (Nemeth *et al.*, 2004a). Mice lacking IL-6 failed to induce hepcidin normally in response to treatment with endotoxin (Nemeth *et al.*, 2004a). Taken together, these studies show that IL-6-mediated induction of hepcidin in inflammation can explain, at least in part, the pathogenesis of the anemia of inflammation.

IL-6 binds to its cell surface receptor to stimulate several distinct signal transduction pathways. Of these, JAK/STAT3 signaling appears to be important for induction of hepcidin expression (Pietrangelo *et al.*, 2007; Verga Falzacappa *et al.*, 2007; Wrighting and Andrews, 2006). An IL-6 responsive, STAT3-binding transcriptional regulatory element was identified in the proximal hepcidin promoter. STAT3 regulation of hepcidin is compelling because while inflammation is one stimulus that can activate the transcription factor, STAT3 can also be activated and repressed by a plethora of other stimuli (Watson, 2001). Thus, it is reasonable to speculate that STAT3 regulation might prove to be a more general mode of modulating hepcidin expression. Conversely, some investigators have suggested that other cytokines may contribute to hepcidin induction in inflammation, though probably to a lesser extent than IL-6 (Lee *et al.*, 2005).

3.4. Hepcidin and the pathogenesis of genetic hemochromatosis

Hemochromatosis is a genetic iron overload disorder characterized by increased dietary iron absorption, increased plasma iron, and deposition of iron in the liver, heart, pancreas, and other tissues. Tissue iron accumulation can lead to liver fibrosis and cirrhosis, heart failure, diabetes, and other endocrinopathies. The disease is caused by mutations in any of five known genes: the classical hemochromatosis gene (HFE) (Feder *et al.*, 1996), TFR2 (Camaschella *et al.*, 2000), hemojuvelin (HJV) (Papanikolaou *et al.*, 2004), ferroportin (as discussed above), and hepcidin itself (Roetto *et al.*, 2003). While mutations in other genes can also lead to iron overload in certain tissues (e.g., ceruloplasmin, transferrin, DMT1), those disorders lack the characteristic features seen in hemochromatosis caused by mutations in the hepcidin, HFE, ferroportin, TFR2, and HJV genes. The pathogenesis of hemochromatosis resulting from ferroportin and hepcidin mutations has already been discussed in the context of known human mutations and animal models. It is now clear that hemochromatosis due to mutations in HFE, TFR2, and HJV genes is invariably associated with inappropriately low hepcidin expression for the amount of total body iron (Bridle *et al.*, 2003; Nemeth *et al.*, 2005; Papanikolaou *et al.*, 2004; Roetto *et al.*, 2003). This implies that all three of those proteins are involved in the regulation of hepcidin expression. In further support of a role for HFE upstream of hepcidin, forced expression of hepcidin in HFE knockout mice not only rescues the iron overload phenotype but also causes anemia (Nicolas *et al.*, 2003). However, the exact cellular functions of HFE, TFR2, and HJV are not fully understood.

HFE encodes an atypical major histocompatibility class I-like protein that forms a heterodimer with β-2-microglobulin but is unable to present small peptide antigens. The large majority of hemochromatosis patients are homozygous for a C282Y missense allele derived from a unique Celtic ancestor. The first clue to its role came when it was discovered that HFE associates with TFR1 to form a stable protein complex (Bennett *et al.*, 2000; Feder *et al.*, 1997, 1998; West *et al.*, 2000). HFE competes with holotransferrin for binding to TFR1. It has been unclear, however, whether the dynamic interactions among these proteins are involved in iron sensing or in modulating cellular iron uptake.

TFR2 is a membrane glycoprotein homologue of TFR1 (Kawabata *et al.*, 1999). It is predominantly expressed in the liver and has been proposed to act as an iron sensor because its cell surface expression increases and is stabilized in response to holotransferrin (Johnson and Enns, 2004; Johnson *et al.*, 2007; Robb and Wessling-Resnick, 2004). In fact, TFR2 has been shown to bind holotransferrin and mediate its internalization, although with decreased efficiency compared to TFR1 (Kawabata *et al.*, 2000).

Despite these studies, the physiological function of TFR2 remains unknown.

Similar to those with mutations in the hepcidin gene, patients with mutations in the HJV gene have severe, early onset (juvenile) hemochromatosis (Papanikolaou *et al.*, 2004). When the gene was first discovered it was found to have two homologues, termed RGM-A and RGM-B (for "repulsive guidance molecules" A and B). Both homologues were predominantly expressed in the central nervous system (Oldekamp *et al.*, 2004), and RGM-A had been implicated in axonal patterning (Monnier *et al.*, 2002). The functions of these proteins were uncertain until it was discovered that RGM-B could act as a bone morphogenetic protein (BMP) coreceptor (Samad *et al.*, 2005). This type of coreceptor enhances the signal of an agonist by increasing receptor sensitivity. It was subsequently shown that HJV also acts as a BMP coreceptor to regulate hepcidin transcription (Babitt *et al.*, 2006). Targeted inactivation of the HJV gene in mice caused marked iron overload in a hemochromatosis distribution pattern (Huang *et al.*, 2005; Niederkofler *et al.*, 2005). In parallel, it was shown that hepatocyte-specific inactivation of the murine gene encoding SMAD4, a key intracellular component of the BMP signaling pathway, resulted in inappropriately low hepcidin expression and an indistinguishable juvenile hemochromatosis-like phenotype (Wang *et al.*, 2005). Finally, administration of BMP to animals induced hepcidin expression *in vivo* (Babitt *et al.*, 2007). Several different BMPs are active in inducing hepcidin expression in cultured cells, but it remains uncertain which are important *in vivo* (Babitt *et al.*, 2006, 2007; Truksa *et al.*, 2006; Wang *et al.*, 2005).

It is not immediately obvious how the BMP coreceptor activity of HJV itself is regulated, but presumably it must respond in some way to information about body iron needs and availability. Part of the answer may relate to a soluble form of HJV, sHJV, which is a proteolytic cleavage product of the mature protein (Kuninger *et al.*, 2006). Treatment of cultured cells with recombinant sHJV can suppress hepcidin gene expression (Lin *et al.*, 2005). In fact, injection of mice with a soluble form of hemojuvelin decreases SMAD signaling and hepcidin expression, resulting in an increase in splenic ferroportin, serum iron, transferrin saturation, and liver iron (Babitt *et al.*, 2007). Furthermore, sHJV shedding can be repressed by treatment with holotransferrin (Silvestri *et al.*, 2007; Zhang *et al.*, 2007). Taken together, these studies suggest that sHJV may be involved in transducing a signal for increased iron absorption, possibly from muscle cells where HJV is abundantly expressed.

The mechanism of action of sHJV remains uncertain. It might function by disrupting a complex between HJV and BMP receptors (or other molecules) to decrease BMP signaling. Alternatively, sHJV might directly bind BMP, sequestering the ligand to block signaling. Other HJV regulatory mechanisms may involve neogenin, a protein primarily known for its functions in the nervous system. Neogenin interacts with related RGM proteins and has been shown to form a complex with HJV (Rodriguez *et al.*, 2007; Zhang *et al.*, 2005). Interaction of neogenin with HJV prevents the liberation of sHJV in response to iron (Zhang *et al.*, 2007).

Loss-of-function mutations in TFR2 and HFE, both in human patients and in animal models, typically result in less severe, later onset iron overload compared with hemochromatosis associated with mutations in hepcidin or HJV (Zoller and Cox, 2005). Moreover, TFR2 and HFE have been shown to interact, and this interaction can be modulated by holotransferrin (Goswami and Andrews, 2006). Based on these observations, it is reasonable to speculate that HJV serves as a major regulator of hepcidin expression while HFE and TFR2 play accessory roles. To date, however, it has been unclear whether these proteins operate in parallel pathways or through a common pathway.

Direct links between ambient iron concentration and changes in hepcidin expression have been difficult to establish *in vitro* because treatment of cultured cells with various forms of iron generally fails to induce hepcidin expression (Muckenthaler *et al.*, 2003; Nemeth *et al.*, 2003; Pigeon *et al.*, 2001). When induction has been observed, the effect has been modest at best (Lin *et al.*, 2007). While this suggests that the mechanism of hepcidin regulation *in vivo* is more complex than can be reproduced *in vitro*, it leaves investigators with few clues to the identities of hepcidin regulators downstream of an iron stimulus. In an alternate approach, Flanagan *et al.* (2007) have developed an *in vivo* bioluminescence imaging system for studying real-time hepcidin expression. This system may aid in pinpointing the region of the hepcidin promoter that is responsive to iron, providing insight into the signaling mechanism involved.

4. Hepcidin and Erythropoeisis

4.1. Hepcidin expression is regulated in response to bone marrow needs

It has long been known that iron absorption is increased in patients with congenital anemias characterized by ineffective erythropoiesis (intramedullary destruction of immature erythrocytes in the bone marrow). Clinically, increased intestinal iron absorption compounds the effects of transfusional iron overload in patients with thalassemia syndromes, sideroblastic anemia, or congenital dyserythropoietic anemias. Finch (1994) proposed the existence of an erythroid regulator of systemic iron homeostasis. The erythron, composed of developing erythroid cells in the bone marrow and circulating erythrocytes, utilizes about 80% of the iron found in the plasma. Anemia results from the inability of the erythroid compartment to receive its full complement of iron. The putative erythroid regulator communicates the iron needs of the erythron to influence changes in intestinal iron absorption.

Because hepcidin is an effective inhibitor of iron absorption, it is reasonable to speculate that the erythroid regulator includes a mechanism to decrease hepcidin production. Accordingly, low hepcidin levels have been reported in mice and patients with thalassemia and other disorders with ineffective erythropoiesis (Adamsky *et al.*, 2004; Breda *et al.*, 2005; Gardenghi *et al.*, 2007; Jenkins *et al.*, 2007; Kattamis *et al.*, 2006; Kearney *et al.*, 2007; Kemna *et al.*, 2007; Roy *et al.*, 2003; Weinstein *et al.*, 2002; Weizer-Stern *et al.*, 2006b). In these disorders, decreased hepcidin expression leads to relief of inhibition of ferroportin, resulting in increased iron release from recycling macrophages and absorptive enterocytes, increasing availability of the metal for erythropoiesis. However, the iron cannot be effectively utilized by the erythron, leading to accumulation and tissue iron overload in the face of anemia.

But what upstream signal shuts off hepcidin production in the service of the erythron? One possibility is that developing erythrocytes secrete a factor that circulates to hepatocytes to negatively regulate hepcidin expression. Supporting this hypothesis, hepcidin expression was downregulated in a hepatocytic cell line after treatment with thalassemic sera (Weizer-Stern *et al.*, 2006a). However, in earlier experiments that predated the discovery of hepcidin, others reported no change in intestinal iron absorption when serum from severely anemic, hyperabsorbing, hypotransferrinemic mice was infused into normal mice (Buys *et al.*, 1991). Those results could support an alternative hypothesis that the erythroid regulator involves decreased action of a factor that normally stimulates hepcidin synthesis. Alternatively, the failure to see increased intestinal absorption *in vivo* might simply mean that an inhibitor of hepcidin expression is short-lived.

Soluble transferrin receptor-1 (sTFR1) could be a compelling candidate for a humoral inducer of hepcidin expression. Plasma levels of sTFR1 increase during iron deficiency and ineffective erythropoiesis, both situations of increased erythron demand for iron (R'Zik and Beguin, 2001). The function of sTFR1 in plasma is unknown, but high sTFR1 levels correlate with low hepcidin levels. However, a recent study showed that transgenic expression of sTFR1 did not have an effect on iron absorption in mice (Flanagan *et al.*, 2006). Other factors secreted by developing erythroid cells could be candidate-negative regulators of hepcidin expression. Elucidation of the upstream portion of the erythroid regulator remains an area of active investigation.

Intestinal iron absorption also increases in response to hypoxia. While iron deficiency anemia can certainly cause hypoxia, it is not yet known whether the erythroid and hypoxia regulators are distinct or overlap at least in part. Hypoxia causes decreased hepcidin expression both *in vivo* and in cell culture models (Nicolas *et al.*, 2002b). Leung *et al.* (2005) correlated hypoxia-induced repression of hepcidin expression with a concomitant increase in intestinal iron absorption. Others have shown that hypoxia increases intestinal iron absorption in mice, but its effects are reversed by administration of turpentine, an inflammatory stimulus that is known to increase hepcidin expression (Raja *et al.*, 1988). Although it is clear that hypoxia reduces hepcidin expression and increases iron absorption, the mechanism by which hypoxia regulates hepcidin has only recently been explored.

Hypoxia-inducible factor (HIF) is a transcription factor that induces and represses a variety of genes involved in responding to the stress associated with hypoxia. HIF regulates erythropoiesis by promoting the expression of erythropoietin (Goldberg *et al.*, 1991). Additionally, HIF regulates iron availability for erythropoiesis through upregulation of HO-1 and transferrin (Lee *et al.*, 1997; Rolfs *et al.*, 1997). The active form of HIF is a heterodimer of one of three distinct α-subunits and one of two β-subunits. Because the β-subunit is constitutively expressed, posttranscriptional regulation of the α-subunit determines the activity of the transcription factor. Under normoxic conditions, iron-dependent prolyl hydroxylases modify HIF-1α (Schofield and Ratcliffe, 2005). Modified HIF-1α readily binds to von Hippel-Lindau (VHL) protein, an E3 ligase that rapidly targets the protein for proteasome-mediated degradation. This repression does not occur under hypoxic conditions. Instead, the two HIF subunits dimerize, translocate into the nucleus, and bind to HIF responsive elements in the promoter regions of target genes. The fact that HIF is involved in regulating genes that increase iron availability and erythropoiesis suggested that repression of hepcidin in hypoxic conditions might be mediated by HIF, either directly or indirectly.

Peyssonnaux *et al.* (2007) tested the hypothesis that HIF can act as a direct repressor of hepcidin expression. They showed that HIF-1α activity was increased in iron-deficient cells. They argued that HIF-1α decreases hepcidin expression in the setting of iron deficiency. Furthermore, they identified functional HIF responsive elements within a putative transcriptional regulatory region upstream of the hepcidin gene (Peyssonnaux *et al.*, 2007). These *in vitro* results were supported by *in vivo* data from a VHL-deficient mouse. VHL deficiency resulted in constitutively active HIF-1α subunits and decreased hepcidin expression that was dependent on the presence of functional HIF-1α subunits (Peyssonnaux *et al.*, 2007). These results were consistent with their model for hepcidin regulation in hypoxia. Another group, however, argued that HIF is not a key mediator of hepcidin regulation, and proposed an alternative mechanism involving reactive oxygen species (Choi *et al.*, 2007). Their data suggested that two positively

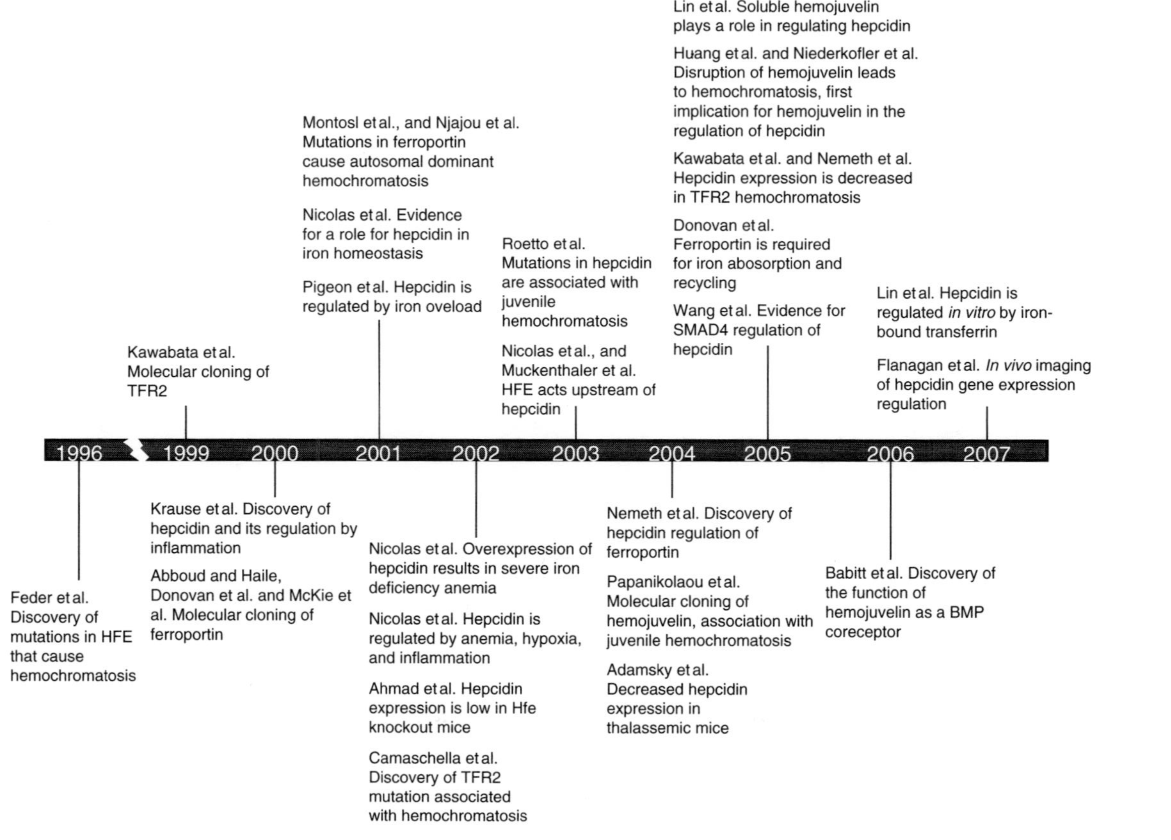
1996
Feder et al. Discovery of mutations in HFE that cause hemochromatosis
1999
Kawabata et al. Molecular cloning of TFR2
2000
Krause et al. Discovery of hepcidin and its regulation by inflammation
Abboud and Haile, Donovan et al. and McKie et al. Molecular cloning of ferroportin
2001
Montosl et al., and Njajou et al. Mutations in ferroportin cause autosomal dominant hemochromatosis
Nicolas et al. Evidence for a role for hepcidin in iron homeostasis
Pigeon et al. Hepcidin is regulated by iron oveload
2002
Nicolas et al. Overexpression of hepcidin results in severe iron deficiency anemia
Nicolas et al. Hepcidin is regulated by anemia, hypoxia, and inflammation
Ahmad et al. Hepcidin expression is low in Hfe knockout mice
Camaschella et al. Discovery of TFR2 mutation associated with hemochromatosis
2003
Roetto et al. Mutations in hepcidin are associated with juvenile hemochromatosis
Nicolas et al., and Muckenthaler et al. HFE acts upstream of hepcidin
2004
Nemeth et al. Discovery of hepcidin regulation of ferroportin
Papanikolaou et al. Molecular cloning of hemojuvelin, association with juvenile hemochromatosis
Adamsky et al. Decreased hepcidin expression in thalassemic mice
2005
Lin et al. Soluble hemojuvelin plays a role in regulating hepcidin
Huang et al. and Niederkofler et al. Disruption of hemojuvelin leads to hemochromatosis, first implication for hemojuvelin in the regulation of hepcidin
Kawabata et al. and Nemeth et al. Hepcidin expression is decreased in TFR2 hemochromatosis
Donovan et al. Ferroportin is required for iron abosorption and recycling
Wang et al. Evidence for SMAD4 regulation of hepcidin
2006
Babitt et al. Discovery of the function of hemojuvelin as a BMP coreceptor
2007
Lin et al. Hepcidin is regulated *in vitro* by iron-bound transferrin
Flanagan et al. *In vivo* imaging of hepcidin gene expression regulation

acting transcription factors—CEBP-α and STAT3—are negatively regulated by hypoxia independent of HIF (Choi *et al.*, 2007; Courselaud *et al.*, 2002; Pietrangelo *et al.*, 2007; Verga Falzacappa *et al.*, 2007; Wrighting and Andrews, 2006).

Although there is still uncertainty about the mechanisms, it is clear that ineffective erythropoiesis and hypoxia have the capacity to negatively regulate hepcidin expression. Because ineffective erythropoiesis can cause hypoxia and hypoxia can influence erythropoiesis, it is difficult to distinguish between the actions of these stimuli on hepcidin expression. Regardless, the net effect of both, individually or in combination, is to increase intestinal iron absorption and macrophage iron recycling to provide more iron for red blood cell production.

5. Conclusions

Over the last 10 years, our understanding of iron metabolism has increased dramatically (Fig. 6.2). Central to this understanding were the discoveries of hepcidin, its mechanism of action, and its role as a hormone regulator of iron homeostasis. Hepcidin indirectly regulates erythropoiesis by controlling iron availability. In response to inflammation or systemic iron overload, hepcidin levels rise to decrease iron egress from enterocytes and tissue macrophages. Conversely, hepcidin levels fall in response to hypoxia, increased erythroid drive, or iron deficiency to release iron from enterocytes and tissue macrophages to meet the needs of maturing erythroid cells. Thus, modulating hepcidin expression serves to maintain iron homeostasis. Appropriate systemic iron homeostasis ensures that erythrocytes develop with an appropriate allowance of iron.

Figure 6.2 Pivotal discoveries in iron homeostasis over the last decade. With the discovery of hepcidin and its role in regulating iron homeostasis, the iron field has developed a hepcidin-centric view of how systemic iron homeostasis is maintained. The discoveries listed on this time line have been instrumental in identifying molecules and mechanisms involved in regulating iron homeostasis. The identification of HFE, TFR2, HJV, and ferroportin and the characterization of their involvement in the etiology of hereditary hemochromatosis have provided the basis for the current understanding of hepcidin regulation and action. The discovery that hepcidin binds ferroportin and targets it for lysosome-mediated degradation has provided a mechanism for how hepcidin mediates iron homeostasis. Identification of regulatory networks involving iron, inflammation, anemia, hypoxia, and bone morphogenetic proteins has provided novel insights into the way hepcidin gene expression may be regulated. Finally, the ability to detect hepcidin in serum, to understand its gene regulation *in vivo*, and to recapitulate iron regulation of hepcidin *in vitro* will undoubtedly propel the field into a future filled with more discovery.

REFERENCES

Abboud, S., and Haile, D. J. (2000). A novel mammalian iron-regulated protein involved in intracellular iron metabolism. *J. Biol. Chem.* **275,** 19906–19912.

Adamsky, K., Weizer, O., Amariglio, N., Breda, L., Harmelin, A., Rivella, S., Rachmilewitz, E., and Rechavi, G. (2004). Decreased hepcidin mRNA expression in thalassemic mice. *Br. J. Haematol.* **124,** 123–124.

Andrews, N. C. (2000). Iron metabolism: Iron deficiency and iron overload. *Annu. Rev. Genomics Hum. Genet.* **1,** 75–98.

Babitt, J. L., Huang, F. W., Wrighting, D. M., Xia, Y., Sidis, Y., Samad, T. A., Campagna, J. A., Chung, R. T., Schneyer, A. L., Woolf, C. J., Andrews, N. C., and Lin, H. Y. (2006). Bone morphogenetic protein signaling by hemojuvelin regulates hepcidin expression. *Nat. Genet.* **38,** 531–539.

Babitt, J. L., Huang, F. W., Xia, Y., Sidis, Y., Andrews, N. C., and Lin, H. Y. (2007). Modulation of bone morphogenetic protein signaling *in vivo* regulates systemic iron balance. *J. Clin. Invest.* **117,** 1933–1939.

Bennett, M. J., Lebron, J. A., and Bjorkman, P. J. (2000). Crystal structure of the hereditary haemochromatosis protein HFE complexed with transferrin receptor. *Nature* **403,** 46–53.

Beutler, E., Gelbart, T., Lee, P., Trevino, R., Fernandez, M. A., and Fairbanks, V. F. (2000). Molecular characterization of a case of atransferrinemia. *Blood* **96,** 4071–4074.

Borregaard, N., and Cowland, J. B. (2006). Neutrophil gelatinase-associated lipocalin, a siderophore-binding eukaryotic protein. *Biometals* **19,** 211–215.

Bothwell, T., Charlton, R., Cook, J., and Finch, C. (1979). "Iron Metabolism." p. 276. Blackwell, Oxford, UK.

Bratosin, D., Mazurier, J., Tissier, J. P., Estaquier, J., Huart, J. J., Ameisen, J. C., Aminoff, D., and Montreuil, J. (1998). Cellular and molecular mechanisms of senescent erythrocyte phagocytosis by macrophages. A review. *Biochimie* **80,** 173–195.

Breda, L., Gardenghi, S., Guy, E., Rachmilewitz, E. A., Weizer-Stern, O., Adamsky, K., Amariglio, N., Rechavi, G., Giardina, P. J., Grady, R. W., and Rivella, S. (2005). Exploring the role of hepcidin, an antimicrobial and iron regulatory peptide, in increased iron absorption in beta-thalassemia. *Ann. NY Acad. Sci.* **1054,** 417–422.

Bridle, K. R., Frazer, D. M., Wilkins, S. J., Dixon, J. L., Purdie, D. M., Crawford, D. H., Subramaniam, V. N., Powell, L. W., Anderson, G. J., and Ramm, G. A. (2003). Disrupted hepcidin regulation in HFE-associated haemochromatosis and the liver as a regulator of body iron homoeostasis. *Lancet* **361,** 669–673.

Britton, R. S., Fleming, R. E., Parkkila, S., Waheed, A., Sly, W. S., and Bacon, B. R. (2002). Pathogenesis of hereditary hemochromatosis: Genetics and beyond. *Semin. Gastrointest. Dis.* **13,** 68–79.

Buys, S. S., Martin, C. B., Eldridge, M., Kushner, J. P., and Kaplan, J. (1991). Iron absorption in hypotransferrinemic mice. *Blood* **78,** 3288–3290.

Camaschella, C., Roetto, A., Cali, A., De Gobbi, M., Garozzo, G., Carella, M., Majorano, N., Totaro, A., and Gasparini, P. (2000). The gene TFR2 is mutated in a new type of haemochromatosis mapping to 7q22. *Nat. Genet.* **25,** 14–15.

Canonne-Hergaux, F., Zhang, A. S., Ponka, P., and Gros, P. (2001). Characterization of the iron transporter DMT1 (NRAMP2/DCT1) in red blood cells of normal and anemic mk/mk mice. *Blood* **98,** 3823–3830.

Canonne-Hergaux, F., Donovan, A., Delaby, C., Wang, H. J., and Gros, P. (2006). Comparative studies of duodenal and macrophage ferroportin proteins. *Am. J. Physiol. Gastrointest. Liver Physiol.* **290,** G156–G163.

Chan, L.-N. L., and Gerhardt, E. M. (1992). Transferrin receptor gene is hyperexpressed and transcriptionally regulated in differentiating erythroid cells. *J. Biol. Chem.* **267,** 8254–8259.

Choi, S. O., Cho, Y. S., Kim, H. L., and Park, J. W. (2007). ROS mediate the hypoxic repression of the hepcidin gene by inhibiting C/EBPalpha and STAT-3. *Biochem. Biophys. Res. Commun.* **356,** 312–317.

Courselaud, B., Pigeon, C., Inoue, Y., Inoue, J., Gonzalez, F. J., Leroyer, P., Gilot, D., Boudjema, K., Guguen-Guillouzo, C., Brissot, P., Loreal, O., and Ilyin, G. (2002). C/EBPalpha regulates hepatic transcription of hepcidin, an antimicrobial peptide and regulator of iron metabolism. Cross-talk between C/EBP pathway and iron metabolism. *J. Biol. Chem.* **277,** 41163–41170.

Dautry-Varsat, A., Ciechanover, A., and Lodish, H. F. (1983). pH and the recycling of transferrin during receptor-mediated endocytosis. *Proc. Natl. Acad. Sci. USA* **80,** 2258–2262.

De Domenico, I., Ward, D. M., Nemeth, E., Vaughn, M. B., Musci, G., Ganz, T., and Kaplan, J. (2005). The molecular basis of ferroportin-linked hemochromatosis. *Proc. Natl. Acad. Sci. USA* **102,** 8955–8960.

De Domenico, I., Ward, D. M., Musci, G., and Kaplan, J. (2006). Iron overload due to mutations in ferroportin. *Haematologica* **91,** 92–95.

De Domenico, I., Ward, D. M., di Patti, M. C., Jeong, S. Y., David, S., Musci, G., and Kaplan, J. (2007a). Ferroxidase activity is required for the stability of cell surface ferroportin in cells expressing GPI-ceruloplasmin. *EMBO J.* **26,** 2823–2831.

De Domenico, I., Ward, D. M., Langelier, C., Vaughn, M. B., Nemeth, E., Sundquist, W. I., Ganz, T., Musci, G., and Kaplan, J. (2007b). The molecular mechanism of hepcidin-mediated ferroportin down-regulation. *Mol. Biol. Cell* **18,** 2569–2578.

Delaby, C., Pilard, N., Goncalves, A. S., Beaumont, C., and Canonne-Hergaux, F. (2005). Presence of the iron exporter ferroportin at the plasma membrane of macrophages is enhanced by iron loading and down-regulated by hepcidin. *Blood* **106,** 3979–3984.

Donovan, A., Brownlie, A., Zhou, Y., Shepard, J., Pratt, S. J., Moynihan, J., Paw, B. H., Drejer, A., Barut, B., Zapata, A., Law, T. C., Brugnara, C., *et al.* (2000). Positional cloning of zebrafish ferroportin1 identifies a conserved vertebrate iron exporter. *Nature* **403,** 776–781.

Donovan, A., Lima, C. A., Pinkus, J. L., Pinkus, G. S., Zon, L. I., Robine, S., and Andrews, N. C. (2005). The iron exporter ferroportin/Slc40a1 is essential for iron homeostasis. *Cell Metab.* **1,** 191–200.

Drakesmith, H., Schimanski, L. M., Ormerod, E., Merryweather-Clarke, A. T., Viprakasit, V., Edwards, J. P., Sweetland, E., Bastin, J. M., Cowley, D., Chinthammitr, Y., Robson, K. J., and Townsend, A. R. (2005). Resistance to hepcidin is conferred by hemochromatosis-associated mutations of ferroportin. *Blood* **106,** 1092–1097.

Feder, J. N., Gnirke, A., Thomas, W., Tsuchihashi, Z., Ruddy, D. A., Basava, A., Dormishian, F., Domingo, R., Jr., Ellis, M. C., Fullan, A., Hinton, L. M., Jones, N. L., *et al.* (1996). A novel MHC class I-like gene is mutated in patients with hereditary haemochromatosis. *Nat. Genet.* **13,** 399–408.

Feder, J. N., Tsuchihashi, Z., Irrinki, A., Lee, V. K., Mapa, F. A., Morikang, E., Prass, C. E., Starnes, S. M., Wolff, R. K., Parkkila, S., Sly, W. S., and Schatzman, R. C. (1997). The hemochromatosis founder mutation in HLA-H disrupts beta2-microglobulin interaction and cell surface expression. *J. Biol. Chem.* **272,** 14025–14028.

Feder, J. N., Penny, D. M., Irrinki, A., Lee, V. K., Lebron, J. A., Watson, N., Tsuchihashi, Z., Sigal, E., Bjorkman, P. J., and Schatzman, R. C. (1998). The hemochromatosis gene product complexes with the transferrin receptor and lowers its affinity for ligand binding. *Proc. Natl. Acad. Sci. USA* **95,** 1472–1477.

Finch, C. (1994). Regulators of iron balance in humans. *Blood* **84,** 1697–1702.

Flanagan, J. M., Peng, H., Wang, L., Gelbart, T., Lee, P., Johnson Sasu, B., and Beutler, E. (2006). Soluble transferrin receptor-1 levels in mice do not affect iron absorption. *Acta Haematol.* **116,** 249–254.

Flanagan, J. M., Truksa, J., Peng, H., Lee, P., and Beutler, E. (2007). *In vivo* imaging of hepcidin promoter stimulation by iron and inflammation. *Blood Cells Mol. Dis.* **38,** 253–257.

Fleming, M. D., Trenor, C. C., 3rd, Su, M. A., Foernzler, D., Beier, D. R., Dietrich, W. F., and Andrews, N. C. (1997). Microcytic anaemia mice have a mutation in Nramp2, a candidate iron transporter gene. *Nat. Genet.* **16,** 383–386.

Fleming, M. D., Romano, M. A., Su, M. A., Garrick, L. M., Garrick, M. D., and Andrews, N. C. (1998). Nramp2 is mutated in the anemic Belgrade (b) rat: Evidence of a role for Nramp2 in endosomal iron transport. *Proc. Natl. Acad. Sci. USA* **95,** 1148–1153.

Gardenghi, S., Marongiu, M. F., Ramos, P., Guy, E., Breda, L., Chadburn, A., Liu, Y., Amariglio, N., Rechavi, G., Rachmilewitz, E. A., Breur, W., Cabantchik, Z. I., *et al.* (2007). Ineffective erythropoiesis in {beta}-thalassemia is characterized by increased iron absorption mediated by down-regulation of hepcidin and up-regulation of ferroportin. *Blood* **109**(11), 5027–5035.

Goldberg, M. A., Gaut, C. C., and Bunn, H. F. (1991). Erythropoietin mRNA levels are governed by both the rate of gene transcription and posttranscriptional events. *Blood* **77,** 271–277.

Goswami, T., and Andrews, N. C. (2006). Hereditary hemochromatosis protein, HFE, interaction with transferrin receptor 2 suggests a molecular mechanism for mammalian iron sensing. *J. Biol. Chem.* **281,** 28494–28498.

Gunshin, H., Mackenzie, B., Berger, U. V., Gunshin, Y., Romero, M. F., Boron, W. F., Nussberger, S., Gollan, J. L., and Hediger, M. A. (1997). Cloning and characterization of a mammalian proton-coupled metal-ion transporter. *Nature* **388,** 482–488.

Gunshin, H., Fujiwara, Y., Custodio, A. O., Direnzo, C., Robine, S., and Andrews, N. C. (2005a). Slc11a2 is required for intestinal iron absorption and erythropoiesis but dispensable in placenta and liver. *J. Clin. Invest.* **115,** 1258–1266.

Gunshin, H., Starr, C. N., Direnzo, C., Fleming, M. D., Jin, J., Greer, E. L., Sellers, V. M., Galica, S. M., and Andrews, N. C. (2005b). Cybrd1 (duodenal cytochrome b) is not necessary for dietary iron absorption in mice. *Blood* **106**(8), 2879–2883.

Gutteridge, J. M. C., Rowley, D. A., and Halliwell, B. (1981). Superoxide-dependent formation of hydroxyl radicals in the presence of iron salts. *Biochem. J.* **199,** 263–265.

Gutteridge, J. M. C., Rowley, D. A., Griffiths, E., and Halliwell, B. (1985). Low-molecular-weight iron complexes and oxygen radical reactions in idiopathic haemochromatosis. *Clin. Sci.* **68,** 463–467.

Hamill, R. L., Woods, J. C., and Cook, B. A. (1991). Congenital atransferrinemia: A case report and review of the literature. *Am. J. Clin. Pathol.* **96,** 215–218.

Harris, Z. L., Takahashi, Y., Miyajima, H., Serizawa, M., MacGillivray, R. T., and Gitlin, J. D. (1995). Aceruloplasminemia: Molecular characterization of this disorder of iron metabolism. *Proc. Natl. Acad. Sci. USA* **92,** 2539–2543.

Harris, Z. L., Durley, A. P., Man, T. K., and Gitlin, J. D. (1999). Targeted gene disruption reveals an essential role for ceruloplasmin in cellular iron efflux. *Proc. Natl. Acad. Sci. USA* **96,** 10812–10817.

Huang, F. W., Pinkus, J. L., Pinkus, G. S., Fleming, M. D., and Andrews, N. C. (2005). A mouse model of juvenile hemochromatosis. *J. Clin. Invest.* **115,** 2187–2191.

Hunter, H. N., Fulton, D. B., Ganz, T., and Vogel, H. J. (2002). The solution structure of human hepcidin, a peptide hormone with antimicrobial activity that is involved in iron uptake and hereditary hemochromatosis. *J. Biol. Chem.* **277,** 37597–37603.

Iolascon, A., d'Apolito, M., Servedio, V., Cimmino, F., Piga, A., and Camaschella, C. (2006). Microcytic anemia and hepatic iron overload in a child with compound heterozygous mutations in DMT1 (SCL11A2). *Blood* **107,** 349–354.

Jabado, N., Canonne-Hergaux, F., Gruenheid, S., Picard, V., and Gros, P. (2002). Iron transporter Nramp2/DMT-1 is associated with the membrane of phagosomes in macrophages and Sertoli cells. *Blood* **100,** 2617–2622.

Jenkins, Z. A., Hagar, W., Bowlus, C. L., Johansson, H. E., Harmatz, P., Vichinsky, E. P., and Theil, E. C. (2007). Iron homeostasis during transfusional iron overload in beta-thalassemia and sickle cell disease: Changes in iron regulatory protein, hepcidin, and ferritin expression. *Pediatr. Hematol. Oncol.* **24,** 237–243.

Johnson, M. B., and Enns, C. A. (2004). Regulation of transferrin receptor 2 by transferrin: Diferric transferrin regulates transferrin receptor 2 protein stability. *Blood* **104**(13), 4287–4293.

Johnson, M. B., Chen, J., Murchison, N., Green, F. A., and Enns, C. A. (2007). Transferrin receptor 2: Evidence for ligand-induced stabilization and redirection to a recycling pathway. *Mol. Biol. Cell* **18,** 743–754.

Kattamis, A., Papassotiriou, I., Palaiologou, D., Apostolakou, F., Galani, A., Ladis, V., Sakellaropoulos, N., and Papanikolaou, G. (2006). The effects of erythropoetic activity and iron burden on hepcidin expression in patients with thalassemia major. *Haematologica* **91,** 809–812.

Kawabata, H., Yang, R., Hirama, T., Vuong, P. T., Kawano, S., Gombart, A. F., and Koeffler, H. P. (1999). Molecular cloning of transferrin receptor 2. A new member of the transferrin receptor-like family. *J. Biol. Chem.* **274,** 20826–20832.

Kawabata, H., Germain, R. S., Vuong, P. T., Nakamaki, T., Said, J. W., and Koeffler, H. P. (2000). Transferrin receptor 2-{alpha} supports cell growth both in iron-chelated cultured cells and *in vivo*. *J. Biol. Chem.* **275,** 16618–16625.

Kearney, S. L., Nemeth, E., Neufeld, E. J., Thapa, D., Ganz, T., Weinstein, D. A., and Cunningham, M. J. (2007). Urinary hepcidin in congenital chronic anemias. *Pediatr. Blood Cancer* **48,** 57–63.

Kemna, E., Pickkers, P., Nemeth, E., van der Hoeven, H., and Swinkels, D. (2005). Time-course analysis of hepcidin, serum iron, and plasma cytokine levels in humans injected with LPS. *Blood* **106,** 1864–1866.

Kemna, E. H., Tjalsma, H., Podust, V. N., and Swinkels, D. W. (2007). Mass spectrometry-based hepcidin measurements in serum and urine: Analytical aspects and clinical implications. *Clin. Chem.* **53,** 620–628.

Klausner, R. D., Ashwell, G., van Renswoude, J., Harford, J. B., and Bridges, K. R. (1983). Binding of apotransferrin to K562 cells: Explanation of the transferrin cycle. *Proc. Natl. Acad. Sci. USA* **80,** 2263–2266.

Knutson, M., and Wessling-Resnick, M. (2003). Iron metabolism in the reticuloendothelial system. *Crit. Rev. Biochem. Mol. Biol.* **38,** 61–88.

Kozyraki, R., Fyfe, J., Verroust, P. J., Jacobsen, C., Dautry-Varsat, A., Gburek, J., Willnow, T. E., Christensen, E. I., and Moestrup, S. K. (2001). Megalin-dependent cubilin-mediated endocytosis is a major pathway for the apical uptake of transferrin in polarized epithelia. *Proc. Natl. Acad. Sci. USA* **98,** 12491–12496.

Krause, A., Neitz, S., Magert, H. J., Schulz, A., Forssmann, W. G., Schulz-Knappe, P., and Adermann, K. (2000). LEAP-1, a novel highly disulfide-bonded human peptide, exhibits antimicrobial activity. *FEBS Lett.* **480,** 147–150.

Kristiansen, M., Graversen, J. H., Jacobsen, C., Sonne, O., Hoffman, H. J., Law, S. K., and Moestrup, S. K. (2001). Identification of the haemoglobin scavenger receptor. *Nature* **409,** 198–201.

Kuninger, D., Kuns-Hashimoto, R., Kuzmickas, R., and Rotwein, P. (2006). Complex biosynthesis of the muscle-enriched iron regulator RGMc. *J. Cell Sci.* **119,** 3273–3283.

Lee, P., Peng, H., Gelbart, T., and Beutler, E. (2004). The IL-6- and lipopolysaccharide-induced transcription of hepcidin in HFE-, transferrin receptor 2-, and beta 2-microglobulin-deficient hepatocytes. *Proc. Natl. Acad. Sci. USA* **101,** 9263–9265.

Lee, P., Peng, H., Gelbart, T., Wang, L., and Beutler, E. (2005). Regulation of hepcidin transcription by interleukin-1 and interleukin-6. *Proc. Natl. Acad. Sci. USA* **102,** 1906–1910.

Lee, P. J., Jiang, B. H., Chin, B. Y., Iyer, N. V., Alam, J., Semenza, G. L., and Choi, A. M. (1997). Hypoxia-inducible factor-1 mediates transcriptional activation of the heme oxygenase-1 gene in response to hypoxia. *J. Biol. Chem.* **272,** 5375–5381.

Leung, P. S., Srai, S. K., Mascarenhas, M., Churchill, L. J., and Debnam, E. S. (2005). Increased duodenal iron uptake and transfer in a rat model of chronic hypoxia is accompanied by reduced hepcidin expression. *Gut* **54,** 1391–1395.

Levy, J. E., Jin, O., Fujiwara, Y., Kuo, F., and Andrews, N. C. (1999). Transferrin receptor is necessary for development of erythrocytes and the nervous system. *Nat. Genet.* **21,** 396–399.

Lin, L., Goldberg, Y. P., and Ganz, T. (2005). Competitive regulation of hepcidin mRNA by soluble and cell-associated hemojuvelin. *Blood* **106**(8), 2884–2889.

Lin, L., Valore, E. V., Nemeth, E., Goodnough, J. B., Gabayan, V., and Ganz, T. (2007). Iron-transferrin regulates hepcidin synthesis in primary hepatocyte culture through hemojuvelin and BMP2/4. *Blood* **110**(6), 2182–2189.

Liu, X. B., Yang, F., and Haile, D. J. (2005). Functional consequences of ferroportin 1 mutations. *Blood Cells Mol. Dis.* **35,** 33–46.

McKie, A. T., Marciani, P., Rolfs, A., Brennan, K., Wehr, K., Barrow, D., Miret, S., Bomford, A., Peters, T. J., Farzaneh, F., Hediger, M. A., Hentz, M. W., *et al.* (2000). A novel duodenal iron-regulated transporter, IREG1, implicated in the basolateral transfer of iron to the circulation. *Mol. Cell* **5,** 299–309.

McKie, A. T., Barrow, D., Latunde-Dada, G. O., Rolfs, A., Sager, G., Mudaly, E., Mudaly, M., Richardson, C., Barlow, D., Bomford, A., Peters, T. J., Raja, K. B., *et al.* (2001). An iron-regulated ferric reductase associated with the absorption of dietary iron. *Science* **291,** 1755–1759.

Mims, M. P., Guan, Y., Pospisilova, D., Priwitzerova, M., Indrak, K., Ponka, P., Divoky, V., and Prchal, J. T. (2004). Identification of a human mutation of DMT1 in a patient with microcytic anemia and iron overload. *Blood* **105**(3), 1337–1342.

Monnier, P. P., Sierra, A., Macchi, P., Deitinghoff, L., Andersen, J. S., Mann, M., Flad, M., Hornberger, M. R., Stahl, B., Bonhoeffer, F., and Mueller, B. K. (2002). RGM is a repulsive guidance molecule for retinal axons. *Nature* **419,** 392–395.

Muckenthaler, M., Roy, C. N., Custodio, A. O., Minana, B., deGraaf, J., Montross, L. K., Andrews, N. C., and Hentze, M. W. (2003). Regulatory defects in liver and intestine implicate abnormal hepcidin and Cybrd1 expression in mouse hemochromatosis. *Nat. Genet.* **34,** 102–107.

Muir, A., and Hopfer, U. (1985). Regional specificity of iron uptake by small intestinal brush-border membranes from normal and iron-deficient mice. *Am. J. Physiol.* **248** (3Pt1), G376–G379.

Nemeth, E., Valore, E. V., Territo, M., Schiller, G., Lichtenstein, A., and Ganz, T. (2003). Hepcidin, a putative mediator of anemia of inflammation, is a type II acute-phase protein. *Blood* **101,** 2461–2463.

Nemeth, E., Rivera, S., Gabayan, V., Keller, C., Taudorf, S., Pedersen, B. K., and Ganz, T. (2004a). IL-6 mediates hypoferremia of inflammation by inducing the synthesis of the iron regulatory hormone hepcidin. *J. Clin. Invest.* **113,** 1271–1276.

Nemeth, E., Tuttle, M. S., Powelson, J., Vaughn, M. B., Donovan, A., Ward, D. M., Ganz, T., and Kaplan, J. (2004b). Hepcidin regulates cellular iron efflux by binding to ferroportin and inducing its internalization. *Science* **306,** 2090–2093.

Nemeth, E., Roetto, A., Garozzo, G., Ganz, T., and Camaschella, C. (2005). Hepcidin is decreased in TFR2 hemochromatosis. *Blood* **105,** 1803–1806.

Nemeth, E., Preza, G. C., Jung, C. L., Kaplan, J., Waring, A. J., and Ganz, T. (2006). The N-terminus of hepcidin is essential for its interaction with ferroportin: Structure-function study. *Blood* **107,** 328–333.

Nicolas, G., Bennoun, M., Devaux, I., Beaumont, C., Grandchamp, B., Kahn, A., and Vaulont, S. (2001). Lack of hepcidin gene expression and severe tissue iron overload in upstream stimulatory factor 2 (USF2) knockout mice. *Proc. Natl. Acad. Sci. USA* **98,** 8780–8785.

Nicolas, G., Bennoun, M., Porteu, A., Mativet, S., Beaumont, C., Grandchamp, B., Sirito, M., Sawadogo, M., Kahn, A., and Vaulont, S. (2002a). Severe iron deficiency anemia in transgenic mice expressing liver hepcidin. *Proc. Natl. Acad. Sci. USA* **99,** 4596–4601.

Nicolas, G., Chauvet, C., Viatte, L., Danan, J. L., Bigard, X., Devaux, I., Beaumont, C., Kahn, A., and Vaulont, S. (2002b). The gene encoding the iron regulatory peptide hepcidin is regulated by anemia, hypoxia, and inflammation. *J. Clin. Invest.* **110,** 1037–1044.

Nicolas, G., Viatte, L., Lou, D. Q., Bennoun, M., Beaumont, C., Kahn, A., Andrews, N. C., and Vaulont, S. (2003). Constitutive hepcidin expression prevents iron overload in a mouse model of hemochromatosis. *Nat. Genet.* **34,** 97–101.

Niederkofler, V., Salie, R., and Arber, S. (2005). Hemojuvelin is essential for dietary iron sensing, and its mutation leads to severe iron overload. *J. Clin. Invest.* **115,** 2180–2186.

Oldekamp, J., Kramer, N., Alvarez-Bolado, G., and Skutella, T. (2004). Expression pattern of the repulsive guidance molecules RGM A, B and C during mouse development. *Gene Expr. Patterns* **4,** 283–288.

Oudit, G. Y., Sun, H., Trivieri, M. G., Koch, S. E., Dawood, F., Ackerley, C., Yazdanpanah, M., Wilson, G. J., Schwartz, A., Liu, P. P., and Backx, P. H. (2003). L-type Ca(2+) channels provide a major pathway for iron entry into cardiomyocytes in iron-overload cardiomyopathy. *Nat. Med.* **9,** 1187–1194.

Papanikolaou, G., Samuels, M. E., Ludwig, E. H., MacDonald, M. L., Franchini, P. L., Dube, M. P., Andres, L., MacFarlane, J., Sakellaropoulos, N., Politou, M., Nemeth, E., Thompson, J., *et al.* (2004). Mutations in HFE2 cause iron overload in chromosome 1q-linked juvenile hemochromatosis. *Nat. Genet.* **36,** 77–82.

Park, C. H., Valore, E. V., Waring, A. J., and Ganz, T. (2001). Hepcidin, a urinary antimicrobial peptide synthesized in the liver. *J. Biol. Chem.* **276,** 7806–7810.

Peyssonnaux, C., Zinkernagel, A. S., Schuepbach, R. A., Rankin, E., Vaulont, S., Haase, V. H., Nizet, V., and Johnson, R. S. (2007). Regulation of iron homeostasis by the hypoxia-inducible transcription factors (HIFs). *J. Clin. Invest.* **117,** 1926–1932.

Pietrangelo, A., Dierssen, U., Valli, L., Garuti, C., Rump, A., Corradini, E., Ernst, M., Klein, C., and Trautwein, C. (2007). STAT3 is required for IL-6-gp130-dependent activation of hepcidin *in vivo. Gastroenterology* **132,** 294–300.

Pigeon, C., Ilyin, G., Courselaud, B., Leroyer, P., Turlin, B., Brissot, P., and Loreal, O. (2001). A new mouse liver-specific gene, encoding a protein homologous to human antimicrobial peptide hepcidin, is overexpressed during iron overload. *J. Biol. Chem.* **276,** 7811–7819.

Poss, K. D., and Tonegawa, S. (1997). Heme oxygenase 1 is required for mammalian iron reutilization. *Proc. Natl. Acad. Sci. USA* **94,** 10919–10924.

Qiu, A., Jansen, M., Sakaris, A., Min, S., Chattopadhyay, S., Tsai, E., Sandoval, C., Zhao, R., Akabas, M., and Goldman, I. (2007). Identification of an intestinal folate transporter and the molecular basis for hereditary folate malabsorption. *Cell* **127**(5), 917–928.

R'Zik, S., and Beguin, Y. (2001). Serum soluble transferrin receptor concentration is an accurate estimate of the mass of tissue receptors. *Exp. Hematol.* **29,** 677–685.

Raja, K. B., Simpson, R. J., Pippard, M. J., and Peters, T. J. (1988). *In vivo* studies on the relationship between intestinal (Fe3+) absorption, hypoxia and erythropoiesis in the mouse. *Br. J. Haematol.* **68,** 373–378.

Rivera, S., Liu, L., Nemeth, E., Gabayan, V., Sorensen, O. E., and Ganz, T. (2005a). Hepcidin excess induces the sequestration of iron and exacerbates tumor-associated anemia. *Blood* **105,** 1797–1802.

Rivera, S., Nemeth, E., Gabayan, V., Lopez, M. A., Farshidi, D., and Ganz, T. (2005b). Synthetic hepcidin causes rapid dose-dependent hypoferremia and is concentrated in ferroportin-containing organs. *Blood* **106,** 2196–2199.

Robb, A., and Wessling-Resnick, M. (2004). Regulation of transferrin receptor 2 protein levels by transferrin. *Blood* **104,** 4294–4299.

Rodriguez, A., Pan, P., and Parkkila, S. (2007). Expression studies of neogenin and its ligand hemojuvelin in mouse tissues. *J. Histochem. Cytochem.* **55,** 85–96.

Roetto, A., Papanikolaou, G., Politou, M., Alberti, F., Girelli, D., Christakis, J., Loukopoulos, D., and Camaschella, C. (2003). Mutant antimicrobial peptide hepcidin is associated with severe juvenile hemochromatosis. *Nat. Genet.* **33,** 21–22.

Rolfs, A., Kvietikova, I., Gassmann, M., and Wenger, R. H. (1997). Oxygen-regulated transferring expression is mediated by hypoxia-inducible factor-1. *J. Biol. Chem.* **272,** 20055–20062.

Roy, C. N., Weinstein, D. A., and Andrews, N. C. (2003). 2002 E. Mead Johnson Award for Research in Pediatrics Lecture: The molecular biology of the anemia of chronic disease: A hypothesis. *Pediatr. Res.* **53,** 507–512.

Roy, C. N., Custodio, A. O., de Graaf, J., Schneider, S., Akpan, I., Montross, L. K., Sanchez, M., Gaudino, A., Hentze, M. W., Andrews, N. C., and Muckenthaler, M. U. (2004). An Hfe-dependent pathway mediates hyposideremia in response to lipopolysaccharide-induced inflammation in mice. *Nat. Genet.* **36,** 481–485.

Roy, C. N., Mak, H. H., Akpan, I., Losyev, G., Zurakowski, D., and Andrews, N. C. (2007). Hepcidin antimicrobial peptide transgenic mice exhibit features of the anemia of inflammation. *Blood* **109,** 4038–4044.

Sacher, A., Cohen, A., and Nelson, N. (2001). Properties of the mammalian and yeast metal-ion transporters DCT1 and Smf1p expressed in Xenopus laevis oocytes. *J. Exp. Biol.* **204,** 1053–1061.

Samad, T. A., Rebbapragada, A., Bell, E., Zhang, Y., Sidis, Y., Jeong, S. J., Campagna, J. A., Perusini, S., Fabrizio, D. A., Schneyer, A. L., Lin, H. Y., Brivanlou, A. H., *et al.* (2005). DRAGON: A bone morphogenetic protein co-receptor. *J. Biol. Chem.* **280**(14), 14122–14129.

Schimanski, L. M., Drakesmith, H., Merryweather-Clarke, A. T., Viprakasit, V., Edwards, J. P., Sweetland, E., Bastin, J. M., Cowley, D., Chinthammitr, Y., Robson, K. J., and Townsend, A. R. (2005). *In vitro* functional analysis of human ferroportin (FPN) and hemochromatosis-associated FPN mutations. *Blood* **105**(10), 4096–4102.

Schofield, C. J., and Ratcliffe, P. J. (2005). Signalling hypoxia by HIF hydroxylases. *Biochem. Biophys. Res. Commun.* **338,** 617–626.

Sham, R. L., Phatak, P. D., West, C., Lee, P., Andrews, C., and Beutler, E. (2005). Autosomal dominant hereditary hemochromatosis associated with a novel ferroportin mutation and unique clinical features. *Blood Cells Mol. Dis.* **34,** 157–161.

Shaw, G. C., Cope, J. J., Li, L., Corson, K., Hersey, C., Ackermann, G. E., Gwynn, B., Lambert, A. J., Wingert, R. A., Traver, D., Trede, N. S., Barut, B. A., *et al.* (2006). Mitoferrin is essential for erythroid iron assimilation. *Nature* **440,** 96–100.

Shayeghi, M., Latunde-Dada, G. O., Oakhill, J. S., Laftah, A. H., Takeuchi, K., Halliday, N., Khan, Y., Warley, A., McCann, F. E., Hider, R. C., Frazer, D. M.,

Anderson, G. J., *et al.* (2005). Identification of an intestinal heme transporter. *Cell* **122,** 789–801.

Sheftel, A. D., Zhang, A. S., Brown, C., Shirihai, O. S., and Ponka, P. (2007). Direct interorganellar transfer of iron from endosome to mitochondrion. *Blood* **110,** 125–132.

Shike, H., Lauth, X., Westerman, M. E., Ostland, V. E., Carlberg, J. M., Van Olst, J. C., Shimizu, C., Bulet, P., and Burns, J. C. (2002). Bass hepcidin is a novel antimicrobial peptide induced by bacterial challenge. *Eur. J. Biochem.* **269,** 2232–2237.

Silvestri, L., Pagani, A., Fazi, C., Gerardi, G., Levi, S., Arosio, P., and Camaschella, C. (2007). Defective targeting of hemojuvelin to plasma membrane is a common pathogenetic mechanism in juvenile hemochromatosis. *Blood* **109,** 4503–4510.

Su, M. A., Trenor, C. C., Fleming, J. C., Fleming, M. D., and Andrews, N. C. (1998). The G185R mutation disrupts function of the iron transporter Nramp2. *Blood* **92,** 2157–2163.

Townsend, A., and Drakesmith, H. (2002). Role of HFE in iron metabolism, hereditary haemochromatosis, anaemia of chronic disease, and secondary iron overload. *Lancet* **359,** 786–790.

Trenor, C. C., 3rd, Campagna, D. R., Sellers, V. M., Andrews, N. C., and Fleming, M. D. (2000). The molecular defect in hypotransferrinemic mice. *Blood* **96,** 1113–1118.

Truksa, J., Peng, H., Lee, P., and Beutler, E. (2006). Bone morphogenetic proteins 2, 4, and 9 stimulate murine hepcidin 1 expression independently of Hfe, transferrin receptor 2 (Tfr2), and IL-6. *Proc. Natl. Acad. Sci. USA* **103,** 10289–10293.

Verga Falzacappa, M. V., Vujic Spasic, M., Kessler, R., Stolte, J., Hentze, M. W., and Muckenthaler, M. U. (2007). STAT3 mediates hepatic hepcidin expression and its inflammatory stimulation. *Blood* **109,** 353–358.

Viatte, L., Lesbordes-Brion, J. C., Lou, D. Q., Bennoun, M., Nicolas, G., Kahn, A., Canonne-Hergaux, F., and Vaulont, S. (2005). Deregulation of proteins involved in iron metabolism in hepcidin-deficient mice. *Blood* **105**(12), 4861–4864.

Vulpe, C. D., Kuo, Y. M., Murphy, T. L., Cowley, L., Askwith, C., Libina, N., Gitschier, J., and Anderson, G. J. (1999). Hephaestin, a ceruloplasmin homologue implicated in intestinal iron transport, is defective in the sla mouse. *Nat. Genet.* **21,** 195–199.

Wallace, D. F., Dixon, J. L., Ramm, G. A., Anderson, G. J., Powell, L. W., and Subramaniam, V. N. (2007). A novel mutation in ferroportin implicated in iron overload. *J. Hepatol.* **46,** 921–926.

Wang, R. H., Li, C., Xu, X., Zheng, Y., Xiao, C., Zerfas, P., Cooperman, S., Eckhaus, M., Rouault, T., Mishra, L., and Deng, C. X. (2005). A role of SMAD4 in iron metabolism through the positive regulation of hepcidin expression. *Cell Metab.* **2,** 399–409.

Watson, C. J. (2001). Stat transcription factors in mammary gland development and tumorigenesis. *J. Mammary Gland Biol. Neoplasia* **6,** 115–127.

Weinberg, E. D. (1978). Iron and infection. *Microbiol. Rev.* **45**(1), 45–66.

Weinstein, D. A., and Wolfsdorf, J. I. (2002). Effect of continuous glucose therapy with uncooked cornstarch on the long-term clinical course of type 1a glycogen storage disease. *Eur. J. Pediatr.* **161,** S35–S39.

Weinstein, D. A., Roy, C. N., Fleming, M. D., Loda, M. F., Wolfsdorf, J. I., and Andrews, N. C. (2002). Inappropriate expression of hepcidin is associated with iron refractory anemia: Implications for the anemia of chronic disease. *Blood* **100,** 3776–3781.

Weiss, G., and Goodnough, L. T. (2005). Anemia of chronic disease. *N. Engl. J. Med.* **352,** 1011–1023.

Weizer-Stern, O., Adamsky, K., Amariglio, N., Levin, C., Koren, A., Breuer, W., Rachmilewitz, E., Breda, L., Rivella, S., Cabantchik, Z. I., and Rechavi, G. (2006a). Downregulation of hepcidin and haemojuvelin expression in the hepatocyte cell-line HepG2 induced by thalassaemic sera. *Br. J. Haematol.* **135,** 129–138.

Weizer-Stern, O., Adamsky, K., Amariglio, N., Rachmilewitz, E., Breda, L., Rivella, S., and Rechavi, G. (2006b). mRNA expression of iron regulatory genes in beta-thalassemia intermedia and beta-thalassemia major mouse models. *Am. J. Hematol.* **81,** 479–483.

West, A. P., Jr., Bennett, M. J., Sellers, V. M., Andrews, N. C., Enns, C. A., and Bjorkman, P. J. (2000). Comparison of the interactions of transferrin receptor and transferrin receptor 2 with transferrin and the hereditary hemochromatosis protein HFE. *J. Biol. Chem.* **275,** 38135–38138.

Wolfsdorf, J. I., and Crigler, J. F. (1999). Effect of continuous glucose therapy begun in infancy on the long-term clinical course of patients with type I glycogen storage disease. *J. Pediatr. Gastroenterol. Nutr.* **29,** 136–143.

Wrighting, D. M., and Andrews, N. C. (2006). Interleukin-6 induces hepcidin expression through STAT3. *Blood* **108,** 3204–3209.

Xu, H., Jin, J., DeFelice, L. J., Andrews, N. C., and Clapham, D. E. (2004). A spontaneous, recurrent mutation in divalent metal transporter-1 exposes a calcium entry pathway. *PLoS Biol.* **2,** E50.

Yachie, A., Niida, Y., Wada, T., Igarashi, N., Kaneda, H., Toma, T., Ohta, K., Kasahara, Y., and Koizumi, S. (1999). Oxidative stress causes enhanced endothelial cell injury in human heme oxygenase-1 deficiency. *J. Clin. Invest.* **103,** 129–135.

Yang, J., Goetz, D., Li, J. Y., Wang, W., Mori, K., Setlik, D., Du, T., Erdjument-Bromage, H., Tempst, P., Strong, R., and Barasch, J. (2002). An iron delivery pathway mediated by a lipocalin. *Mol. Cell* **10,** 1045–1056.

Zhang, A. S., West, A. P., Jr., Wyman, A. E., Bjorkman, P. J., and Enns, C. A. (2005). Interaction of hemojuvelin with neogenin results in iron accumulation in human embryonic kidney 293 cells. *J. Biol. Chem.* **280,** 33885–33894.

Zhang, A. S., Anderson, S. A., Meyers, K. R., Hernandez, C., Eisenstein, R. S., and Enns, C. A. (2007). Evidence that inhibition of hemojuvelin shedding in response to iron is mediated through neogenin. *J. Biol. Chem.* **282,** 12547–12556.

Zohn, I. E., De Domenico, I., Pollock, A., Ward, D. M., Goodman, J. F., Liang, X., Sanchez, A. J., Niswander, L., and Kaplan, J. (2007). The flatiron mutation in mouse ferroportin acts as a dominant negative to cause ferroportin disease. *Blood* **109,** 4174–4180.

Zoller, H., and Cox, T. M. (2005). Hemochromatosis: Genetic testing and clinical practice. *Clin. Gastroenterol. Hepatol.* **3,** 945–958.

Zoller, H., McFarlane, I., Theurl, I., Stadlmann, S., Nemeth, E., Oxley, D., Ganz, T., Halsall, D. J., Cox, T. M., and Vogel, W. (2005). Primary iron overload with inappropriate hepcidin expression in V162del ferroportin disease. *Hepatology* **42,** 466–472.

CHAPTER SEVEN

Effects of Nitric Oxide on Red Blood Cell Development and Phenotype

Vladan P. Čokić* *and* Alan N. Schechter†

Contents

Abstract

Nitric oxide (NO) is a diffusible free radical generated primarily by NO synthases (NOS), isoenzymes that convert the L-arginine and molecular oxygen to citrulline and NO in cells. Endothelial cells as well as macrophages, components of hematopoietic microenvironment and potent NO producers, play an active role

* Laboratory of Experimental Hematology, Institute for Medical Research, Belgrade, Serbia
† Molecular Medicine Branch, NIDDK, National Institutes of Health, Bethesda, Maryland 20892

Current Topics in Developmental Biology, Volume 82
ISSN 0070-2153, DOI: 10.1016/S0070-2153(07)00007-5

in the modulation of human hematopoietic cell growth and differentiation. A role of NO in erythroid cell differentiation has been postulated based on demonstration that NO inhibits growth, differentiation, and hemoglobinization of erythroid primary cells. Endothelial NOS (eNOS) mRNA and protein levels, as well as bioactivity, decrease during erythroid differentiation, concomitantly with the elevation of hemoglobin levels. Human red blood cells (RBCs) have been reported to contain some eNOS activity; NO appears to affect RBC's deformability. Generally, NO activates cellular soluble guanylyl cyclase (sGC) to produce a second messenger molecule cGMP. NO increases cGMP, γ-globin, and HbF levels in human erythroid cells whereas inhibition of sGC prevents NO-induced increase in γ-globin gene expression. Activation of sGC increases γ-globin gene expression in primary human erythroblasts. High cAMP levels continuously decrease in contrast to steady but low levels of cGMP during erythroid differentiation. The activation of the cAMP pathway has also been reported to induce expression of the γ-globin gene in human erythroid cells. NO is hydrophobic and accumulates in lipid membranes, and most autoxidation to nitrite *in vivo* occurs there. The reaction of NO with deoxyhemoglobin produces nitrosylhemoglobin (HbFe(II)NO), while that with oxyhemoglobin produced methemoglobin and nitrate. Nitrite can also react with deoxyhemoglobin to produce NO. This reaction as well as the postulated formation of a thiol-NO derivative of hemoglobin (SNO-Hb) appears to be major mechanisms for the preservation and transport of NO bioactivity by red cells—making NO act as a "hormone." Thus, RBCs and hemoglobin molecules are essential factors in regulating the bioactivity of NO throughout the mammalian body and may be important in the pathophysiology of several circulatory diseases and be the basis for new therapeutic approaches to these diseases.

1. Nitric Oxide

Nitric oxide (NO) is reported to have been first prepared in about 1620 by the Belgian scientist Jan Baptist van Helmont (*Encyclopedia Britannica*, 15th edn., vol. 8, p.726). In the 1840s, Walter Crum devised a method for preparing pure NO by shaking together nitric acid, concentrated sulfuric acid, and mercury, and in 1908, Fritz Haber described the synthesis of NO in an electric arc. NO is a colorless gas with good solubility in water. NO is one of the simplest odd-electron species in which the presence of the unpaired electron gives it paramagnetic properties and makes it reactive with many atoms in biological molecules and other free radicals. The biological half-life of NO is generally considered to be about 1–10 s (Ignarro *et al.*, 1993).

Nitroglycerine has been used for over a century to treat coronary heart disease, and it has long been suggested that humans synthesize oxides of nitrogen (see for example Mitchell *et al.*, 1916). Murad discovered in 1977 that nitroglycerin and similar substances release NO, which relaxes smooth muscle cells. The existence of an *endothelium-derived relaxing factor* (EDRF) was first postulated by Furchgott and Zawadzki (1980). EDRF was later identified as NO gas synthesized from L-arginine by the action of nitric oxide synthase (NOS, Palmer *et al.*, 1988). In addition, involvement of Ca^{2+} in NO-mediated signal transduction was demonstrated in blood vessels (Rees *et al.*, 1989) and neurons (Garthwaite *et al.*, 1988). At a scientific conference in 1986, Furchgott and Ignarro presented their conclusions that NO transmits signals in the human organism. NO was named as a molecule of the year by the journal *Science* in 1992. The winners of the 1998 Nobel Medicine Prize, scientists who won for discovering new properties of NO, are Robert F. Furchgott, Louis J. Ignaro, and Ferid Murad.

NO mediates vasodilatation, causes inhibition of platelet aggregation, has a role as an important effector molecule of the host defense system (septic shock, acute and chronic inflammation, destruction of invading bacteria, viruses, and tumor cells), and functions in neuronal transmission. As both NO and O_2 are hydrophobic, they accumulate in lipid membranes, and most autoxidation to nitrite *in vivo* occurs there (Liu *et al.*, 1998b). NO reacts rapidly with oxyhemoglobin or oxymyoglobin to form nitrate. The rapid rate of this reaction, together with the high hemoglobin (Hb) concentration in blood and the rapid diffusion of NO, makes nitrate the major endpoint of NO metabolism *in vivo* (Liu *et al.*, 2005). It has been determined that the rate constant of NO consumption by 5×10^6 red blood cells (RB)/mL at 37 °C is 0.54 s^{-1}. The half-life of NO calculated from this rate constant is 1.3 s (Cassoly and Gibson, 1975).

1.1. Chemistry of NO

NO is a small, highly reactive, diffusible free radical. NO is a gas at room and body temperature, making it highly diffusible within the vasculature. In the gaseous phase, reaction of NO with oxygen results in the formation of nitrogen dioxide (NO_2). In aqueous aerobic solutions (e.g., supernatant medium), NO_2 decomposes to give mainly nitrite and to a smaller extent nitrate. Assuming that enough NO is available, NO_2 reacts with it to form dinitrogen trioxide adding to NO's nitrosative action (Liu *et al.*, 2005). NO reacts with deoxy- and (as noted above) oxyhemoglobin (deoxyHb and oxyHb, respectively) at a very high rate to form nitrosyl hemoglobin (HbNO) and methemoglobin (metHb), respectively (Cassoly and Gibson, 1975; Eich *et al.*, 1996). NO can react rapidly in the intracellular environment to form nitrite and nitrate, *S*-nitrosothiols, or peroxynitrite. NO has been shown to bind rapidly, and with high affinity, to ferrous iron (Fe^{2+}). As a consequence of this, NO can bind easily to three forms of iron: free, within iron–sulfur centers, and within hemoproteins (Thomsen *et al.*, 1995).

1.2. Synthesis of NO

NO is generated by NOS, a group of evolutionarily conserved cytosolic or membrane-bound isoenzymes that convert the nonaromatic amino acid L-arginine, and molecular oxygen, to citrulline and NO in mammalian and nonmammalian animals (including protozoa and insects) as well as in plants (Klessig *et al.*, 2000; Nappi *et al.*, 2000; Stuehr, 1999). The neuronal NOS (nNOS), inducible NOS (iNOS), and endothelial NOS (eNOS) genes were localized to human chromosome 12q24.2, 17cen–q12, and 7q35–q36, respectively. The human eNOS comprises 26 exons that span 21 kb. The 28-exon nNOS gene on human chromosome 12q24.2 extends over 100 kb. Human chromosome 17cen-q11.2 houses the 26-exon, 37 kb iNOS gene (Chartrain *et al.*, 1994; Hall *et al.*, 1994; Robinson *et al.*, 1994). Cloning of the human and bovine eNOS shows that these proteins have approximately 60% sequence identity with nNOS and 50% with iNOS. A sequence similarity between iNOS and nNOS is 53%, and between iNOS and eNOS is 51%. Murine iNOS and human iNOS share a similarity of some 80% (Taylor and Geller, 2000).

1.2.1. NOS isoforms

All NOS isoforms are homodimeric enzymes that require the same cosubstrates [molecular oxygen, nicotinamide adenine dinucleotide phosphate (NADPH)] and cofactors [flavin mononucleotide (FMN), flavine adenine dinucleotide (FAD), tetrahydrobiopterin (H_4B), heme, Ca^{2+}/calmodulin (CaM), and, possibly, also Zn^{2+} ions, Stuehr, 1999]. NOS isoforms oxidize L-arginine in a process that consumes NADPH (catalyzes the NADPH-dependent oxidation) and produces stoichiometric amounts of L-citrulline and NO. NOS isozymes are divided into two functional classes, constitutive and inducible.

Constitutive NOSs are present in unstimulated cells, and enzyme activity is stimulated by increased intracellular calcium levels and interaction with CaM. Constitutive isoforms are given below:

1. *Neuronal NOS* (nNOS or NOS I). nNOS first isolated from rat cerebellum (Bredt and Snyder, 1990). nNOS is expressed in: central nervous system, sympathetic ganglia, adrenal glands, certain areas of the spinal cord, skeletal muscle, pancreatic islets, macula densa cells of the kidney, epithelial cells of lung, uterus, and stomach.
2. *Endothelial NOS* (eNOS or NOS III). eNOS first isolated from cultured bovine aortic endothelial cells (Forstermann *et al.*, 1991). eNOS has been purified from arterial and venous endothelial cells, kidney tubular epithelial cells, and the syncytiotrophoblast of human placenta.

3. *Mitochondrial NOS* (mtNOS). An additional Ca-dependent mtNOS also synthesizes NO-inhibiting oxidative phosphorylation, that is mitochondrial energy producing metabolic process, and protects mitochondria from oxygen radicals. Mitochondrial membrane possess an electrogenic uniporter transporting Ca into mitochondria (stimulation of mtNOS), while the Na^{+}/Ca^{2+} exchanger removes Ca from mitochondria. Mitochondrial disorders with low mtNOS activity participate in accelerated aging and age-related diseases. mtNOS has the same cofactor and substrate requirements as other constitutive NOS. mtNOS has been identified as the α-isoform of nNOS, acylated at a Thr or Ser residue, and phosphorylated at the C-terminal end. Endogenous NO reversibly inhibits oxygen consumption and ATP synthesis by competitive inhibition of cytochrome oxidase. NO is the first molecule that fulfills the requirement for a cytochrome oxidase activity modulator: it is a competitive inhibitor, produced endogenously at a fair rate near the target site, at concentrations high enough to exhibit an inhibitory effect on cytochrome oxidase (Giulivi, 1998; Tatoyan and Giulivi, 1998).

Two different constitutively present isoforms of NOS (eNOS and nNOS) are responsible for biosynthesis of the relatively small but physiologically very important amounts of NO. Such small amounts of NO are completely noncytotoxic. nNOS, in the CNS, is activated by increases in postsynaptic intracellular Ca^{2+} levels mediated by glutamate and the *N*-methyl-d-aspartate receptor, a glutamate receptor subtype. Glutamate release is stimulated by a signal from the presynaptic neuron. In brain, nNOS is mainly a soluble enzyme that migrates with a molecular mass of 150–160 kDa in SDS-PAGE. In skeletal muscle, nNOS is linked to plasma membrane through a PDZ–PDZ domain interaction (Brenman *et al.*, 1996). The endoplasmic reticulum localization was suggested for the major portion of the particulate nNOS fraction in rat cerebellum (Hecker *et al.*, 1994). The enzyme can be phosphorylated at serine and threonine residues by Ca^{2+}/CaM-dependent protein kinase II and protein kinases A, C, and G.

Myristylation at the N-terminal glycine (glycine-2 before removal of the initiation methionine by a specific aminopeptidase) is mainly responsible for the membrane association of the eNOS enzyme; palmitoylation may also contribute (Shaul *et al.*, 1996). eNOS can undergo serine phosphorylation in response to bradykinin resulting in a translocation from the particulate to the cytosolic fraction. The presence of a consensus sequence for N-terminal myristoylation, absent in the other two isoforms, accounts for the observation that more than 90% of the enzymatic activity remains associated with the particulate fraction. The terminal amino acid in the mature eNOS expressed in mammalian cells is a glycine due to posttranslational processing of the terminal methionine. Both human and bovine endothelial proteins are proline-rich and contain a consensus target sequence for myristoylation by acyltransferases. It has been proposed that caveolin protein members function as scaffolding proteins to target signaling proteins to the caveolae, vesicular organelles located at or near the plasma membrane (Galbiati *et al.*, 1998). Caveolin-1 targets eNOS to the caveolae and inhibits eNOS activity via a direct interaction with the enzyme (Govers *et al.*, 2002). Other groups have proposed a dynamic cycling of eNOS from the Golgi apparatus to the plasma membrane thereby regulating eNOS activity (Sessa *et al.*, 1995). In cultured aortic endothelial cells and intact blood vessels, eNOS is primarily expressed in the Golgi region of the cells (Garcia-Gardena *et al.*, 1996). eNOS is involved in the regulation of blood pressure, organ blood flow distribution, the inhibition of the adhesion and activation of platelets and polymorphonuclear granulocytes.

iNOSs or NOS II with a tightly bound CaM are present in cells after incubation with cytokines and/or lipopolysaccharide (LPS), and enzyme activity is relatively insensitive to intracellular calcium levels. iNOS was first isolated from activated rat peritoneal macrophages (Yui *et al.*, 1991). iNOS is prominent in a wide array of cells and tissues: macrophages (MacMicking *et al.*, 1997), chondrocytes (Charles *et al.*, 1993), Kupffer cells, hepatocytes (Geller *et al.*, 1993), neutrophils (Sethi and Dikshit, 2000), pulmonary epithelium (Asano *et al.*, 1994), colonic epithelium (Perner *et al.*, 2002), vasculature (Hickey *et al.*, 2001), and various neoplastic diseases. The main switch for activity of iNOS is the level of its mRNA.

An inducible isoform of NOS, which is a distinct gene product from the constitutive isoforms, is responsible for high-output production of NO. Generation of large amounts of NO is associated with cytotoxicity and pathophysiology. A CaM consensus sequence has also been found in the iNOS. This was unexpected because this isoform had demonstrated no dependence on Ca^{2+} or CaM. Nathan and coworkers report that the macrophage iNOS contains CaM that is tightly bound and requires very low levels of Ca^{2+} for activation. These results suggest that iNOS is also regulated by Ca^{2+} and CaM but shows no apparent requirement because of an apparently high affinity for the Ca^{2+}/CaM complex. iNOS has a denatured molecular mass of 125–135 kDa and like nNOS it is a predominantly soluble enzyme. Once expressed, no regulatory mechanisms (except for product inhibition by NO) are known to regulate the activity of iNOS. Examples include induction of NOS in vascular smooth muscle resulting in profound vasorelaxation and hypotension, the characteristics of endotoxin shock; induction of NOS in brain microglia and astrocytes thereby resulting in production of large amounts of NO that permeate nearby oligodendrocytes resulting in impaired myelin production, the hallmark feature of multiple sclerosis; induction of NOS in joint tissues resulting in acute and chronic inflammation, characteristic signs of arthritic disease. nNOS and eNOS produce low NO concentrations for neurotransmission, insulin release, penile erection, vasorelaxation, oxygen detection, and memory storage, whereas cytokine-iNOS produces larger NO concentrations to counter pathogens and coordinate the T-cell response [through the production of peroxynitrite (ONOO–)].

1.2.2. NOS structure

All NOS isoforms catalyze a five-electron oxidation of guanidino nitrogen of the amino acid L-arginine to NO. This process involves the oxidation of NADPH and the reduction of molecular oxygen. All NOS isoforms are homodimers, the dimer being the catalytically active form. Each monomer comprises three distinct domains:

1. *The reductase domain:* a C-terminal electron-supplying reductase domain (NOS_{RED}, residues 531–1144 for iNOS). NOS_{RED} is homologous to cytochrome P450 reductase and binds FMN, FAD, and NADPH. Consensus sequences for the reduced form of NADPH, FAD, and FMN binding are located within the C-terminal half of NOS.
2. *The oxygenase domain:* a catalytic N-terminal oxygenase domain (NOS_{OX}, residues 1–498 for iNOS). NOS_{OX} binds heme, H_4B, and substrate L-arginine. NOS oxygenase domain is formed by one continuous fold made up of several overlapping or winged ß sheets. Thus, the structural elements responsible for binding L-arginine, H_4B, and iron containing heme group (iron protoporphyrin IX) are located throughout the oxygenase domain, rather than being arranged in a linear series of subdomains along polypeptide sequence. The heme iron is predominantly five-coordinated and has high spin in its ferric form in NOS. It is bound to the protein through a cysteine thiolate axial ligation, with proximal substrate L-arginine binding and an overall ligand environment similar to the heme iron in the cytochrome P450s. The FAD- and FMN-containing reductase domain transfers the electrons from NADPH to heme iron, enabling it to activate oxygen. Heme-activated O_2 first oxidizes L-arginine to hydroxyl-L-arginine (NOH-Arg), and this intermediate is then oxidized to L-citrulline and NO by a second molecule of O_2 that is bound to the same heme cofactor. These reactions occur on the oxygenase domain and require three electrons from NADPH. Approximately 1.5 mol NADPH is oxidized per mole of NO formed (Fig. 7.1).
3. *CaM binding:* the domains are linked by a Ca^{2+}/CaM-binding region. In eNOS and nNOS, Ca^{2+}/CaM binding aligns the domains and stimulates electron flow, accounting for enzyme activation. The intervening CaM-binding region (residues 499–530 for iNOS) regulates reduction of NOS_{OX} by NOS_{RED}. The CaM-free form of nNOS was found to load NADPH-derived electrons exclusively into its flavins, and could transfer electrons to its heme iron only upon CaM binding. iNOS also appears to require its tightly bound CaM to carry out heme iron reduction.

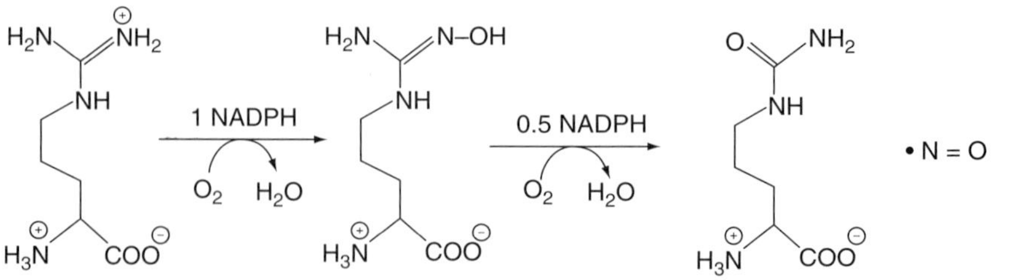

Figure 7.1 The synthesis of nitric oxide by eNOS. The FAD- and FMN-containing reductase domain transfers the electrons from NADPH to the enzyme's heme iron, enabling it to activate O_2. Heme-activated O_2 oxidizes L-arginine to hydroxyl-L-arginine (NOH-Arg), and this intermediate is then oxidized to L-citrulline and NO by a second molecule of O_2 that is bound to the same heme cofactor. These reactions occur on the oxygenase domain and require three electrons from NADPH. Approximately 1.5 mol NADPH oxidized per 1 mol NO formed.

$$\text{L-Arginine} + 2O_2 + 1.5\ \text{NADPH} \rightarrow \text{L-Citrulline} + \text{NO} + 1.5\ \text{NADP}^+ + 2H_2O$$

Most NOS inhibitors described to date bind to the oxygenase domain of NOS and interact with the guanidinium region of the L-arginine-binding site within NOS.

1.2.3. Arginine–citrulline cycle

Citrulline, the by-product of NO synthesis, is recycled to arginine through the intermediate formation of arginosuccinate. The relevant metabolic enzymes for this recycling, arginosuccinate synthase and arginosuccinase, are highly active in brain, endothelium, and other tissues enriched in NOS. The formation of citrulline from arginine via the urea cycle is largely restricted to liver and kidney, as ornithine transcarbomoylase and carbamoyl phosphate are not present in most cell types, including neurons and endothelial cells (Fig. 7.2).

1.3. Degradation of NOS in human cells

iNOS, eNOS, and nNOS protein levels are increased in the presence of the proteasome inhibitors (Bender *et al.*, 2000; Cokic *et al.*, 2007b; Musial and Eissa, 2001). Proteasomes are large multi-subunit complexes, localized in the nucleus and cytosol that selectively degrade intracellular proteins. The ubiquitin–proteasome pathway has been implicated in the destruction of proteins that participate in cell cycle progression, transcription control,

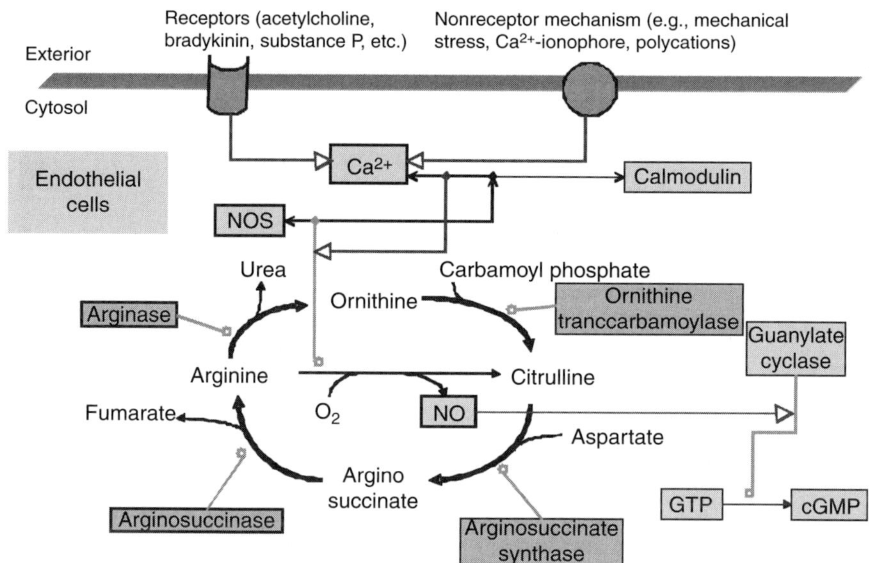

Figure 7.2 Nitric oxide synthesis in endothelial cells. The diagram shows how signal transduction is transmitted to activate the NOS via calcium, the arginine–citrulline cycle, and the effects of NO on conversion of GTP to cGMP.

signal transduction, and metabolic regulation. A protein marked for degradation is covalently attached to multiple molecules of ubiquitin, which escorts it for rapid hydrolysis to the multicomponent enzymatic complex known as the 26S proteasome. The proteolytic core of this complex, the 20S proteasome, contains multiple peptidase activities and functions as the catalytic machine (Magill *et al.*, 2003). NO upregulates p21 protein expression through the prevention of p21 protein degradation by the ubiquitin–proteasome pathway and NO-induced accumulation of tumor suppressor protein p53 is also mediated by inhibition of the proteasome (Glockzin *et al.*, 1999; Kibbe *et al.*, 2000). Proteasome inhibitors completely blocked tumor necrosis factor (TNF)- and interleukin (IL)-1β-stimulated cell surface expression of VCAM-1 and ICAM-1 in endothelial cells (Cobb *et al.*, 1996). NO donors, as well as proteasome inhibitors, inhibited cytokine-induced ICAM-1 and -VCAM-1 expression (Cobb *et al.*, 1996; De Caterina *et al.*, 1995). The proteasome inhibition also significantly enhanced endothelial-dependent vasorelaxation of rat aortic rings (Stangl *et al.*, 2004). In addition, the proteasome inhibitors increased NO production in endothelial cells (Cokic *et al.*, 2007b). The proteasome inhibitor bortezomib decreases proliferation, induces apoptosis, enhances the activity of chemotherapy and radiation, and reverses chemoresistance in a variety of hematologic and solid malignancy models *in vitro* and *in vivo* (Hamilton *et al.*, 2005; Hideshima *et al.*, 2001).

1.4. NO–cGMP pathway

1.4.1. Guanylyl cyclase and a second messenger molecule cGMP

cGMP acts as an intracellular signal molecule in the regulation of various cellular events. cGMP is recognized as a second messenger molecule involved in smooth muscle relaxation (Lincoln, 1989), inhibition of platelet aggregation (Mellion *et al.*, 1981), and neurotransmission (Garthwaite *et al.*, 1988). cGMP applies its effects through cGMP-dependent protein kinases, cGMP-regulated phosphodiesterases, and cGMP-regulated ion channels. Most of the cGMP effects have been shown to be mediated by the cGMP-dependent kinase.

Although cGMP-forming activity was first described in 1969 (Hardman and Sutherland, 1969), it took until the mid-1970s to find out that there are two different types of guanylyl cyclase (GC; Chrisman *et al.*, 1975), which were subsequently found to differ not only in their cellular localization. Two major types of cGMP-forming enzymes are the membrane-spanning peptide-regulated guanylyl-cyclases and the NO-sensitive guanylyl-cyclases (NO-sensitive GCs) found mainly in the cytosol. NO-sensitive GC consists of two distinct subunits, α and β. The enzyme ($\alpha_1\beta_1$) was purified from lung tissue and both subunits were cloned and sequenced 20 years ago (Koesling *et al.*, 1988). Two additional subunits, α_2 and β_2, were obtained by homology screening (Harteneck *et al.*, 1991). The α_2-subunit was able to form a catalytically active and highly NO-sensitive GC upon coexpression with the β_1-subunit in COS cells (Yuen *et al.*, 1990) and $\alpha_2\beta_1$-heterodimers have been shown to exist physiologically (Russwurm *et al.*, 1998). The enzyme contains a prosthetic ferrous heme group to which NO binds with high affinity (Gerzer *et al.*, 1981). Formation of NO–heme complex and a subsequent conformational change are responsible for the up to 200-fold increase in the catalytic rate (Stone and Marletta, 1995). NO-sensitive GC, similar to other nucleotide-converting enzymes, requires Mg^{2+} as cofactor for catalysis. The NO-sensitive GCas well as the adenylyl cyclases contain two different catalytic, whereas the peptide-regulated GCs contain two identical catalytic domains. Each subunit of NO-sensitive GC can be divided into three parts: a C-terminal catalytic domain, a domain commonly referred to as dimerization domain, and an N-terminal regulatory domain involved in heme binding. The C-terminal catalytic domains are highly conserved among the subunits of NO-sensitive GC and show homologies to the respective regions in the peptide receptor GCs and the adenylyl cyclases. The N-terminal parts of the subunits bind and coordinate with the heme and are therefore commonly designated as regulatory domains. The enzyme binds heme in a ratio of one mole per mole heterodimer (Brandish *et al.*, 1998) The enzyme contains a five-coordinated ferrous heme iron with a histidine, as the axial ligand, at the fifth coordinating position (Stone and Marletta, 1994). The histidine 1 residue (His-105) of the β_1-subunit was identified as the proximal heme ligand (Wedel *et al.*, 1994). The prosthetic heme group mediates the NO-induced stimulation subsequent to binding NO. Activation of NO-sensitive GC is initiated by NO binding to the sixth coordinating position of the heme iron that leads to breakage of the histidine–iron bond yielding a five-coordinated nitrosyl–heme complex (Gerzer *et al.*, 1981). The change in heme conformation is transduced to the catalytic cGMP forming domain resulting in the up to 200-fold activation of the enzyme. By replacing the heme, the heme precursor protoporphyrin IX stimulates the enzyme NO independently indicating that protoporphyrin IX, due to the lack of the central iron, is able to mimic the conformation of NO-bound heme (Ignarro *et al.*, 1982). The finding is compatible with the assumption that the release of the histidine–iron bond is required for stimulation of GC. Several models exist to explain the activation of NO-sensitive GC, the simplest is thought to occur in two steps: binding of NO to the sixth coordination position of the heme results in a six-coordinated NO-Fe^{2+}–His complex. The subsequent breakage of the histidine-to-iron bond leads to the formation of a five-coordinated nitrosyl–heme complex (Gerzer *et al.*, 1981). The opening of the histidine-to-iron bond is considered to initiate a conformational change resulting in the activation of the enzyme, up to 200-fold. In support of this simple model, protoporphyrin IX activates NO-sensitive GC independently of NO (Ignarro *et al.*, 1982).

Generally, dissociation of NO from the heme group is considered to trigger deactivation of GC. Other five-coordinated heme proteins exhibit very slow NO dissociation rates (Kharitonov *et al.*, 1997); with respect to NO-sensitive GC, this would imply that the enzyme is not able to immediately respond to changing NO concentrations. The half-life of the NO–GC complex of 2 min in the absence of substrate has been confirmed (Brandish *et al.*, 1998). NO-sensitive GC is catalytically active only as a heterodimer (Budworth *et al.*, 1999; Harteneck *et al.*, 1990; Kamisaki *et al.*, 1986). Besides the interaction of the $\alpha_2\beta_1$-isoform with PDZ-containing proteins, translocation of the $\alpha_1\beta_1$-isoform to the plasma membrane in response to elevated calcium concentrations has been reported. Furthermore, the membrane-associated enzyme appeared to display a higher NO sensitivity. The tissue distribution of the NO-sensitive GC has been studied on RNA and protein levels. Northern blot analysis revealed a broad distribution of both the α_1- and β_1-subunits, which is in accordance with the notion of the $\alpha_1\beta_1$-isoform being the predominant isoform. The α_2-subunit was only detected in brain, placenta, and uterus (Budworth *et al.*, 1999). In rat brain, a widespread distribution of the α_1-, α_2- and β_1-subunits was demonstrated by RT-PCR and *in situ* hybridization; some regions predominantly expressed either the α_1-subunit or the α_2-subunit explaining mismatches between the α_1 and β_1 distributions observed earlier (Furuyama *et al.*, 1993; Gibb and Garthwaite, 2001). The major occurrence of $\alpha_1\beta_1$ in brain is compatible with the concept that this isoform represents the neuronal isoform of NO-sensitive GC and plays a specific role in synaptic transmission. This is supported by the finding that this isoform is targeted to synaptic membranes. On the other hand, the $\alpha_2\beta_1$-isoform is most prominent in vascularized tissues and therefore may represent the vascular form of the enzyme. Thus, the differential subcellular and tissue localization of the two GC-isoforms may reflect the association to the neuronal and eNOS, respectively.

2. NO Influence on Cell Differentiation

2.1. Role of NO in hematopoietic cell differentiation

2.1.1. NO paracrine effects via stromal cells

Bone marrow stromal cells serve hematopoietic microenvironments where different blood cells are controlled in their growth and differentiation. The different components of human bone marrow stroma (fibroblasts, fat cells, macrophages, and endothelial cells) may provide the preferable microenvironment for a rapid expansion of the lineage-restricted progenitor cells (Gordon *et al.*, 1983). Numerous growth factors, cytokines, and chemokines are secreted by human $CD34^+$ cells, myeloblasts, erythroblasts, and megakaryoblasts and regulate normal hematopoiesis in an autocrine/paracrine manner (Majka *et al.*, 2001). G-protein signaling induces mobilization of hematopoietic stem/progenitor cells in mice (Papayannopoulou *et al.*, 2003). eNOS is regulated in endothelial cells by reversible and inhibitory interactions with G–protein-coupled receptors (Marrero *et al.*, 1999). Mice deficient in eNOS demonstrate a defect in progenitor cell mobilization. These findings indicate that eNOS, expressed by bone marrow stromal cells, influences recruitment of stem and progenitor cells (Aicher *et al.*, 2003). The G–protein-coupled, extracellular calcium-sensing receptor (CaR) is expressed in bone marrow-derived cells, including stromal cells. CaR agonists stimulate phosphorylation of ERK1/2 and p38 MAPK (Yamaguchi *et al.*, 2000). Furthermore, inhibition of ERK activity induced erythroid differentiation, and it acted synergistically with hydroxyurea, a γ-globin inducer, on Hb synthesis, whereas inhibition of p38 activity inhibited induction of Hb production by hydroxyurea in human erythroleukemic K562 cells (Park *et al.*, 2001).

2.1.2. Hematopoietic stem cells

Since its inception in the 1960s, the hematopoietic stem cells research community has accumulated a tremendous amount of knowledge regarding stem cell systems, as well as developing a large number of techniques with uses for both basic research and clinical application (Metcalf, 2001).

Hematopoietic stem cells are pluripotent cells able to give rise to at least 10 different functional cell types (neutrophil, monocytes/macrophages, basophils, eosinophils, RBCs, platelet, mast cells, dendritic cells, B and T lymphocytes; Abramson *et al.*, 1977). Two types of stem cells have been defined. The long-term repopulating cells are capable of producing all blood cell types for the entire life span of the individual and of generating progeny that display similar potentiality on secondary transplant. The short-term repopulating cells reconstitute myeloid and/or lymphoid compartments for a short period (Harrison and Zhong, 1992; Morrison and Weissman, 1994). The process by which stem cells give rise to terminally differentiated cells occurs through a variety of committed progenitor cells, often overlapping in their hemopoietic capacity (Dexter *et al.*, 1984).

2.1.3. Erythropoietin and a role of NO in erythroid cell differentiation

Erythropoietin (Epo) acts primarily on apoptosis to decrease the rate of cell death in erythroid progenitor cells in the bone marrow (Koury and Bondurant, 1990; Lin *et al.*, 1985). Epo is predominantly synthesized and secreted by tubular and juxtatubular capillary endothelial and interstitial cells of the kidney (Fisher *et al.*, 1996; Mujais *et al.*, 1999), whereas approximately 10–15% of the total amount Epo comes from extrarenal sources and is predominantly produced by hepatocytes and Kupffer cells of the liver (Eckardt *et al.*, 1994). Epo acts on the later stages of development of erythroid progenitor cells, primarily on colony-forming unit erythroid (CFU-E) to induce these cells to proliferate and mature through the normoblast into reticulocytes and mature RBCs (Gregory and Eaves, 1974). Epo acts synergistically with SCF, GM-CSF, IL-3, IL-4, IL-9, and IGF-1 to cause maturation and proliferation from the stage of the burst-forming unit erythroid (BFU-E) and CFU-E to the normoblast stage of erythroid cell development (Wu *et al.*, 1995). The Epo receptor is apparently expressed primarily on erythroid cells between the CFU-E and the pronormoblast stage of erythroid cell development (Sawada *et al.*, 1990). Epo binding to the receptor changes the conformation of the Epo receptor, which is necessary for JAK2 activation by a mechanism of self-dimerization (Constantinescu *et al.*, 2001). It has been recently reported that an increase in eNOS expression as well as in NO and cGMP production occurs in response to Epo during hypoxia in endothelial cells (Beleslin-Cokic *et al.*, 2004).

A role of NO in erythroid cell differentiation has been postulated based on demonstration that exogenous NO inhibits growth of erythroid primary cells and colony cultures (Maciejewski *et al.*, 1995). NO levels decreased significantly erythroid differentiation in hemin-induced and control cells; the decrease was, however, more in the hemin-induced group (Kucukkaya *et al.*, 2006). NO is able to inhibit erythroid cell differentiation induced by some (butyric acid and the anthracycline antitumor drugs aclarubicin and doxorubicin), but not all (hemin) agents. Also, erythroid cell hemoglobinization was inhibited by NO donors (Chenais *et al.*, 1999). NO decreased CFU-E and colony-forming unit-granulocyte macrophage (CFU-GM) formation derived from mononuclear cells isolated from human bone marrow. However, NO increased CFU-GM and decreased CFU-E formation derived from $CD34^+$ cells (Shami and Weinberg, 1996). Although NO increased intracellular levels of cGMP in bone marrow cells, addition of a membrane permeable cGMP analogue did not reproduce the previous mentioned NO effects on bone marrow colonies (Shami and Weinberg, 1996).

2.2. NO function in endothelial cell differentiation

2.2.1. Endothelial cell as a component of bone marrow stroma

The endothelial cells as well as macrophages, normal components of bone marrow stroma, play an active role in the modulation of human hematopoietic stem cell growth. The murine endothelial cell lines stimulate the proliferation and differentiation of erythroid precursors, where close cell contact is necessary for erythropoiesis (Ohneda and Bautch, 1997). A media, conditioned by human endothelial cell cultures, also stimulates the growth of human multipotent and committed erythroid (BFU-E, CFU-E) progenitors (Yamaguchi *et al.*, 1996). It has been shown that human umbilical vein endothelia cells (HUVEC), as well as human bone marrow endothelial cell line (TrHBMEC), can also support hematopoietic progenitor cells as a stromal microenvironment and induce formation of erythroid colonies (Yamaguchi *et al.*, 1996). No preferential adhesion of hematopoietic progenitor cells to HBMEC compared to HUVEC cells was observed (Rood *et al.*, 1999). NO produced by endothelial stromal cells should have a paracrine effect on erythroid cells in the bone marrow, as well as HbF induction. The erythroblast–macrophage contact in the erythroblastic islands promotes proliferation and terminal maturation of erythroid cells leading to their enucleation (Wilson and Tavassoli, 1994), whereas the growth and composition of granulocytic and erythroid colonies were unaffected in fibroblast-deficient human marrow cells (Abboud *et al.*, 1986). Hb synthesis of K562 cells was increased after coculture with HUVEC and most monolayers of bone marrow-derived macrophages, as well as with cell-free culture media conditioned by blood monocyte-derived macrophages (Zuhrie *et al.*, 1988). Further, induction of NO production in macrophages by hydroxyurea has been reported, which would also amplify effects on hematopoietic cells (Pai *et al.*, 1997).

2.2.2. Control of eNOS activity in endothelial cells

Several reports suggest that cytokines and interleukins, once thought to be explicit for the hematopoietic system, particularly granulocyte-macrophage colony-stimulating factor (GM-CSF), IL-3, IL-4, IL-6, and IL-8, are capable of affecting metabolism and function of endothelial cells (Bussolino *et al.*, 1989, 1991). G-CSF, GM-CSF, vascular endothelial growth factor, sphingosine-1-phosphate, and shear stress are known to activate endothelial NO production through the phosphatidylinositol 3′-kinase (PI3K)/protein kinase B (PKB/Akt) pathway and to induce endothelial cell proliferation and migration (Boo *et al.*, 2002; Dimmeler *et al.*, 1999; Michell *et al.*, 2001; van der Zee *et al.*, 1997). eNOS is rapidly activated and phosphorylated on both Ser-1177 and Thr-495 in the presence of cGMP-dependent protein kinase II and the catalytic subunit of PKA in endothelial cells (Fig. 7.3). These processes are more prominent in the presence of Ca^{2+}/CaM. PKA signaling acts by increasing phosphorylation of Ser-1177 and dephosphorylation of

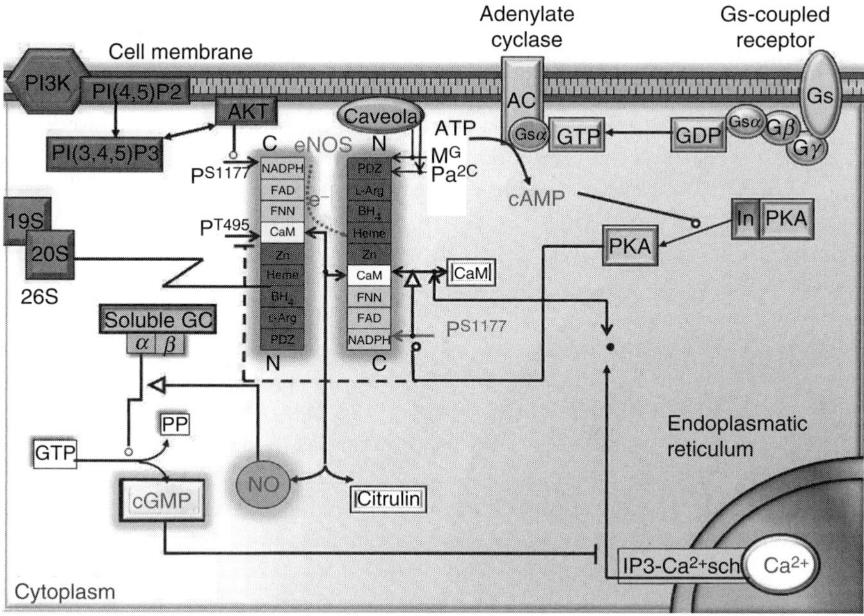

Figure 7.3 Activation of eNOS enzyme. eNOS enzyme is a homodimer composed of two identical subunits that contain the reductase (yellow), oxygenase (green), and calmodulin (CaM)-binding domains. cAMP-dependent protein kinase A (PKA) signaling acts by increasing phosphorylation of residue Ser-1177 and dephosphorylation of Thr-495 to activate eNOS. cAMP-mediated eNOS activation depends on Ca^{2+} release from internal stores. Ca^{2+}/CaM binding aligns the domains and stimulates electron flow, accounting for enzyme activation. eNOS oxidizes L-arginine in a process that consumes NADPH and produces L-citrulline and NO. NO activates soluble GC to produce cGMP. An elevated cGMP level attenuates the store-operated Ca^{2+} entry. The phosphatidylinositol 3′-kinase (PI3K)/protein kinase B (PKB/Akt) pathway can also activate eNOS. Caveolin-1 targets eNOS to the caveolae, vesicular organelles located at the plasma membrane. Myristylation at the N-terminal glycine (glycine-2 before removal of the initiation methionine by a specific aminopeptidase) is mainly responsible for the membrane association of the eNOS enzyme; palmitoylation may also contribute. 26S proteasome are large multi-subunit complexes that selectively degrade eNOS protein. The proteolytic core of this complex is the 20S proteasome. (See Color Insert.)

Thr-495 to activate eNOS (Boo *et al.*, 2002). Shear stress stimulates phosphorylation of bovine eNOS at the corresponding serine in a PKA-dependent, but PKB/Akt-independent manner, whereas NO production is regulated by the mechanisms dependent on both PKA and PKB/Akt (Boo *et al.*, 2002). The cAMP elevation activates the L-arginine/NO system and induces vasorelaxation in rabbit femoral artery *in vivo* and human umbilical vein (Ferro *et al.*, 1999; Xu *et al.*, 2000), as well as inhibition of platelet adhesion to endothelium (Queen *et al.*, 2000). A relaxant response of rat aorta to cAMP-mediated vasodilators is regulated by NO production in endothelium and subsequent increase in cGMP in vascular smooth muscle cells (Toyoshima *et al.*, 1998). cAMP-mediated eNOS activation is associated with phosphorylation of residue Ser-1177 in HUVEC in a PI3K)-independent manner (Kaufmann *et al.*, 2003). It has been reported that Akt is transiently activated in bradykinin-stimulated BAEC downstream from PI3K and that a rapid increase in NO production by bradykinin is mediated by the PKA-dependent phosphorylation of eNOS at Ser-1179 (Bae *et al.*, 2003). cGMP responses to bradykinin were greater in cells with increased cAMP levels than in control cells with basal levels of cAMP. The transient rises of cGMP levels induced by bradykinin and endothelin, which both cause release of Ca^{2+} from internal stores, were similarly enhanced by activation of adenylyl cyclase. Therefore, cAMP seems to enhance NO formation that depends on Ca^{2+} release from internal stores (Reiser, 1992). An elevated cGMP level attenuated the store-operated Ca^{2+} entry in vascular endothelial cells (Kwan *et al.*, 2000). The cGMP-mediated $[Ca^{2+}]_i$-reducing mechanisms may operate as a negative reaction to protect endothelial cells from damaging effect of excessive $[Ca^{2+}]_i$. The main targets of cGMP are phosphodiesterases, resulting in interference with the cAMP-signaling pathway (Vaandrager and de Jonge, 1996).

2.3. NO induction of differentiation in different cell types

NO is a negative regulator of proliferation of neuroblasts. Its function in neuronal differentiation basically consists in permitting neuroblasts escape from the proliferative condition, and supporting their differentiation to a neuronal phenotype (Contestabile and Ciani, 2004). NO, as a molecule involved in myoblast proliferation and fusion, also enhanced myoblast differentiation (Long *et al.*, 2006). NO–cGMP pathway stimulates the proliferation and osteoblastic differentiation of primary mouse bone marrow-derived mesenchymal stem cells and osteoblasts (Hikiji *et al.*, 1997). At low concentrations, NO has a selective enhancing effect on the induction and differentiation of Th1 but not Th2 cells. NO, at high concentrations, inhibits the differentiation and development of Th1 cells via the suppression of IL-12 synthesis (Huang *et al.*, 1998). NO synergizes with TNF-α and LPS to trigger a modified dendritic cells maturation program, resulting in an enhanced ability to stimulate T-lymphocyte proliferation. NO's action on maturing dendritic cells depends on activation of guanylate cyclase and generation of cGMP (Paolucci *et al.*, 2003).

2.4. NO effects on cancer cells

Both NO gas and NO generated by activated macrophage lysates inhibit tumor cell ribonucleotide reductase. NO-dependent cytostasis begins with a rapid and reversible inhibition of ribonucleotide reductase, progressively reinforced by other, long-lasting antiproliferative effects (Lepoivre *et al.*, 1994).

2.4.1. NO and leukemic cells

NO decreased the viability and the growth of freshly isolated acute non-lymphocytic leukemia cells *in vitro* (Shami *et al.*, 1995). It was also reported that iNOS-transfected erythroleukemic cells grew more slowly than control cells (Rafferty *et al.*, 1996). The host defense against human erythroleukemic cells is, in part, based on the production of NO by activated macrophages (Kwon *et al.*, 1991). Stimulated alveolar rat macrophages release reactive nitrogen intermediates, in the form of NO, which are cytostatic to murine leukemia L1210 cells (Huot *et al.*, 1992). After maturation of human monocytes to macrophages, reactive nitrogen intermediates are employed in mediating macrophage cytotoxicity toward human erythroleukemic K562 cells (Martin and Edwards, 1993). However, cells of the blood vessels can also participate in antitumor defense responses. They produce NO constitutively (endothelial cells) and after stimulation by proinflammatory cytokines (endothelial cells and vascular smooth muscle cells). Treatment with IFN-γ and TNF-α induced death of human erythroleukemic K562 cells cocultured with vascular smooth muscle cells or endothelial cells. K562 cells did not produce any appreciable levels of NO, but they were targeted by reactive nitrogen intermediates released from the cytokine-stimulated vascular cells that showed formation of nonheme iron–nitrosyl complexes in the tumor cells (Geng *et al.*, 1996).

2.4.2. Apoptotic properties of NO

High NO concentrations promote apoptosis in most cases, while low NO concentrations can result in resistance to apoptosis. NO induces biochemical characteristics of apoptosis in macrophages (Albina *et al.*, 1993), thymocytes (Fehsel *et al.*, 1995), pancreatic islets (McDaniel *et al.*, 1997), certain neurons (Heneka *et al.*, 1998), and tumor cells (Cui *et al.*, 1994). The factors affecting cell-specific sensitivity to NO-mediated apoptosis can be associated with the redox state within the cells, activation of the apoptotic signaling cascade (such as caspases, Kim *et al.*, 2000b), the mitochondrial cytochrome *c* release (Brown and Borutaite, 1999), or regulation of cell survival and apoptotic gene expression (Tamatani *et al.*, 1998). The induction of apoptosis often requires exposure to high levels of exogenous NO donors (Messmer *et al.*, 1995), which may overwhelm the natural protective mechanism of cells. Furthermore, the threshold of the NO level triggering apoptosis is different from one cell to the other. The cytosolic cytochrome *c* activates the caspase-dependent apoptotic signal cascade, resulting in the degradation of the inhibitor of caspase-activated DNase (CAD), activation of CAD, and DNA fragmentation (Sakahira *et al.*, 1998). NO binds to cytochrome *c* oxidase in the mitochondrial electron transfer chain (Podcroso *et al.*, 1996). NO produces peroxynitrite ($ONOO^-$) in response to a rapid reaction with superoxide (O_2^-), a powerful oxidant with sufficient stability to diffuse through cells to react with targets (Blaise *et al.*, 2005). Enhanced formation of $ONOO^-$ induces DNA damage. In response to DNA damage, p53, a key player in the caspase-independent pathway, is upregulated and thereby activates the DNA repair enzyme poly-ADP ribose polymerase. The amount of energy necessary for DNA repair may stimulate the cell to initiate apoptosis (Borst and Rottenberg, 2004). Peroxynitrite induces both the nitration of the tyrosine residue in proteins and the apoptotic cell death in thymocytes (Salgo *et al.*, 1995), neuronal cells (Bonfoco *et al.*, 1995), and HL-60 cells (Lin *et al.*, 1995). In the case of apoptosis promotion, the physiological relevant concentration of peroxynitrite induces apoptosis in HL-60 human leukemia cells, but fails to affect normal human endothelial and mononuclear cells (Lin *et al.*, 1995). The sustained activation of JNK/SAPK and p38 MAPK contributes to NO-mediated apoptosis by activation of caspase-3 through the release of mitochondrial cytochrome *c* into the cytosol (Assefa *et al.*, 2000; Tournier *et al.*, 2000).

2.4.3. Antiapototic properties of NO

Although NO promotes apoptosis in some cells, it becomes apparent that NO displays antiapototic properties in other cell types. These include hepatocytes (Kim *et al.*, 1997a), human B lymphocytes (Mannick *et al.*, 1994), endothelial cells (Dimmelder *et al.*, 1997), splenocytes (Genaro *et al.*, 1995), and eosinophils (Beauvais *et al.*, 1995). The intracellular elevation of cGMP activates PKG and in turn decreases the cellular Ca^{2+} concentration, which is one of the key signals of apoptosis. The interference of the NO–cGMP pathway with the apoptotic signal transduction is controversial. Undoubtedly, cGMP production by NO can prevent apoptosis in some cell types including hepatocytes (Kim *et al.*, 1997b), embryonic motor neurons (Estevez *et al.*, 1998), B lymphocytes (Genaro *et al.*, 1995), eosinophils (Beauvais *et al.*, 1995), and ovarian follicles (Chun *et al.*, 1995). NO also counteracts the reactive oxygen species generated by proapoptotic ceramides and, in turn, inhibits the assembly of Apaf-1 and pro-caspase-9, which are essential for apoptosis (Zech *et al.*, 2003). Through a cGMP-independent mechanism, NO can directly inhibit the activity of downstream caspase-3 by *S*-nitrosylation of the enzyme, resulting in suppression of apoptotic activity. NO oxidative products (NO_2, $ONOO^-$, HNO_2, and NOx) may deaminate, crosslink, and oxidize DNA bases.

2.4.4. NO effects on angiogenesis during tumor growth

NO has been shown to stimulate angiogenesis during tumor growth through its stimulation of proliferative and migratory function of endothelial cells (Jenkins *et al.*, 1995). Within a tumor microenvironment, NO can be produced by cancer cells and/or endothelial cells in the vasculature of the tumor. Also, macrophages and stromal cells within the tumor can produce NO. All NOS isoforms have been found to be expressed in many different cancers, *in vitro* and/or *in vivo* (Barreiro Arcos *et al.*, 2003; Cobbs *et al.*, 1995; Jenkins *et al.*, 1994; Klotz *et al.*, 1998; Lee *et al.*, 2003; Mortensen *et al.*, 1999; Park *et al.*, 2003). In many different types of cancer, NOS expression has been positively correlated with tumor progression. These include human gastric carcinoma tissue, where total NOS activity was 75% higher than in normal tissue (Wang *et al.*, 2005). Moreover, iNOS is dominant because 81% of cancer tissues showed iNOS expression compared with 20% of normal and iNOS expression was positively correlated with lymph node metastasis and clinical stage (Wang *et al.*, 2005). Increased iNOS protein expression was associated with an increase in tumor growth in primary breast cancer tissue (Loibl *et al.*, 2005). In addition to breast cancer, iNOS has also been shown to be markedly expressed in approximately 60% of human adenomas and in 20–25% of colon carcinomas, while expression was either low or absent in the surrounding normal tissues (Ambs *et al.*, 1998b). In human ovarian cancer, iNOS activity has been localized in tumor cells and not found in normal tissue (Thomsen *et al.*, 1994). Other tumors that have demonstrated iNOS gene expression are brain (Cobbs *et al.*, 1995), esophagus (Wilson *et al.*, 1998), lung (Ambs *et al.*, 1998a), prostate (Klotz *et al.*, 1998), bladder (Swana *et al.*, 1999), pancreatic (Hajri *et al.*, 1998), and Kaposi's sarcoma (Weninger *et al.*, 1998). eNOS was shown to be positively correlated with breast cancer progression through sequential activation of sGC and mitogen-activated protein kinase (MAPK, Jadeski *et al.*, 2003).

3. NO Interaction with RBCs

3.1. NO production in RBCs

3.1.1. NOs levels in RBC

It has been reported that human RBCs contain iNOS and eNOS as well as CaM, suggesting that RBCs may synthesize their own NO (Jubelin and Gierman, 1996). This notion was supported by the observation that RBCs have an active eNOS protein (Chen and Mehta, 1998). However, it was later reported that RBCs possess iNOS and eNOS, but the proteins are without catalytic activity (Kang *et al.*, 2000). Recent studies revealed eNOS protein in the cytoplasm and in the internal side of RBC membranes according to its activity, apparently serving essential functions for RBC deformability and, indirectly, platelet aggregation (Bor-Kucukatay *et al.*, 2003; Kleinbongard *et al.*, 2006). This eNOS is regulated by L-arginine, calcium, and phosphorylation via PI3K (Kleinbongard *et al.*, 2006). Human erythroid progenitor and precursor cells also contain eNOS mRNA and protein, and demonstrate eNOS activity (Cokic *et al.*, 2007a). The eNOS mRNA and protein levels, as well as its activity, descend during erythroid differentiation, parallel with the elevation of Hb levels. The continuous decline of eNOS presence and activity, throughout erythroid differentiation, suggests that such activity may be a residual of eNOS production at an earlier stage (Cokic *et al.*, 2007a).

NO produced *in vitro* by iNOS in RBCs can convert Hb contained in the RBCs to *S*-nitrosohemoglobin (Mamone *et al.*, 1999). The half-life of NO in the presence of 2.1×10^6 RBCs/ml is 4.2 s. Estimated half-life of NO in whole blood is 1.8 ms (Liu *et al.*, 1998a). Relaxations induced by endogenous, endothelium-derived, NO were more inhibited by fetal than by adult RBCs suggesting that fetal RBCs have a higher NO scavenging effect than those present in adult blood (Calatayud *et al.*, 1998). Low concentrations of NO were reported to augment the oxygen affinity of sickle RBCs *in vitro* and *in vivo* without significant metHb production (Head *et al.*, 1997), but this was later shown to be incorrect (Gladwin *et al.*, 1999).

3.1.2. External NO consumption rate by RBC

It is now recognized that the NO consumption rate by RBC is much slower than that expected based on the *in vitro* reaction rate of NO with free Hb (Vaughn *et al.*, 1998). The *in vivo* quenching of NO by RBC is first reduced by the RBC-free zone created by the flow field (Liao *et al.*, 1999; Vaughn *et al.*, 1998). However, even without the RBC-free zone, the NO consumption by RBC is still about 500–1000 times slower than the NO reaction rate with free Hb (Liao *et al.*, 1999). NO produced from the endothelium must go through four steps to react with the RBC-enclosed Hb: (1) diffusion through the RBC-free region near the vessel wall, created by intravascular flow (Liu *et al.*, 1998a), to the bulk solution, (2) diffusion from the bulk solution to the RBC surface, (3) diffusion across RBC membrane, and (4) diffusion and reaction inside RBC cytosol. The first two are affected by extracellular factors, whereas the last two steps are affected by intracellular components intrinsic to RBC itself (Liu *et al.*, 1998a; Vaughn *et al.*, 2000). The difference in NO consumption by RBCs and free Hb is also hematocrit dependent (Tsoukias and Popel, 2002).

The RBC membrane contains a spectrin–actin protein skeleton connected to the integral membrane proteins: band 3 and glycophorin C, via the bridging proteins: ankyrin, and protein 4.1. The band 3 anion exchanger is the major integral protein of the RBC membrane and exists as dimers (60%) and tetramers (40%). The tetrameric form of band 3 binds ankyrin and protein 4.2 constituting the major attachment site of the RBC membrane to the underlying cytoskeleton (Michaely and Bennett, 1995; Rybicki *et al.*, 1996). The N-terminal cytoplasmic domain of band 3 binds to the 2,3-disphosphoglycerate binding site of deoxyHb with high affinity (Fig. 7.4). Liao and coworkers proposed a mechanism for the modulation of

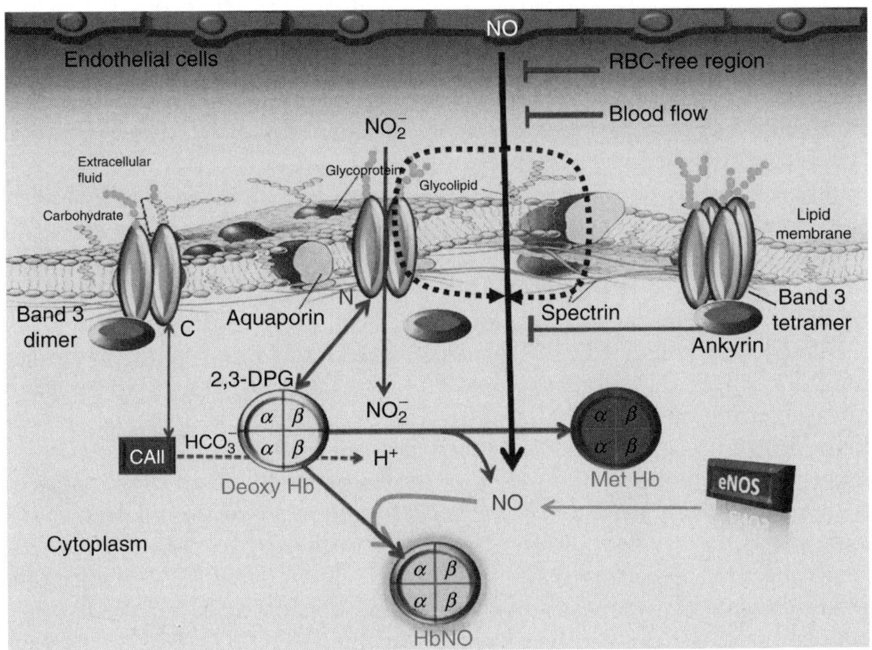

Figure 7.4 Nitric oxide uptake through the membrane skeleton of RBC. On the cytosolic side of the RBC membrane, there is a cytoskeletal network of proteins, including spectrin and band 3 tetramers association with ankyrin, which significantly slow the diffusion of NO. In addition to diffusing through the lipid membrane, it is thought that NO and nitrite ions may use band 3 and/or aquaporin to pass through the RBC membrane. The band 3 protein binds both deoxyhemoglobin and carbonic anhydrase II (CAII, forms the H^+ necessary for nitrite reduction by deoxyhemoglobin), and thus may channel nitrite. The products of deoxyhemoglobin (deoxyHb, $HbFe^{II}$) and nitrite (NO_2^-) reaction are methemoglobin (metHb, $HbFe^{III}$) and NO, which rapidly reacts with deoxyhemoglobin to form iron-nitrosyl-hemoglobin (HbNO, $HbFe^{II}NO$). The RBC-free region, near the vessel wall, and laminar intravascular flow limit NO delivery from the endothelium to RBC-enclosed Hb, in contrast to extracellular Hb, which is more rapidly exposed to NO and appears to destroy its activity to a much greater extent than intracellular Hb. 2,3-disphosphoglycerate (2,3-DPG). (See Color Insert.)

NO uptake by RBCs in which HbNO binds band 3 tetramers in the cytoskeleton and thereby displaces ankyrin. Kim-Shapiro *et al.* suggests that the binding of deoxyHb to band 3 may be primarily responsible for distraction to the cytoskeleton barrier and changes in RBC NO permeability (Huang *et al.*, 2007).

3.2. Interactions between NO and hemoglobin

The first hemoglobins (Hbs) evolved before the presence of oxygen in atmosphere. Moreover, the principal function of the primordial prokaryotic (Hbs) appears to be utilization of NO equivalents to supply protection from nitrosative stress. Several of the invertebrate (Hbs) come out to enclose structural elements considered to promote their redox activity. There are microorganism (Hbs) that additionally function to control nitrosative stress, plant (Hbs) that regulate NO levels, and the Hb from *Ascaris* that uses NO to modulate O_2 levels. These observations have lead to the proposition that (Hbs) have evolved, in part, to handle NO and that their use as oxygen transport proteins is a relatively recent evolutionary development (Minning *et al.*, 1999).

3.2.1. Reaction of NO with oxyHb and deoxyHb

As discussed earlier in this chapter, it has been known for about a century that NO reacts with oxyHb to form metHb and nitrate and in the last several decades it was realized that NO reacts with deoxyHb to form nitrosylhemoglobin (HbFe(II)NO), with the iron bound to the ferrous heme iron instead of oxygen. In view of the slow dissociation rate of this species and the stability of nitrate ions, it was generally believed that both of these reactions were effectively irreversible from a physiological perspective and that the main effect of Hb on NO was to destroy its bioactivity.

In the last 10 years, two additional mechanisms of NO interaction with Hb both intracellular and normally small amounts of the extracellular forms circulating in the plasma have been postulated. Stamler and his coworkers have presented a series of papers suggesting that NO can react with the highly conserved β-93 cysteine thiols of each Hb tetramer to form *S*-nitrosyl Hb (SNO-Hb) and that the stability of this species is allosterically linked to oxygen binding at the heme, that is, it is more stable in the oxyHb form (R state) than in the deoxyHb form (T state, Gow and Stamler, 1998; Jia *et al.*, 1996; Stamler *et al.*, 1997). Such a mechanism could form the basis for a model in which blood flow (and other properties dependent on NO concentration) was regulated in a homeostatic fashion: NO would be carried (and not destroyed) by Hb from the lungs and other oxygenated tissues to the regions of low pO_2 where the NO would dissociate from the SNO-Hb and lead to vascular dilatation, increased blood flow, and increased oxygen delivery. Various molecular mechanisms for these reactions were proposed but work by other groups have largely disproved each aspect of this model and ultimately this now controversial proposal rests on a unique way (photolysis-based assay) in which the original group has quantitated SNO-Hb and obtained much higher values for its formation than have been obtained with several other assays (Gladwin *et al.*, 2000a, 2002; Xu *et al.*, 2003).

This work did lead to extensive studies on NO–Hb interactions by other research groups and an alternate model of how Hb may effectively transport NO bioactivity in the blood and thus allow NO to act as a hormone. Reduction of nitrite by deoxyHb was first reported by Brooks (1937) and later studied by Doyle *et al.* (1981). The primary reaction of nitrite with deoxyHb produces NO and MetHb. The NO can then bind to available deoxyHb, forming HbNO. This model involves the postulate that nitrite ions are the major storage source of NO in the body (Cosby *et al.*, 2003; Dejam *et al.*, 2005). Although bacteria and other organisms can reduce nitrite to NO, this reaction was thought to occur to a very limited extent under ordinary conditions in mammalian tissues. However, in the last few years, data have been presented that suggest that deoxyHb proteins, especially Hb, can reduce nitrite to NO (Grubina *et al.*, 2007). This hypothesis is based on *in vitro* studies that show that partially deoxygenated RBCs can dilate vascular rings when nitrite ions are present (Cosby *et al.*, 2003), inhalation and infusion administration of nitrite to experimental animals resulting in vasodilatation and NO production (Duranski *et al.*, 2005; Hunter *et al.*, 2004; Pluta *et al.*, 2005), and studies in human volunteers where nitrite infusions at slightly superphysiological doses cause measurable peripheral vasodilatation (Gladwin *et al.*, 2000b). These studies, which have also been extended to demonstrations of potential therapeutic value of nitrite administration in preventing ischemia–reperfusion injury in models of coronary artery obstruction (Duranski *et al.*, 2005) and subarachnoid hemorrhage (Pluta *et al.*, 2005), have raised the possibility of using nitrite administration in the therapy of various circulatory diseases. Studies are in progress on the mechanism of production of nitrite in the human body, in addition to that which occurs due to dietary ingestion, the control of nitrite levels in human RBCs (Lauer *et al.*, 2001; Nagababu *et al.*, 2003), and the effects of storage of blood on nitrite levels. In addition, there has recently been growing recognition of the fact that extracellular Hb—markedly increased in acute hemolysis as well as in chronic hemolytic anemias (such as sickle-cell disease and thalassemia)—has different properties with regard to NO and nitrite metabolism than intracellular Hb (Reiter *et al.*, 2002; Rother *et al.*, 2005). In particular, whereas intracellular is partitioned from NO by cell-poor plasma streaming, unstirred layers, and the RBC membrane itself, extracellular Hb is much more efficient in reacting with both NO and nitrite. Indeed, it has been proposed that many of the manifestations of hemolytic diseases may be due to the destruction of NO by the increased cell-free Hb levels in these syndromes (Minneci *et al.*, 2005; Rother *et al.*, 2005). Clearly the RBC and Hb molecule are crucial factors in regulating the bioactivity of NO and its storage form, nitrite ions. Nitrite ions increase blood flow in the human circulation as well as vasodilation of rat aortic rings. Formation of both NO gas and NO-modified Hb results from the nitrite reductase activity of deoxyHb and deoxygenated RBCs levels (Fig. 7.5).

4. Effects of NO on Hemoglobin Expression

4.1. NO interaction with cyclic nucleotides

High cAMP levels continuously decrease in contrast to steady low levels of cGMP during erythroid differentiation (Cokic *et al.*, 2007a). NO increases cGMP levels in human erythroid progenitor cells. NO also increased levels of HbF protein and γ-globin mRNA in the primary erythroid cells, as well as in erythroleukemic cells. Soluble GC inhibitors prevented γ-globin induction by the NO donor in human erythroid progenitor cells (Cokic *et al.*, 2003). However, the ability to induce HbF is reduced upon full erythroid maturation, likely secondary to reduced gene expression with pyknosis of the nuclei and the increase in intracellular Hb, which is a potent NO scavenger. NO increases intracellular cGMP levels, however it reduces cAMP levels in erythroid progenitor cells (Cokic *et al.*, 2007a). Furthermore, sGC activators or analogs increased γ-globin gene expression in primary human erythroblasts (Ikuta *et al.*, 2001). It has been described that guanine, guanosine and guanine ribonucleotides are inducers of erythroid differentiation of erythroleukemic cells, associated with a larger γ-globin mRNA accumulation (Osti *et al.*, 1997).

4.2. Signaling pathways related to hemoglobin induction

There are several transcription factors regulated by the NO/cGMP signaling pathway that might participate in Hb switching. The enhancer activity and inducibility of the γ-globin promoter region are both dependent upon the synergistic action of proteins bound to the tandem activator protein-1 (AP-1), which is a heterodimeric protein consisting of c-fos and jun subunits (Moi and Kan, 1990; Safaya *et al.*, 1994). The NO–cGMP pathway is known to increase both c-fos and jun mRNA levels and promotes AP-1 binding to DNA (Haby *et al.*, 1994; Idriss *et al.*, 1999). JunB is a member of the AP-1 family of transcription factors that binds to a specific DNA sequence to activate transcription of target genes. Expression of junB induced erythroid differentiation of an erythroleukemic cell line. In both murine and human primary erythroid progenitors and precursors, elevated

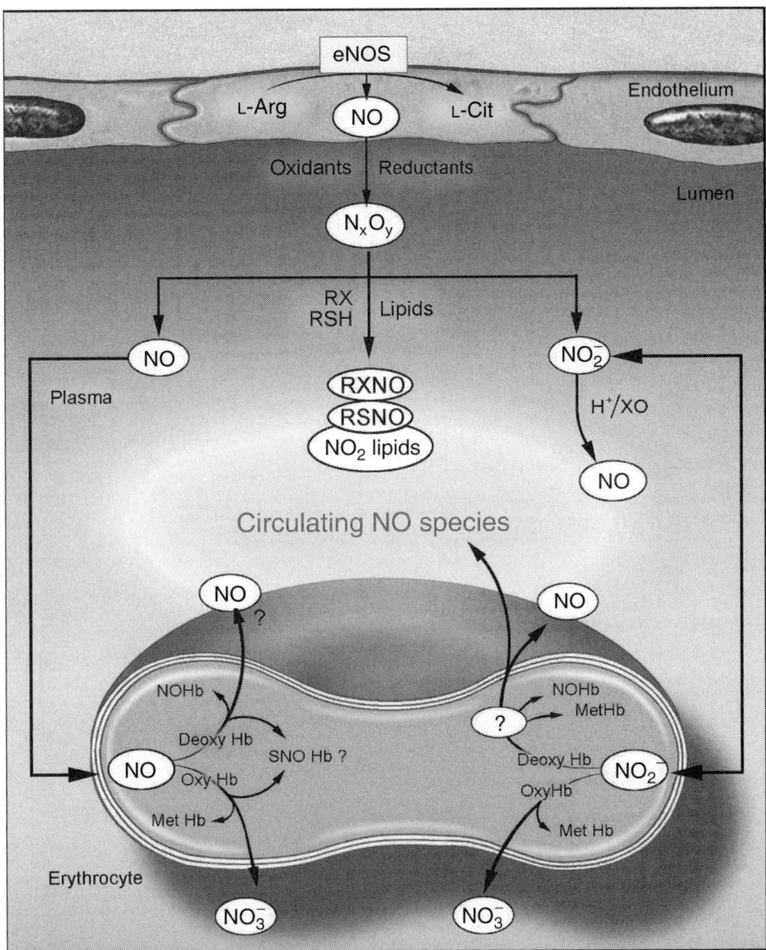

Figure 7.5 A model of intravascular metabolism of nitric oxide. NO produced by eNOS may diffuse into the vascular lumen as well as the underlying smooth muscle. The majority of this NO enters the erythrocyte and reacts with oxyhemoglobin (oxyHb) to form nitrate (NO_3^-); a minor portion may escape the Hb scavenger and react with plasma constituents to form nitros(yl)ated species (RXNO), including nitrosothiols (RSNO), nitrated lipids (NO_2 lipids), and nitrite (NO_2^-). Each of these species is capable of transducing NO bioactivity far from its location of formation. Nitrite may diffuse into the erythrocytes where it appears in a higher concentration than in plasma. In the erythrocyte, nitrite reacts with deoxyhemoglobin (deoxyHb) to form NO and methemoglobin (metHb) and other NO adducts. NO can then diffuse out of the erythrocyte either directly or via an intermediate NO metabolite. The question mark circled in white refers to the possibility of an intermediate during nitrite bioactivation. NOHb, iron nitrosylhemoglobin; SNOHb, nitrosohemoglobin; L-Arg, L-arginine; L-Cit, L-citrulline; N_xO_y: higher N oxides. This figure was originally published in *Blood* (Dejam *et al.*, 2005) and is reproduced with permission. © The American Society of Hematology.

junB expression is observed during differentiation of the cells in the presence of Epo (Jacobs-Helber *et al.*, 2002). The cellular AP-1 activity is also regulated by the phosphorylation of c-fos and jun proteins by the MAPK. MAPK are activated by phosphorylation by upstream MAPK kinases (MEKs) that in turn are activated in response to growth factors, cytokines, or various forms of cellular stress (including NO). There have been previous reports of increased phosphorylation of p38 MAPK in human neutrophils (Browning *et al.*, 1999) and other cell types (Browning *et al.*, 2000; Kim *et al.*, 2000a) by both NO and cGMP. The inhibition of ERK1/2 and p38 also caused a significant downregulation in the expression and production of IL-8 where the phosphorylation of ERK1/2 was essential for the activation of AP-1 (Kumar *et al.*, 2003). Previous studies showed that p38 MAPK is involved in inducing IL-8 gene transcription via AP-1 activation in human-vascular smooth muscle cells (Jung *et al.*, 2002). In correlation with that, it was reported that proteasome inhibitors through upstream signaling molecules ERK and JNK induce AP-1 activation and IL-8 gene expression (Wu *et al.*, 2002). In addition, the Sp1-binding CCACCC motif is also thought to be critical for high activity of the γ-globin promoter (Fischer *et al.*, 1993). It was reported that NO increased the activity of the TNF-α promoter by targeting the Sp1-binding site (Wang *et al.*, 1999). Thus, NO could potentially affect γ-globin expression through Sp1 as well as AP-1.

5. Role of NO During Malaria Pathogenesis

During the course of malaria pathogenesis, there are two possible mechanisms of hemolysis. First, *Plasmodium* parasites rupture RBCs after completing their development within them, thus releasing progeny merozoites. The rupture will also release the remaining contents of the RBC, which includes Hb. Second, there are more complex, immune system-mediated mechanisms of RBC lysis leading to malaria anemia (McDevitt *et al.*, 2004). Complement-mediated erythrophagocytosis has long been postulated as a major mechanism of malaria anemia but the other nonphagocytic mechanisms, such as T-cell-mediated and TNF-mediated cytotoxicity, might also be involved (Saxena and Chandrasekhar, 2000). Several hypotheses are postulated regarding the role of NO during malaria pathogenesis. In first one, NO derived from iNOS-expressing monocytes or macrophages kills *Plasmodium* as part of innate immunity (Stevenson and Riley, 2004). The second hypothesis suggests that NO is overproduced during *Plasmodium* infection and that NO has a function in pathogenesis. For example, it has been demonstrated that iNOS-derived overproduction of NO in the brain might disrupt the regulatory role of NO in the CNS, leading to the impaired consciousness of cerebral malaria (Clark and Cowden, 2003). Several lines of evidence argue against the hypothesis that NO is overproduced during malaria and that the excess NO functions systemically to mediate severe malaria pathogenesis. First, the presence of hypoargininemia in malaria patients suggests that NO production might be limited in malaria (Lopansri *et al.*, 2003). Human RBCs contain arginase; the rupture of RBCs by the parasite will increase plasma arginase levels, which is a mechanism of hypoargininemia in sickle-cell patients (Rother *et al.*, 2005). Finally, mononuclear cells iNOS protein and mRNA levels correlate inversely with disease severity (Anstey *et al.*, 1996; Chiwakata *et al.*, 2000). Collectively, these observations suggest that NO is not overproduced and that iNOS-derived NO might be valuable. The increased production of NO might be beneficial, either by killing the parasite or by preventing the development of disease. There are two main candidates for NO scavenging: Hb and superoxide. Free Hb released during the asexual cycle of blood-stage *Plasmodium* might quench NO, thus having an important role in limiting NO bioavailability during malaria (Reiter *et al.*, 2002). iNOS activities have been reported to correlate inversely with Hb levels (Keller *et al.*, 2004). Free Hb is a powerful *in vivo* scavenger of NO, leading to vasoconstriction and impaired microvascular blood perfusion, which are major determinants of tissue and organism survival. The efficiency of free Hb at scavenging NO is almost 1000-fold that of Hb packaged in RBCs (Liu *et al.*, 1998a).

Malaria is associated with genetic blood disorder sickle-cell anemia in individuals who are homozygous for the S gene. Sickle-cell anemia is an autosomal recessive genetic disorder caused by a mutation in the hemoglobin-β gene found on chromosome 11. The β-subunit has the amino acid valine at position 6 instead of the glutamic acid that is normally present. The presence of two defective genes (SS) is needed for sickle-cell anemia, whereas one defective gene is necessary for sickle-cell trait. The gene causing sickling provides a classic example of overdominance; heterozygotes are healthier than either homozygote when malaria is present (Allison, 1954). Protective effects of the S gene are associated with reduced parasitemia and clinical symptoms (Lell *et al.*, 1999).

6. Effects of Fetal Hemoglobin Inducers on NO–cGMP Pathway

6.1. Effects of butyric acid and 5-azacytidine

Sodium butyrate activated p38 MAPK and increased γ-globin mRNA levels in human primary erythroid progenitors (Pace *et al.*, 2003). Butyric acid, a cytostatic agent used to increase HbF in genetic diseases of Hb, increased NO production in endothelial cells (Cokic *et al.*, 2006). This is very consistent with the known vasorelaxation effect of butyrate. However, in contrast to hydroxyurea induction of NO, it has been reported that butyrate-induced vasorelaxation was unaffected by inhibition of NOS, although it was abolished by cAMP stimulation (Aaronson *et al.*, 1996). Moreover, butyric acid suppressed eNOS protein and mRNA levels in HUVEC (Aaronson *et al.*, 1996; Urbich *et al.*, 2002). 5-Azacytidine, an inhibitor of DNA methyltransferase activity, increased eNOS mRNA levels in nonendothelial cell lines (human aortic vascular smooth muscle cells and several human carcinoma cell lines). In contrast, the expression of eNOS mRNA levels in HUVEC was not increased by treatment with 5-azacytidine (Chan *et al.*, 2004). These observations indicate that cytostatic agents do not generally induce NO or eNOS activity, as was demonstrated by the lack of NO production by 5-azacytidine (Cokic *et al.*, 2006). Cytostatic agents are used in the treatment of myeloproliferative disorders as well as to increase HbF in genetic diseases of Hb.

6.2. Effects of hydroxyurea

6.2.1. In human endothelial cells

Hydroxyurea, a HbF inducer, increased NO production through eNOS phosphorylation and therefore eNOS activity in endothelial cells (Fig. 7.6, Cokic *et al.*, 2006). Hydroxyurea stimulation of NO production is regulated by the mechanisms dependent on both PKA and PI3K-induced stimulation of PKB/Akt (Cokic *et al.*, 2006). Involvement of eNOS, the major eNOS enzyme, in hydroxyurea induction of NO is supported by finding of phosphorylation of eNOS at Ser-1177, which represents the activation site of the enzyme. Phosphorylation of Ser-1177 by hydroxyurea was completely PKA, and partially PKB/Akt dependent. Subsequently hydroxyurea induction of eNOS is accompanied by rise of the intracellular calcium concentration (Cokic *et al.*, 2006). Further, cAMP and cGMP levels are amplified in HUVEC during incubation with hydroxyurea, which is in accordance with reported results that hydroxyurea induced cGMP levels in erythroid progenitor cells (Cokic *et al.*, 2003). Furthermore, cGMP production demonstrated eNOS dependence during stimulation by hydroxyurea. Phosphorylation of eNOS and induction of NO levels, as well as of cGMP levels, suggest that eNOS contributes to hydroxyurea effects on endothelial cells (Cokic *et al.*, 2006). Beyond short-term phosphorylation of eNOS protein, the effects of hydroxyurea on endothelial cells appear to be also mediated by long-term posttranscriptional increases in eNOS protein levels by inhibiting protein degradation (Cokic *et al.*, 2007b). Incubation of aortic rings with hydroxyurea weakly enhanced endothelial-dependent vasorelaxation (Chalupsky *et al.*, 2004).

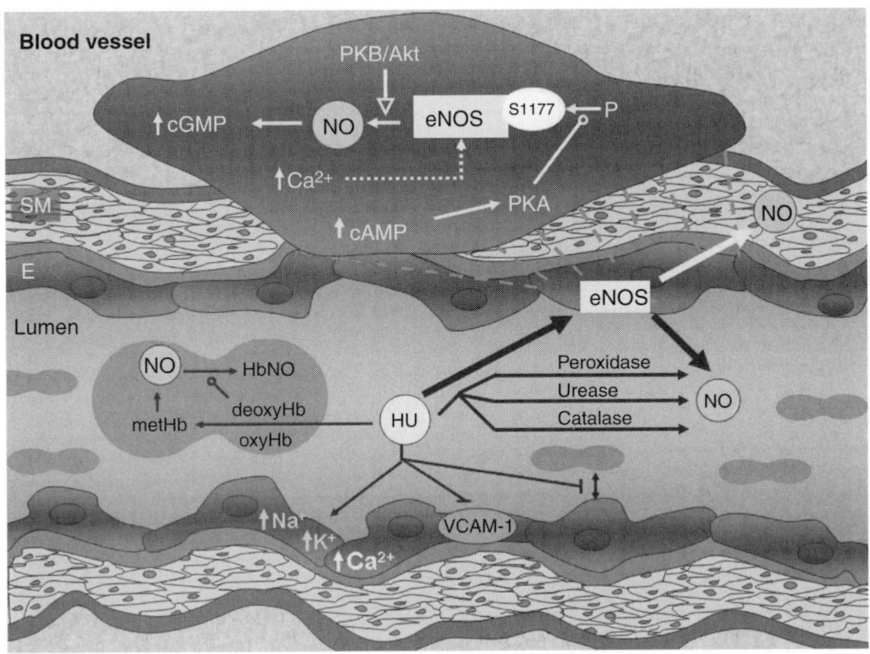

Figure 7.6 Hydroxyurea effects on blood vessels. Hydroxyurea oxidizes oxy- and deoxyhemoglobin to methemoglobin (metHb), which may react directly with other molecules of hydroxurea (or via hydroxylamine) to form nitrosylhemoglobin (HbNO) which can slowly release NO. Hydroxyurea may also chemically or enzymatically (peroxidase, urease, catalase) decompose to NO, as well as directly effect intracellular levels of cations, production of VCAM-1, RBC adhesion, etc. Our data show that hydroxyurea also stimulates the phosphorylation and thus activation of endothelial NOS (eNOS) with resultant production of NO. This effect is modulated by increased intracellular calcium levels and increases in cAMP that stimulates phosphorylation at Ser-1177 by PKA. NO production is regulated by both PKA and PKB/Akt. The increase in eNOS activity presumably leads to increases in cGMP in endothelial and surrounding cells, which result in many of the pleiotropic effects of endothelial produced NO. SMm smooth muscle cells; E, endothelial cells. This figure was originally published in *Blood* (Cokic *et al.*, 2006) and is reproduced with permission. © The American Society of Hematology. (See Color Insert.)

6.2.2. In human erythroid cells

Hydroxyurea, a drug which changes the phenotype of human erythroid cells, increased intracellular cGMP levels as well as cAMP levels in human erythroid progenitor cells (Cokic *et al.*, 2007a). Moreover, the cAMP-dependent pathway was activated in erythroleukemic cells in which the cGMP-dependent pathway was stimulated by hemin, which is also a γ-globin inducer as hydroxyurea (Inoue *et al.*, 2004). The cAMP-dependent pathway, which is independent of the MAPK pathways, appears to play a negative role in γ-globin gene expression in K562 erythroleukemia cells (Inoue *et al.*, 2004). In contrast to these results, there have been two reports of cAMP apparently increasing HbF levels. Ikuta's group (Kuroyanagi *et al.*, 2006) has reported that expression of the γ-globin gene is induced upon activation of the cAMP pathway, while Dover and co-workers (Keefer *et al.*, 2006) found that adenylate cyclase inhibition markedly decreased HbF induction by hydroxyurea in human erythroid cells. It has been demonstrated that hydroxyurea stimulated cAMP production in erythroid progenitor cells (Cokic *et al.*, 2007a), whereas Dover and co-workers (Keefer *et al.*, 2006) reported that hydroxyurea failed to significantly stimulate adenylate cyclase activity in erythroid precursor cells culture. Further work is necessary to clarify the role of the cAMP pathway, while the importance of the cGMP pathway now seems clearly established.

6.2.3. In patients with sickle-cell anemia

Induction of HbF of similar magnitude in the cells of normal individuals (1.3- to 3.5-fold) and sickle-cell patients (2- to 5-fold) by hydroxyurea was reported using a similar erythroid cell culture system (Moi and Kan, 1990). The increase in HbF levels in erythroid progenitor cells treated with hydroxyurea *in vitro* is comparable to the rise of peripheral blood HbF production (*in vivo*) following hydroxyurea therapy in patients with sickle-cell anemia (Yang *et al.*, 1997). Hydroxyurea has little suppressive effect on effective erythropoiesis but increases RBC survival (life span from 18.6 ± 11 days to 70 ± 21 days) as a result of decreased hemolysis in patients with sickle-cell anemia (Ballas *et al.*, 1999), presumably due to reduced intracellular HbS polymerization as a result of increased intracellular HbF levels. These selective effects prolong the period of increase of HbF to reach a therapeutically significant level in peripheral blood of sickle-cell patients and could account for the fact that maximal increases in HbF levels may take months to occur (Rodgers *et al.*, 1993). Indeed maximal increments in F-reticulocytes are attained at 10–11 days after the start of hydroxyurea treatment in patients with sickle-cell anemia (Veith *et al.*, 1985), suggesting a rapid effect on the fundamental mechanisms responsible for determining globin gene phenotype. Measurements of F-cells also show maximal increases in 14–21 days, confirming the disparity between cellular responses and HbF levels (Charache *et al.*, 1987).

cGMP levels were found to be significantly higher in RBCs of patients with sickle-cell disease (who also have small baseline elevations of HbF) and were further increased with hydroxyurea therapy, and cGMP levels correlated with HbF levels in hydroxyurea treated sickle-cell patients (Conran *et al.*, 2004). Chronic hydroxyurea therapy also significantly increased NO, cGMP, and HbF levels in the blood of patients with sickle-cell anemia (Nahavandi *et al.*, 2002). NOS activity has been reported as higher in RBCs of sickle-cell disease patients, on hydroxyurea therapy than in untreated patients (Iyamu *et al.*, 2005). This is in accordance with results that L-arginine alone does not increase serum NOx production in steady-state patients; however, it does when given together with hydroxyurea (Morris *et al.*, 2003). *In vitro* NOx production by RBCs (normal and sickle) is increased by treatment with hydroxyurea, but is not decreased by NOS inhibition (Nahavandi *et al.*, 2006). However, it has been also shown that hydroxyurea increased NO production via induction of eNOS activity in endothelial cells (Cokic *et al.*, 2006). Inhibition of NOS partially reversed the hydroxyurea effects on HbF synthesis in BFU-E colonies (Haynes *et al.*, 2004), but did not decrease NOx production in RBCs during incubation with hydroxyurea (Nahavandi *et al.*, 2006). However, the inhibition of eNOS reduced hydroxyurea-stimulated NO production in endothelial cells (Cokic *et al.*, 2006).

7. Summary: NO and Erythropoiesis

Human erythroid precursors contain an active eNOS protein. The eNOS mRNA and protein levels, as well as their activities, decrease during erythroid differentiation, corresponding with the rise of Hb levels. The continuous decline of eNOS presence and activity, throughout erythroid differentiation, suggests that any activity in mature erythroid cells may be a residual of eNOS production at an earlier stage. Human RBCs contain iNOS and eNOS as well as CaM, and RBCs may synthesize their own NO via an active eNOS protein. The eNOS protein is present in the cytoplasm and in the internal side of RBC membranes, as measured by activity, and may affect RBC deformability, platelet aggregation, and other properties.

cAMP production constantly declines in distinction to stable cGMP levels during erythroid differentiation. NO increases cGMP levels and decreases cAMP levels, in human erythroid progenitor cells. NO also augments levels of HbF protein and γ-globin mRNA in primary erythroid cells during their differentiation, as well as γ-globin mRNA in erythroleukemic cells. Activation of sGC increases γ-globin gene expression in primary human erythroblasts. Soluble GC inhibitors prevented γ-globin stimulation by NO in human erythroid progenitor cells.

These findings open the gate to novel therapeutic strategies based on NO delivery by NO donors, arginine salt administration, or increasing NO synthase expression or activity, or based on the amplification of NO signal transduction with phosphodiesterase inhibitors. Modulators of the eNOS–cGMP pathway may result in synergistic regimens that increase responsiveness and reduce morbidity linked to cytotoxic therapy.

REFERENCES

Aaronson, P. I., McKinnon, W., and Poston, L. (1996). Mechanism of butyrate-induced vasorelaxation of rat mesenteric resistance artery. *Br. J. Pharmacol.* **117,** 365–371.

Abboud, C. N., Duerst, R. E., Frantz, C. N., Ryan, D. H., Liesveld, J. L., and Brennan, J. K. (1986). Lysis of human fibroblast colony-forming cells and endothelial cells by monoclonal antibody (6–19) and complement. *Blood* **68,** 1196–1200.

Abramson, S., Miller, R. G., and Phillips, R. A. (1977). The identification in adult bone marrow of pluripotent and restricted stem cells of the myeloid and lymphoid systems. *J. Exp. Med.* **145,** 1567–1579.

Aicher, A., Heeschen, C., Mildner-Rihm, C., Urbich, C., Ihling, C., Technau-Ihling, K., Zeiher, A. M., and Dimmeler, S. (2003). Essential role of endothelial nitric oxide synthase for mobilization of stem and progenitor cells. *Nat. Med.* **9,** 1370–1376.

Albina, J. E., Cui, S., Mateo, R. B., and Reichner, J. S. (1993). Nitric oxide-mediated apoptosis in murine peritoneal macrophages. *J. Immunol.* **150,** 5080–5085.

Allison, A. C. (1954). Protection afforded by sickle-cell trait against subtertian malarial infection. *Br. Med. J.* **1,** 290–294.

Ambs, S., Bennett, W. P., Merriam, W. G., Ogunfusika, M. O., Oser, S. M., Khan, M. A., Jones, R. T., and Harris, C. C. (1998a). Vascular endothelial growth factor and nitric oxide synthase expression in human lung cancer and the relation to p53. *Br. J. Cancer* **78,** 233–239.

Ambs, S., Merriam, W. G., Bennett, W. P., Felley-Bosco, E., Ogunfusika, M. O., Oser, S. M., Klein, S., Shields, P. G., Billiar, T. R., and Harris, C. C. (1998b). Frequent nitric oxide synthase-2 expression in human colon adenomas: Implication for tumor angiogenesis and colon cancer progression. *Cancer Res.* **58,** 334–341.

Anstey, N. M., Weinberg, J. B., Hassanali, M. Y., Mwaikambo, E. D., Manyenga, D., Misukonis, M. A., Arnelle, D. R., Hollis, D., McDonald, M. I., and Granger, D. L. (1996). Nitric oxide in Tanzanian children with malaria: Inverse relationship between malaria severity and nitric oxide production/nitric oxide synthase type 2 expression. *J. Exp. Med.* **184,** 557–567.

Asano, K., Chee, C. B., Gaston, B., Lilly, C. M., Gerard, C., Drazen, J. M., and Stamler, J. S. (1994). Constitutive and inducible nitric oxide synthase gene expression, regulation, and activity in human lung epithelial cells. *Proc. Natl. Acad. Sci. USA* **91,** 10089–10093.

Assefa, Z., Vantieghem, A., Garmyn, M., Declercq, W., Vandenabeele, P., Vandenheede, J. R., Bouillon, R., Merlevede, W., and Agostinis, P. (2000). p38 mitogen-activated protein kinase regulates a novel, caspase-independent pathway for the mitochondrial cytochrome c release in ultraviolet B radiation induced apoptosis. *J. Biol. Chem.* **275,** 21416–21421.

Bae, S. W., Kim, H. S., Cha, Y. N., Park, Y. S., Jo, S. A., and Jo, I. (2003). Rapid increase in endothelial nitric oxide production by bradykinin is mediated by protein kinase A signaling pathway. *Biochem. Biophys. Res. Commun.* **306,** 981–987.

Ballas, S. K., Marcolina, M. J., Dover, G. J., and Barton, F. B. (1999). Erythropoietic activity in patients with sickle cell anaemia before and after treatment with hydroxyurea. *Br. J. Haematol.* **105,** 491–496.

Barreiro Arcos, M. L., Gorelik, G., Klecha, A., Goren, N., Cerquetti, C., and Cremaschi, G. A. (2003). Inducible nitric oxide synthase-mediated proliferation of a Tl ymphoma cell line. *Nitric Oxide* **8,** 111–118.

Beauvais, F., Michel, L., and Dubertret, L. (1995). The nitric oxide donors, azide and hydroxylamine, inhibit the programmed cell death of cytokine-deprived human eosinophils. *FEBS Lett.* **361,** 229–232.

Beleslin-Cokic, B. B., Cokic, V. P., Yu, X., Weksler, B. B., Schechter, A. N., and Noguchi, C. T. (2004). Erythropoietin and hypoxia stimulate erythropoietin receptor and nitric oxide production by endothelial cells. *Blood* **104,** 2073–2080.

Bender, A. T., Demady, D. R., and Osawa, Y. (2000). Ubiquitination of neuronal nitric-oxide synthase *in vitro* and *in vivo*. *J. Biol. Chem.* **275,** 17407–17411.

Blaise, G. A., Gauvin, D., Gangal, M., and Authier, S. (2005). Nitric oxide, cell signaling and cell death. *Toxicology* **208,** 177–192.

Bonfoco, E., Krainc, D., Ankarcrona, M., Nicotera, P., and Lipton, S. A. (1995). Apoptosis and necrosis: Two distinct events induced, respectively, by mild and intense insults with N-methyl-D-aspartate or nitric oxide/superoxide in cortical cell cultures. *Proc. Natl. Acad. Sci. USA* **92,** 7162–7166.

Boo, Y. C., Sorescu, G., Boyd, N., Shiojima, I., Walsh, K., Du, J., and Jo, H. (2002). Shear stress stimulates phosphorylation of endothelial nitric-oxide synthase at Ser1179 by Akt-independent mechanisms: Role of protein kinase A. *J. Biol. Chem.* **277,** 3388–3396.

Bor-Kucukatay, M., Wenby, R. B., Meiselman, H. J., and Baskurt, O. K. (2003). Effects of nitric oxide on red blood cell deformability. *Am. J. Physiol. Heart Circ. Physiol.* **284,** H1577–H1584.

Borst, P., and Rottenberg, S. (2004). Cancer cell death by programmed necrosis? *Drug Resist. Update* **7,** 321–324.

Brandish, P. E., Buechler, W., and Marletta, M. A. (1998). Regeneration of the ferrous heme of soluble guanylate cyclase from the nitric oxide complex: Acceleration by thiols and oxyhemoglobin. *Biochemistry* **37,** 16898–16907.

Bredt, D. S., and Snyder, S. H. (1990). Isolation of nitric oxide synthetase, a calmodulin-requiring enzyme. *Proc. Natl. Acad. Sci. USA* **87,** 682–685.

Brenman, J. E., Chao, D. S., Gee, S. H., McGee, A. W., Craven, S. E., Santillano, D. R., Wu, Z., Huang, F., Xia, H., Peters, M. F., Froehner, S. C., and Bredt, D. S. (1996). Interaction of nitric oxide synthase with the postsynaptic density protein PSD-95 and alpha1-syntrophin mediated by PDZ domains. *Cell* **84,** 757–767.

Brooks, J. (1937). The action of nitrite on haemoglobin in the absence of oxygen. *Proc. R. Soc. Lond. B Biol. Sci.* **123,** 368–382.

Brown, G. C., and Borutaite, V. (1999). Nitric oxide, cytochrome c and mitochondria. *Biochem. Soc. Symp.* **66,** 17–25.

Browning, D. D., Windes, N. D., and Ye, R. D. (1999). Activation of p38 mitogen-activated protein kinase by lipopolysaccharide in human neutrophils requires nitric oxide-dependent cGMP accumulation. *J. Biol. Chem.* **274,** 537–542.

Browning, D. D., McShane, M. P., Marty, C., and Ye, R. D. (2000). Nitric oxide activation of p38 mitogen-activated protein kinase in 293T fibroblasts requires cGMP-dependent protein kinase. *J. Biol. Chem.* **275,** 2811–2816.

Budworth, J., Meillerais, S., Charles, I., and Powell, K. (1999). Tissue distribution of the human soluble guanylate cyclases. *Biochem. Biophys. Res. Commun.* **263,** 696–701.

Bussolino, F., Wang, J. M., Defilippi, P., Turrini, F., Sanavio, F., Edgell, C. J., Aglietta, M., Arese, P., and Mantovani, A. (1989). Granulocyte- and granulocyte-macrophage-colony stimulating factors induce human endothelial cells to migrate and proliferate. *Nature* **337,** 471–473.

Bussolino, F., Ziche, M., Wang, J. M., Alessi, D., Morbidelli, L., Cremona, O., Bosia, A., Marchisio, P. C., and Mantovani, A. (1991). *In vitro* and *in vivo* activation of endothelial cells by colony-stimulating factors. *J. Clin. Invest.* **87,** 986–995.

Calatayud, S., Beltran, B., Brines, J., Moncada, S., and Esplugues, J. V. (1998). Foetal erythrocytes exhibit an increased ability to scavenge for nitric oxide. *Eur. J. Pharmacol.* **347,** 363–366.

Cassoly, R., and Gibson, Q. (1975). Conformation, co-operativity and ligand binding in human hemoglobin. *J. Mol. Biol.* **91,** 301–313.

Chalupsky, K., Lobysheva, I., Nepveu, F., Gadea, I., Beranova, P., Entlicher, G., Stoclet, J. C., and Muller, B. (2004). Relaxant effect of oxime derivatives in isolated rat aorta: Role of nitric oxide (NO) formation in smooth muscle. *Biochem. Pharmacol.* **67,** 1203–1214.

Chan, Y., Fish, J. E., D'Abreo, C., Lin, S., Robb, G. B., Teichert, A. M., Karantzoulis-Fegaras, F., Keightley, A., Steer, B. M., and Marsden, P. A. (2004). The cell-specific expression of endothelial nitric-oxide synthase: A role for DNA methylation. *J. Biol. Chem.* **279,** 35087–35100.

Charache, S., Dover, G. J., Moyer, M. A., and Moore, J. W. (1987). Hydroxyurea-induced augmentation of fetal hemoglobin production in patients with sickle cell anemia. *Blood* **69,** 109–116.

Charles, I. G., Palmer, R. M., Hickery, M. S., Bayliss, M. T., Chubb, A. P., Hall, V. S., Moss, D. W., and Moncada, S. (1993). Cloning, characterization, and expression of a cDNA encoding an inducible nitric oxide synthase from the human chondrocyte. *Proc. Natl. Acad. Sci. USA* **90,** 11419–11423.

Chartrain, N. A., Geller, D. A., Koty, P. P., Sitrin, N. F., Nussler, A. K., Hoffman, E. P., Billiar, T. R., Hutchinson, N. I., and Mudgett, J. S. (1994). Molecular cloning, structure, and chromosomal localization of the human inducible nitric oxide synthase gene. *J. Biol. Chem.* **269,** 6765–6772.

Chen, L. Y., and Mehta, J. L. (1998). Evidence for the presence of L-arginine-nitric oxide pathway in human red blood cells: Relevance in the effects of red blood cells on platelet function. *J. Cardiovasc. Pharmacol.* **32,** 57–61.

Chenais, B., Molle, I., and Jeannesson, P. (1999). Inhibitory effect of nitric oxide on chemically induced differentiation of human leukemic K562 cells. *Biochem. Pharmacol.* **58,** 773–778.

Chiwakata, C. B., Hemmer, C. J., and Dietrich, M. (2000). High levels of inducible nitric oxide synthase mRNA are associated with increased monocyte counts in blood and have a beneficial role in Plasmodium falciparum malaria. *Infect. Immun.* **68,** 394–399.

Chrisman, T. D., Garbers, D. L., Parks, M. A., and Hardman, J. G. (1975). Characterization of particulate and soluble guanylate cyclases from rat lung. *J. Biol. Chem.* **250,** 374–381.

Chun, S. Y., Eisenhauer, K. M., Kubo, M., and Hsueh, A. J. (1995). Interleukin-1 beta suppresses apoptosis in rat ovarian follicles by increasing nitric oxide production. *Endocrinology* **136,** 3120–3127.

Clark, I. A., and Cowden, W. B. (2003). The pathophysiology of falciparum malaria. *Pharmacol. Ther.* **99,** 221–260.

Cobb, R. R., Felts, K. A., Parry, G. C., and Mackman, N. (1996). Proteasome inhibitors block VCAM-1 and ICAM-1 gene expression in endothelial cells without affecting nuclear translocation of nuclear factor-kappa B. *Eur. J. Immunol.* **26,** 839–845.

Cobbs, C. S., Brenman, J. E., Aldape, K. D., Bredt, D. S., and Israel, M. A. (1995). Expression of nitric oxide synthase in human central nervous system tumors. *Cancer Res.* **55,** 727–730.

Cokic, V. P., Smith, R. D., Beleslin-Cokic, B. B., Njoroge, J. M., Miller, J. L., Gladwin, M. T., and Schechter, A. N. (2003). Hydroxyurea induces fetal hemoglobin by the nitric oxide-dependent activation of soluble guanylyl cyclase. *J. Clin. Invest.* **111,** 231–239.

Cokic, V. P., Beleslin-Cokic, B. B., Tomic, M., Stojilkovic, S. S., Noguchi, C. T., and Schechter, A. N. (2006). Hydroxyurea induces the eNOS–cGMP pathway in endothelial cells. *Blood* **108,** 184–191.

Cokic, V. P., Andric, S. A., Stojilkovic, S. S., Noguchi, C. T., and Schechter, A. N. (2007a). Hydroxyurea nitrosylates and activates soluble guanylyl cyclase in human erythroid cells. *Blood.* Prepublished Novembar 9; DOI 10.1182/blood-2007-05-088732.

Cokic, V. P., Beleslin-Cokic, B. B., Noguchi, C. T., and Schechter, A. N. (2007b). Hydroxyurea increases eNOS protein levels through inhibition of proteasome activity. *Nitric Oxide* **16,** 371–378.

Conran, N., Oresco-Santos, C., Acosta, H. C., Fattori, A., Saad, S. T., and Costa, F. F. (2004). Increased soluble guanylate cyclase activity in the red blood cells of sickle cell patients. *Br. J. Haematol.* **124,** 547–554.

Constantinescu, J. N., Keren, T., Socolovsky, M., Nam, H., Henis, Y. I., and Lodish, H. F. (2001). Ligand-independent oligomerization of cell-surface erythropoietin receptor is mediated by the transmembrane domain. *Proc. Natl. Acad. Sci. USA* **98,** 4379–4384.

Contestabile, A., and Ciani, E. (2004). Role of nitric oxide in the regulation of neuronal proliferation, survival and differentiation. *Neurochem. Int.* **45,** 903–914.

Cosby, K., Partovi, K. S., Crawford, J. H., Patel, R. P., Reiter, C. D., Martyr, S., Yang, B. K., Waclawiw, M. A., Zalos, G., Xu, X., Huang, K. T., Shields, H., *et al.* (2003). Nitrite reduction to nitric oxide by deoxyhemoglobin vasodilates the human circulation. *Nat. Med.* **9,** 1498–1505.

Cui, S., Reichner, J. S., Mateo, R. B., and Albina, J. E. (1994). Activated murine macrophages induce apoptosis in tumor cells through nitric oxide-dependent or -independent mechanisms. *Cancer Res.* **54,** 2462–2467.

De Caterina, R., Libby, P., Peng, H. B., Thannickal, V. J., Rajavashisth, T. B., Jr., Gimbrone, M. A., Shin, W. S., and Liao, J. K. (1995). Nitric oxide decreases cytokine-induced endothelial activation. Nitric oxide selectively reduces endothelial expression of adhesion molecules and proinflammatory cytokines. *J. Clin. Invest.* **96,** 60–68.

Dejam, A., Hunter, C. J., Pelletier, M. M., Hsu, L. L., Machado, R. F., Shiva, S., Power, G. G., Kelm, M., Gladwin, M. T., and Schechter, A. N. (2005). Erythrocytes are the major intravascular storage sites of nitrite in human blood. *Blood* **106,** 734–739.

Dexter, T. M., Spooncer, E., Schofield, R., Lord, B. I., and Simmons, P. (1984). Haemopoietic stem cells and the problem of self-renewal. *Blood Cells* **10,** 315–339.

Dimmelder, S., Haendeler, J., Nehls, M., and Zeiher, A. M. (1997). Suppression of apoptosis by nitric oxide via inhibition of interleukin-1beta-converting enzyme (ICE)-like and cysteine protease protein (CPP)-32-like proteases. *J. Exp. Med.* **185,** 601–607.

Dimmeler, S., Fleming, I., Fisslthaler, B., Hermann, C., Busse, R., and Zeiher, A. M. (1999). Activation of nitric oxide synthase in endothelial cells by Akt-dependent phosphorylation. *Nature* **399,** 601–605.

Doyle, M. P., Pickering, R. A., Deweert, T. M., Hoekstra, J. W., and Pater, D. (1981). Kinetics and mechanism of the oxidation of human deoxyhemoglobin by nitrites. *J. Biol. Chem.* **256,** 2393–2398.

Duranski, M. R., Greer, J. J. M., Dejam, A., Jaganmohan, S., Hogg, N., Langston, W., Patel, R. P., Yet, S. F., Wang, X. D., Kevil, C. G., Gladwin, M. T., and Lefer, D. J. (2005). Cytoprotective effects of nitrite during *in vivo* ischemia–reperfusion of the heart and liver. *J. Clin. Invest.* **115,** 1232–1240.

Eckardt, K. U., Pugh, C. W., Meier, M., Tan, C. C., Ratcliffe, P. J., and Kurtz, A. (1994). Production of erythropoietin by liver cells *in vivo* and *in vitro*. *Ann. N. Y. Acad. Sci.* **718,** 50–60.

Eich, R. F., Li, T., Lemon, D. D., Doherty, D. H., Curry, S. R., Aitken, J. F., Mathews, A. J., Johnson, K. A., Smith, R. D., Phillips, G. N., and Olson, J. S. (1996). Mechanism of NO-induced oxidation of myoglobin and hemoglobin. *Biochemistry* **35,** 6976–6983.

Estevez, A. G., Spear, N., Thompson, J. A., Cornwell, T. L., Radi, R., Barbeito, L., and Beckman, J. S. (1998). Nitric oxide dependent production of cGMP supports the survival of rat embryonic motor neurons cultured with brain-derived neurotrophic factor. *J. Neurosci.* **18,** 3708–3714.

Fehsel, K., Kroncke, K. D., Meyer, K. L., Huber, H., Wahn, V., and Kolb-Bachofen, V. (1995). Nitric oxide induces apoptosis in mouse thymocytes. *J. Immunol.* **155,** 2858–2865.

Ferro, A., Queen, L. R., Priest, R. M., Xu, B., Ritter, J. M., Poston, L., and Ward, J. P. (1999). Activation of nitric oxide synthase by beta 2-adrenoceptors in human umbilical vein endothelium *in vitro*. *Br. J. Pharmacol.* **126,** 1872–1880.

Fisher, J. W., Koury, S., Ducey, T., and Mendel, S. (1996). Erythropoietin (Epo) production by interstitial cells of hypoxic monkey kidneys. *Br. J. Haematol.* **95,** 27–32.

Fischer, K. D., Haese, A., and Nowock, J. (1993). Cooperation of GATA-1 and Sp1 can result in synergistic transcriptional activation or interference. *J. Biol. Chem.* **268,** 23915–23923.

Forstermann, U., Pollock, J. S., Schmidt, H. H., Heller, M., and Murad, F. (1991). Calmodulin-dependent endothelium-derived relaxing factor/nitric oxide synthase activity is present in the particulate and cytosolic fractions of bovine aortic endothelial cells. *Proc. Natl. Acad. Sci. USA* **88,** 1788–1792.

Furchgott, R. F., and Zawadzki, J. V. (1980). The obligatory role of endothelial cells in the relaxation of arterial smooth muscle by acetylcholine. *Nature* **288,** 373–376.

Furuyama, T., Inagaki, S., and Takagi, H. (1993). Localizations of $\alpha 1$ and $\beta 1$ subunits of soluble guanylate cyclase in the rat brain. *Brain Res. Mol. Brain Res.* **20,** 335–344.

Galbiati, F., Volonte, D., Gil, O., Zanazzi, G., Salzer, J. L., Sargiacomos, M., Scherer, P. E., Engelman, J. A., Schlegel, A., Parenti, M., Okamoto, T., and Lisanti, M. P. (1998). Expression of caveolin-1 and -2 in differentiating PC12 cells and dorsal root ganglion neurons: Caveolin-2 is up-regulated in response to cell injury. *Proc. Natl. Acad. Sci. USA* **95,** 10257–10262.

Garcia-Gardena, G., Fan, R., Stern, D. F., Liu, J., and Sessa, W. C. (1996). Endothelial nitric oxide synthase is regulated by tyrosine phosphorylation and interacts with caveolin-1. *J. Biol. Chem.* **271,** 27237–27240.

Garthwaite, J., Charles, S. L., and Chess-Williams, R. (1988). Endothelium-derived relaxing factor release on activation of NMDA receptors suggests role as intercellular messenger in the brain. *Nature* **336,** 385–388.

Geller, D. A., Lowenstein, C. J., Shapiro, R. A., Nussler, A. K., Di Silvio, M., Wang, S. C., Nakayama, D. K., Simmons, R. L., Snyder, S. H., and Billiar, T. R. (1993). Molecular cloning and expression of inducible nitric oxide synthase from human hepatocytes. *Proc. Natl. Acad. Sci. USA* **90,** 3491–3495.

Genaro, A. M., Hortelano, S., Alvarez, A., Martinez, C., and Bosca, L. (1995). Splenic B lymphocyte programmed cell death is prevented by nitric oxide release through mechanisms involving sustained Bcl-2 levels. *J. Clin. Invest.* **95,** 1884–1890.

Geng, Y. J., Hellstrand, K., Wennmalm, A., and Hansson, G. K. (1996). Apoptotic death of human leukemic cells induced by vascular cells expressing nitric oxide synthase in response to gamma-interferon and tumor necrosis factor-alpha. *Cancer Res.* **56,** 866–874.

Gerzer, R., Böhme, E., Hofmann, F., and Schultz, G. (1981). Soluble guanylate cyclase purified from bovine lung contains heme and copper. *FEBS Lett.* **132,** 71–74.

Gibb, B. J., and Garthwaite, J. (2001). Subunits of the nitric oxide receptor, soluble guanylyl cyclase, expressed in rat brain. *Eur. J. Neurosci.* **13,** 539–544.

Giulivi, C. (1998). Functional implications of nitric oxide produced by mitochondria in mitochondrial metabolism. *Biochem. J.* **332,** 673–679.

Gladwin, M. T., Schechter, A. N., Shelhamer, J. H., Pannell, L. K., Conway, D. A., Hrinczenko, B. W., Nichols, J. S., Pease-Fye, M. E., Noguchi, C. T., Rodgers, G. P., and Ognibene, F. P. (1999). Inhaled nitric oxide augments nitric oxide transport on sickle cell hemoglobin without affecting oxygen affinity. *J. Clin. Invest.* **104,** 937–945.

Gladwin, M. T., Ognibene, F. P., Pannell, L. K., Nichols, J. S., Pease-Fye, M. E., Shelhamer, J. H., and Schechter, A. N. (2000a). Relative role of heme nitrosylation and beta-cysteine 93 nitrosation in the transport and metabolism of nitric oxide by hemoglobin in the human circulation. *Proc. Natl. Acad. Sci. USA* **97,** 9943–9948.

Gladwin, M. T., Shelhamer, J. H., Schechter, A. N., Pease-Fye, M. E., Waclawiw, M. A., Panza, J. A., Ognibene, F. P., and Cannon, R. O. (2000b). Role of circulating nitrite and S-nitrosohemoglobin in the regulation of regional blood flow in humans. *Proc. Natl. Acad. Sci. USA* **97,** 11482–11487.

Gladwin, M. T., Wang, X. D., Reiter, C. D., Yang, B. K., Vivas, E. X., Bonaventura, C., and Schechter, A. N. (2002). S-nitrosohemoglobin is unstable in the reductive erythrocyte environment and lacks O-2/NO-linked allosteric function. *J. Biol. Chem.* **277,** 27818–27828.

Glockzin, S., von Knethen, A., Scheffner, M., and Brune, B. (1999). Activation of the cell death program by nitric oxide involves inhibition of the proteasome. *J. Biol. Chem.* **274,** 19581–19586.

Gordon, M. Y., Kearney, L., and Hibbin, J. A. (1983). Effects of human marrow stromal cells on proliferation by human granulocytic (GM-CFC), erythroid (BFU-E) and mixed (Mix-CFC) colony-forming cells. *Br. J. Haematol.* **53,** 317–325.

Govers, R., van der Sluijs, P., van Donselaar, E., Slot, J. W., and Rabelink, T. J. (2002). Endothelial nitric oxide synthase and its negative regulator caveolin-1 localize to distinct perinuclear organelles. *J. Histochem. Cytochem.* **50,** 779–788.

Gow, A. J., and Stamler, J. S. (1998). Reactions between nitric oxide and haemoglobin under physiological conditions. *Nature* **391,** 169–173.

Gregory, C. J., and Eaves, A. C. (1974). Human marrow cells capable of erythropoietic differentiation *in vitro*: Definition of three erythroid colony responses. *Blood* **49,** 855–864.

Grubina, R., Huang, Z., Shiva, S., Joshi, M. S., Azarov, I., Basu, S., Ringwood, L. A., Jiang, A., Hogg, N., Kim-Shapiro, D. B., and Gladwin, M. T. (2007). Concerted nitric oxide formation and release from the simultaneous reactions of nitrite with deoxy- and oxyhemoglobin. *J. Biol. Chem.* **282,** 12916–12927.

Haby, C., Lisovoski, F., Aunis, D., and Zwiller, J. (1994). Stimulation of the cyclic GMP pathway by NO induces expression of the immediate early genes c-fos and junB in PC12 cells. *J. Neurochem.* **62,** 496–501.

Hajri, A., Metzger, E., Vallat, F., Coffy, S., Flatter, E., Evrard, S., Marescaux, J., and Aprahamian, M. (1998). Role of nitric oxide in pancreatic tumour growth: *In vivo* and *in vitro* studies. *Br. J. Cancer* **78,** 841–849.

Hall, A. V., Antoniou, H., Wang, Y., Cheung, A. H., Arbus, A. M., Olson, S. L., Lu, W. C., Kau, C. L., and Marsden, P. A. (1994). Structural organization of the human neuronal nitric oxide synthase gene (NOS1). *J. Biol. Chem.* **269,** 33082–33090.

Hamilton, A. L., Eder, J. P., Pavlick, A. C., Clark, J. W., Liebes, L., Garcia-Carbonero, R., Chachoua, A., Ryan, D. P., Soma, V., Farrell, K., Kinchla, N., Boyden, J., *et al.* (2005). Proteasome inhibition with bortezomib (PS-341): A phase I study with pharmacodynamic end points using a day 1 and day 4 schedule in a 14-day cycle. *J. Clin. Oncol.* **23,** 6107–6116.

Hardman, J. G., and Sutherland, E. W. (1969). Guanyl cyclase, an enzyme catalyzing the formation of guanosine 3′,5′-monophosphate from guanosine triphosphate. *J. Biol. Chem.* **244,** 6363–6370.

Harrison, D. E., and Zhong, R. K. (1992). The same exhaustible multilineage precursor produces both myeloid and lymphoid cells as early as 3–4 weeks after marrow transplantation. *Proc. Natl. Acad. Sci. USA* **89,** 10134–10138.

Harteneck, C., Koesling, D., Söling, A., Schultz, G., and Böhme, E. (1990). Expression of soluble guanylate cyclase: Catalytic activity requires two subunits. *FEBS Lett.* **272,** 221–223.

Harteneck, C., Wedel, B., Koesling, D., Malkewitz, J., Böhme, E., and Schultz, G. (1991). Molecular cloning and expression of a new α-subunit of soluble guanylyl cyclase. Interchangeability of the α-subunits of the enzyme. *FEBS Lett.* **292,** 217–222.

Haynes, J., Jr., Baliga, B. S., Obiako, B., Ofori-Acquah, S., and Pace, B. (2004). Zileuton induces hemoglobin F synthesis in erythroid progenitors: Role of the L-arginine-nitric oxide signaling pathway. *Blood* **103,** 3945–3950.

Head, C. A., Brugnara, C., Martinez-Ruiz, R., Kacmarek, R. M., Bridges, K. R., Kuter, D., Bloch, K. D., and Zapol, W. M. (1997). Low concentrations of nitric oxide increase oxygen affinity of sickle erythrocytes *in vitro* and *in vivo*. *J. Clin. Invest.* **100,** 1193–1198.

Hecker, M., Mülsch, A., and Busse, R. (1994). Subcellular localization and characterization of neuronal nitric oxide synthase. *J. Neurochem.* **62,** 1524–1529.

Heneka, M. T., Loschmann, P. A., Gleichmann, M., Weller, M., Schulz, J. B., Wullner, U., and Klockgether, T. (1998). Induction of nitric oxide synthase and nitric oxide-mediated apoptosis in neuronal PC12 cells after stimulation with tumor necrosis factor-alpha/lipopolysaccharide. *J. Neurochem.* **71,** 88–94.

Hickey, M. J., Granger, D. N., and Kubes, P. (2001). Inducible nitric oxide synthase (iNOS) and regulation of leucocyte/endothelial cell interactions: Studies in iNOS-deficient mice. *Acta Physiol. Scand.* **173,** 119–126.

Hideshima, T., Richardson, P., Chauhan, D., Palombella, V. J., Elliott, P. J., Adams, J., and Anderson, K. C. (2001). The proteasome inhibitor PS-341 inhibits growth, induces apoptosis, and overcomes drug resistance in human multiple myeloma cells. *Cancer Res.* **61,** 3071–3076.

Hikiji, H., Shin, W. S., Oida, S., Takato, T., Koizumi, T., and Toyo-oka, T. (1997). Direct action of nitric oxide on osteoblastic differentiation. *FEBS Lett.* **30,** 238–242.

Huang, F. P., Niedbala, W., Wei, X. Q., Feng, G. J., Robinson, J. H., Lam, C., and Liew, F. Y. (1998). Nitric oxide regulates Th1 cell development through the inhibition of IL-12 synthesis by macrophages. *Eur. J. Immunol.* **28,** 4062–4070.

Huang, K. T., Huang, Z., and Kim-Shapiro, D. B. (2007). Nitric oxide red blood cell membrane permeability at high and low oxygen tension. *Nitric Oxide* **16,** 209–216.

Hunter, C. J., Dejam, A., Blood, A. B., Shields, H., Kim-Shapiro, D., Machado, R. F., Tarekegn, S., Mulla, N., Hopper, A. O., Schechter, A. N., Power, G. G., and Gladwin, M. T. (2004). Inhaled nebulized nitrite is a hypoxia-sensitive NO-dependent selective pulmonary vasodilator. *Nat. Med.* **10,** 1122–1127.

Huot, A. E., Kruszyna, H., Kruszyna, R., Smith, R. P., and Hacker, M. P. (1992). Formation of nitric oxide hemoglobin in erythrocytes co-cultured with alveolar macrophages taken from bleomycin treated rats. *Biochem. Biophys. Res. Commun.* **182,** 151–157.

Idriss, S. D., Gudi, T., Casteel, D. E., Kharitonov, V. G., Pilz, R. B., and Boss, G. R. (1999). Nitric oxide regulation of gene transcription via soluble guanylate cyclase and type I cGMP-dependent protein kinase. *J. Biol. Chem.* **274,** 9489–9493.

Ignarro, L. J., Wood, K. S., and Wolin, M. S. (1982). Activation of purified soluble guanylate cyclase by protoporphyrin IX. *Proc. Natl. Acad. Sci. USA* **79,** 2870–2873.

Ignarro, L. J., Fukoto, J. M., Griscavage, J. M., Rogers, N. E., and Burns, R. E. (1993). Oxidation of nitric oxide in aqueous solution to nitrite but not nitrate: Comparison with enzymatically formed nitric oxide from L-arginine. *Proc. Natl. Acad. Sci. USA* **90,** 8103–8107.

Ikuta, T., Ausenda, S., and Cappellini, M. D. (2001). Mechanism for fetal globin gene expression: Role of the soluble guanylate cyclase-cGMP-dependent protein kinase pathway. *Proc. Natl. Acad. Sci. USA* **98,** 1847–1852.

Inoue, A., Kuroyanagi, Y., Terui, K., Moi, P., and Ikuta, T. (2004). Negative regulation of gamma-globin gene expression by cyclic AMP-dependent pathway in erythroid cells. *Exp. Hematol.* **32,** 244–253.

Iyamu, E. W., Cecil, R., Parkin, L., Woods, G., Ohene-Frempong, K., and Asakura, T. (2005). Modulation of erythrocyte arginase activity in sickle cell disease patients during hydroxyurea therapy. *Br. J. Haematol.* **131,** 389–394.

Jacobs-Helber, S. M., Abutin, R. M., Tian, C., Bondurant, M., Wickrema, A., and Sawyer, S. T. (2002). Role of JunB in erythroid differentiation. *J. Biol. Chem.* **277,** 4859–4866.

Jadeski, L. C., Chakraborty, C., and Lala, P. K. (2003). Nitric oxide-mediated promotion of mammary tumour cell migration requires sequential activation of nitric oxide synthase, guanylate cyclase and mitogen activated protein kinase. *Int. J. Cancer* **106,** 496–504.

Jenkins, D. C., Charles, I. G., Baylis, S. A., Lelchuk, R., Radomski, M. W., and Moncada, S. (1994). Human colon cancer cell lines show a diverse pattern of nitric oxide synthase gene expression and nitric oxide generation. *Br. J. Cancer* **70,** 847–849.

Jenkins, D. C., Charles, I. G., Thomsen, L. L., Moss, D. W., Holmes, L. S., Baylis, S. A., Rhodes, P., Westmore, K., Emson, P. C., and Moncada, S. (1995). Roles of nitric oxide in tumor growth. *Proc. Natl. Acad. Sci. USA* **92,** 4392–4396.

Jia, L., Bonaventura, C., Bonaventura, J., and Stamler, J. S. (1996). S-nitrosohaemoglobin: A dynamic activity of blood involved in vascular control. *Nature* **380,** 221–226.

Jubelin, B. C., and Gierman, J. L. (1996). Erythrocytes may synthesize their own nitric oxide. *Am. J. Hypertens.* **9,** 1214–1219.

Jung, Y. D., Fan, F., McConkey, D. J., Jean, M. E., Liu, W., Reinmuth, N., Stoeltzing, O., Ahmad, S. A., Parikh, A. A., Mukaida, N., and Ellis, L. M. (2002). Role of P38 MAPK, AP-**1,** and NF-kappaB in interleukin-1beta-induced IL-8 expression in human vascular smooth muscle cells. *Cytokine* **18,** 206–213.

Kamisaki, Y., Saheki, S., Nakane, M., Palmieri, J. A., Kuno, T., Chang, B. Y., Waldman, S. A., and Murad, F. (1986). Soluble guanylate cyclase from rat lung exists as a heterodimer. *J. Biol. Chem.* **261,** 7236–7241.

Kang, E. S., Ford, K., Grokulsky, G., Wang, Y. B., Chiang, T. M., and Acchiardo, S. R. (2000). Normal circulating adult human red blood cells contain inactive NOS proteins. *J. Lab. Clin. Med.* **135,** 444–451.

Kaufmann, J. E., Iezzi, M., and Vischer, U. M. (2003). Desmopressin (DDAVP) induces NO production in human endothelial cells via V2 receptor- and cAMP-mediated signaling. *J. Thromb. Haemost.* **1,** 821–828.

Keefer, J. R., Schneidereith, T. A., Mays, A., Purvis, S. H., Dover, G. J., and Smith, K. D. (2006). Role of cyclic nucleotides in fetal hemoglobin induction in cultured $CD34^+$ cells. *Exp. Hematol.* **34,** 1151–1161.

Keller, C. C., Kremsner, P. G., Hittner, J. B., Misukonis, M. A., Weinberg, J. B., and Perkins, D. J. (2004). Elevated nitric oxide production in children with malarial anemia: Hemozoin-induced nitric oxide synthase type 2 transcripts and nitric oxide in blood mononuclear cells. *Infect. Immun.* **72,** 4868–4873.

Kharitonov, V. G., Sharma, V. S., Magde, D., and Koesling, D. (1997). Kinetics of nitric oxide dissociation from five- and six-coordinate nitrosyl hemes and heme proteins, including soluble guanylate cyclase. *Biochemistry* **36,** 6814–6818.

Kibbe, M. R., Nie, S., Seol, D. W., Kovesdi, I., Lizonova, A., Makaroun, M., Billiar, T. R., and Tzeng, E. (2000). Nitric oxide prevents p21 degradation with the ubiquitin-proteasome pathway in vascular smooth muscle cells. *J. Vasc. Surg.* **31,** 364–374.

Kim, S. O., Xu, Y., Katz, S., and Pelech, S. (2000a). Cyclic GMP-dependent and-independent regulation of MAP kinases by sodium nitroprusside in isolated cardiomyocytes. *Biochim. Biophys. Acta* **1496,** 277–284.

Kim, Y. M., de Vera, M. E., Watkins, S. C., and Billiar, T. R. (1997a). Nitric oxide protects cultured rat hepatocytes from tumor necrosis factor-alpha-induced apoptosis by inducing heat shock protein 70 expression. *J. Biol. Chem.* **272,** 1402–1411.

Kim, Y. M., Talanian, R. V., and Billiar, T. R. (1997b). Nitric oxide inhibits apoptosis by preventing increases in caspase-3-like activity via two distinct mechanisms. *J. Biol. Chem.* **272,** 31138–31148.

Kim, Y. M., Chung, H. T., Simmons, R. L., and Billiar, T. R. (2000b). Cellular non-heme iron content is a determinant of nitric oxide-mediated apoptosis, necrosis, and caspase inhibition. *J. Biol. Chem.* **275,** 10954–10961.

Kleinbongard, P., Schulz, R., Rassaf, T., Lauer, T., Dejam, A., Jax, T., Kumara, I., Gharini, P., Kabanova, S., Ozuyaman, B., Schnurch, H. G., Godecke, A., *et al.* (2006). Red blood cells express a functional endothelial nitric oxide synthase. *Blood* **107,** 2943–2951.

Klessig, D. F., Durner, J., Noad, R., Navarre, D. A., Wendehenne, D., Kumar, D., Zhou, J. M., Shah, J., Zhang, S., Kachroo, P., Trifa, Y., Pontier, D., *et al.* (2000). Nitric oxide and salicylic acid signaling in plant defense. *Proc. Natl. Acad. Sci. USA* **97,** 8849–8855.

Klotz, T., Bloch, W., Volberg, C., Engelmann, U., and Addicks, K. (1998). Selective expression of inducible nitric oxide synthase in human prostate carcinoma. *Cancer* **82,** 1897–1903.

Koesling, D., Herz, J., Gausepohl, H., Niroomand, F., Hinsch, K. D., Mülsch, A., Böhme, E., Schultz, G., and Frank, R. (1988). The primary structure of the 70 kDa subunit of bovine soluble guanylate cyclase. *FEBS Lett.* **239,** 29–34.

Koury, M. J., and Bondurant, M. C. (1990). Erythropoietin retards DNA breakdown and prevents programmed death in erythroid progenitor cells. *Science* **248,** 378–381.

Kucukkaya, B., Ozturk, G., and Yalcintepe, L. (2006). Nitric oxide levels during erythroid differentiation in K562 cell line. *Indian J. Biochem. Biophys.* **43,** 251–253.

Kumar, A., Knox, A. J., and Boriek, A. M. (2003). CCAAT/enhancer-binding protein and activator protein-1 transcription factors regulate the expression of interleukin-8 through the mitogen-activated protein kinase pathways in response to mechanical stretch of human airway smooth muscle cells. *J. Biol. Chem.* **278,** 18868–18876.

Kuroyanagi, Y., Kaneko, Y., Muta, K., Park, B. S., Moi, P., Ausenda, S., Cappellini, M. D., and Ikuta, T. (2006). cAMP differentially regulates gamma-globin gene expression in erythroleukemic cells and primary erythroblasts through c-Myb expression. *Biochem. Biophys. Res. Commun.* **344,** 1038–1047.

Kwan, H. Y., Huang, Y., and Yao, X. (2000). Store-operated calcium entry in vascular endothelial cells is inhibited by cGMP via a protein kinase G-dependent mechanism. *J. Biol. Chem.* **275,** 6758–6763.

Kwon, N. S., Stuehr, D. J., and Nathan, C. F. (1991). Inhibition of tumor cell ribonucleotide reductase by macrophage-derived nitric oxide. *J. Exp. Med.* **174,** 761–767.

Lauer, T., Preik, M., Rassaf, T., Strauer, B. E., Deussen, A., Feelisch, M., and Kelm, M. (2001). Plasma nitrite rather than nitrate reflects regional endothelial nitric oxide synthase activity but lacks intrinsic vasodilator action. *Proc. Natl. Acad. Sci. USA* **98,** 12814–12819.

Lee, T. W., Chen, G. G., Xu, H., Yip, J. H., Chak, E. C., Mok, T. S., and Yim, A. P. (2003). Differential expression of inducible nitric oxide synthase and peroxisome proliferator-activated receptor gamma in non-small cell lung carcinoma. *Eur. J. Cancer* **39,** 1296–1301.

Lell, B., May, J., Schmidt-Ott, R. J., Lehman, L. G., Luckner, D., Grevem, B., Matousek, P., Schmid, D., Herbich, K., Mockenhaupt, F. P., Meyer, C. G., Bienzle, U., *et al.* (1999). The role of red blood cell polymorphisms in resistance and susceptibility to malaria. *Clin. Infect. Dis.* **28,** 794–799.

Lepoivre, M., Flaman, J. M., Bobe, P., Lemaire, G., and Henry, Y. (1994). Quenching of the tyrosyl free radical of ribonucleotide reductase by nitric oxide. Relationship to cytostasis induced in tumor cells by cytotoxic macrophages. *J. Biol. Chem.* **269,** 21891–21897.

Liao, J. C., Hein, T. W., Vaughn, M. W., Huang, K. T., and Kuo, L. (1999). Intravascular flow decreases erythrocyte consumption of nitric oxide. *Proc. Natl. Acad. Sci. USA* **96,** 8757–8761.

Lin, F. K., Suggs, S., Lin, C. H., Browne, J. K., Smailing, R., Egric, J. C., Chen, K. K., Fox, G. M., Martin, F., and Wasser, Z. (1985). Cloning and expression of the human erythropoietin gene. *Proc. Natl. Acad. Sci. USA* **92,** 7850–7884.

Lin, K. T., Xue, J. Y., Nomen, M., Spur, B., and Wong, P. Y. (1995). Peroxynitrite-induced apoptosis in HL-60 cells. *J. Biol. Chem.* **270,** 16487–16490.

Lincoln, T. M. (1989). Cyclic GMP and mechanisms of vasodilation. *Pharmacol. Ther.* **41,** 479–502.

Liu, X., Miller, M. J., Joshi, M. S., Sadowska-Krowicka, H., Clark, D. A., and Lancaster, J. R., Jr. (1998a). Diffusion-limited reaction of free nitric oxide with erythrocytes. *J. Biol. Chem.* **273,** 18709–18713.

Liu, X., Miller, M. J. S., Joshi, M. S., Thomas, D. D., and Lancaster, J. R. (1998b). Accelerated reaction of nitric oxide with oxygen within the hydrophobic interior of biological membranes. *Proc. Natl. Acad. Sci. USA* **95,** 2175–2179.

Liu, X., Liu, Q., Gupta, E., Zorko, N., Brownlee, E., and Zweier, J. L. (2005). Quantitative measurements of NO reaction kinetics with a Clark-type electrode. *Nitric Oxide* **13,** 68–77.

Loibl, S., Buck, A., Strank, C., von Minckwitz, G., Roller, M., Sinn, H. P., Schini-Kerth, V., Solbach, C., Strebhardt, K., and Kaufmann, M. (2005). The role of early expression of inducible nitric oxide synthase in human breast cancer. *Eur. J. Cancer* **41,** 265–271.

Long, J. H. D., Lira, V. A., Soltow, Q. A., Betters, J. L., Sellman, J. E., and Criswell, D. S. (2006). Arginine supplementation induces myoblast fusion via augmentation of nitric oxide production. *J. Muscle Res. Cell. Motil.* **27,** 577–584.

Lopansri, B. K., Anstey, N. M., Weinberg, J. B., Stoddard, G. J., Hobbs, M. R., Levesque, M. C., Mwaikambo, E. D., and Granger, D. L. (2003). Low plasma arginine concentrations in children with cerebral malaria and decreased nitric oxide production. *Lancet* **361,** 676–678.

Maciejewski, J. P., Selleri, C., Sato, T., Cho, H. J., Keefer, L. K., Nathan, C. F., and Young, N. S. (1995). Nitric oxide suppression of human hematopoiesis *in vitro*. Contribution to inhibitory action of interferon-gamma and tumor necrosis factor-alpha. *J. Clin. Invest.* **96,** 1085–1092.

MacMicking, J., Xie, Q. W., and Nathan, C. (1997). Nitric oxide and macrophage function. *Annu. Rev. Immunol.* **15,** 323–350.

Magill, L., Walker, B., and Irvine, A. E. (2003). The proteasome: A novel therapeutic target in haematopoietic malignancy. *Hematology* **8,** 275–283.

Majka, M., Janowska-Wieczorek, A., Ratajczak, J., Ehrenman, K., Pietrzkowski, Z., Kowalska, M. A., Gewirtz, A. M., Emerson, S. G., and Ratajczak, M. Z. (2001). Numerous growth factors, cytokines, and chemokines are secreted by human CD34(+) cells, myeloblasts, erythroblasts, and megakaryoblasts and regulate normal hematopoiesis in an autocrine/paracrine manner. *Blood* **97,** 3075–3085.

Mamone, G., Sannolo, N., Malorni, A., and Ferranti, P. (1999). *In vitro* formation of S-nitrosohemoglobin in red cells by inducible nitric oxide synthase. *FEBS Lett.* **462,** 241–245.

Mannick, J. B., Asano, K., Izumi, K., Kieff, E., and Stamler, J. S. (1994). Nitric oxide produced by human B-lymphocytes inhibits apoptosis and Epstein-Barr virus reactivation. *Cel1* **79,** 1137–1146.

Marrero, M. B., Venema, V. J., Ju, H., He, H., Liang, H., Caldwell, R. B., and Venema, R. C. (1999). Endothelial nitric oxide synthase interactions with G-protein-coupled receptors. *Biochem. J.* **343,** 335–340.

Martin, J. H., and Edwards, S. W. (1993). Changes in mechanisms of monocyte/macrophage-mediated cytotoxicity during culture. Reactive oxygen intermediates are involved in monocyte-mediated cytotoxicity, whereas reactive nitrogen intermediates are employed by macrophages in tumor cell killing. *J. Immunol.* **150,** 3478–3486.

McDaniel, M. L., Corbett, J. A., Kwon, G., and Hill, J. R. (1997). A role for nitric oxide and other inflammatory mediators in cytokine-induced pancreatic beta-cell dysfunction and destruction. *Adv. Exp. Med. Biol.* **426,** 313–319.

McDevitt, M. A., Xie, J., Gordeuk, V., and Bucala, R. (2004). The anemia of malaria infection: Role of inflammatory cytokines. *Curr. Hematol. Rep.* **3,** 97–106.

Mellion, B. T., Ignarro, L. J., Ohlstein, E. H., Pontecorvo, E. G., Hyman, A. L., and Kadowitz, P. J. (1981). Evidence for the inhibitory role of guanosine 3′,5′-monophosphate in ADP-induced human platelet aggregation in the presence of nitric oxide and related vasodilators. *Blood* **57,** 946–955.

Messmer, U. K., Lapetina, E. G., and Brune, B. (1995). Nitric oxide-induced apoptosis in RAW 264.7 macrophages is antagonized by protein kinase C- and protein kinase A activating compounds. *Mol. Pharmacol.* **47,** 757–765.

Metcalf, D. (2001). Some general aspects of hematopoietic cell development. *In* "Hematopoiesis: A Developmental Approach" (L. I. Zon, eds.), pp. 3–14. Oxford University Press, Oxford, UK.

Michaely, P., and Bennett, V. (1995). Mechanism for binding site diversity on ankyrin. Comparison of binding sites on ankyrin for neurofascin and the Cl-/HCO3- anion exchanger. *J. Biol. Chem.* **270,** 31298–31302.

Michell, B. J., Chen, Z., Tiganis, T., Stapleton, D., Katsis, F., Power, D. A., Sim, A. T., and Kemp, B. E. (2001). Coordinated control of endothelial nitric-oxide synthase phosphorylation by protein kinase C and the cAMP-dependent protein kinase. *J. Biol. Chem.* **276,** 17625–17628.

Minneci, P. C., Deans, K. J., Huang, Z., Yuen, P. S. T., Star, R. A., Banks, S. M., Schechter, A. N., Natanson, C., Gladwin, M. T., and Solomon, S. B. (2005). Hemolysis associated endothelial dysfunction mediated by accelerated NO inactivation by decompartmentalized oxyhemoglobin. *J. Clin. Invest.* **115,** 3409–3417.

Minning, D. M., Gow, A. J., Bonaventura, J., Braun, R., Dewhirst, M., Goldberg, D. E., and Stamler, J. S. (1999). Ascaris haemoglobin is a nitric oxide-activated 'deoxygenase'. *Nature* **401,** 497–502.

Mitchell, H. H., Shonle, H. A., and Grindley, H. S. (1916). The origin of the nitrates in the urine. *J. Biol. Chem.* **24,** 461–490.

Moi, P., and Kan, Y. W. (1990). Synergistic enhancement of globin gene expression by activator protein-1-like proteins. *Proc. Natl. Acad. Sci. USA* **87,** 9000–9004.

Morris, C. R., Vichinsky, E. P., van Warmerdam, J., Machado, L., Kepka-Lenhart, D., Jr., Morris, S. M., and Kuypers, F. A. (2003). Hydroxyurea and arginine therapy: Impact on nitric oxide production in sickle cell disease. *J. Pediatr. Hematol. Oncol.* **25,** 629–634.

Morrison, S. J., and Weissman, I. L. (1994). The long-term repopulating subset of haemopoietic stem cells is deterministic and isolatable by phenotype. *Immunity* **1,** 661–673.

Mortensen, K., Skouv, J., Hougaard, D. M., and Larsson, L. I. (1999). Endogenous endothelial cell nitric-oxide synthase modulates apoptosis in cultured breast cancer cells and is transcriptionally regulated by p53. *J. Biol. Chem.* **274,** 37679–37684.

Mujais, S. K., Bery, N., Pullman, T. N., and Goldwasser, E. (1999). Erythropoietin is produced by tubular cells of the rat kidney. *Cell Biochem. Biophys.* **30,** 153–166.

Musial, A., and Eissa, N. T. (2001). Inducible nitric-oxide synthase is regulated by the proteasome degradation pathway. *J. Biol. Chem.* **276,** 24268–24273.

Nagababu, E., Ramasamy, S., Abernethy, D. R., and Rifkind, J. M. (2003). Active nitric oxide produced in the red cell under hypoxic conditions by deoxyhemoglobin-mediated nitrite reduction. *J. Biol. Chem.* **278,** 46349–46356.

Nahavandi, M., Tavakkoli, F., Wyche, M. Q., Perlin, E., Winter, W. P., and Castro, O. (2002). Nitric oxide and cyclic GMP levels in sickle cell patients receiving hydroxyurea. *Br. J. Haematol.* **119,** 855–857.

Nahavandi, M., Tavakkoli, F., Millis, R. M., Wyche, M. Q., Habib, M. J., and Tavakoli, N. (2006). Effects of hydroxyurea and L-arginine on the production of nitric oxide metabolites in cultures of normal and sickle erythrocytes. *Hematology* **11,** 291–294.

Nappi, A. J., Vass, E., Frey, F., and Carton, Y. (2000). Nitric oxide involvement in Drosophila immunity. *Nitric Oxide* **4,** 423–430.

Ohneda, O., and Bautch, V. L. (1997). Murine endothelial cells support fetal liver erythropoiesis and myelopoiesis via distinct interactions. *Br. J. Haematol.* **98,** 798–808.

Osti, F., Corradini, F. G., Hanau, S., Matteuzzi, M., and Gambari, R. (1997). Human leukemia K562 cells: Induction to erythroid differentiation by guanine, guanosine and guanine nucleotides. *Haematologica* **82,** 395–401.

Pace, B. S., Qian, X. H., Sangerman, J., Ofori-Acquah, S. F., Baliga, B. S., Han, J., and Critz, S. D. (2003). p38 MAP kinase activation mediates gamma-globin gene induction in erythroid progenitors. *Exp. Hematol.* **31,** 1089–1096.

Pai, K., Shrivastava, A., Kumar, R., Khetarpal, S., Sarmah, B., Gupta, P., and Sodhi, A. (1997). Activation of P388D1 macrophage cell line by chemotherapeutic drugs. *Life Sci.* **60,** 1239–1248.

Palmer, R. M., Ashton, D. S., and Moncada, S. (1988). Vascular endothelial cells synthesize nitric oxide from L-arginine. *Nature* **333,** 664–666.

Paolucci, C., Burastero, S. E., Rovere-Querini, P., De Palma, C., Falcone, S., Perrotta, C., Capobianco, A., Manfredi, A. A., and Clementi, E. (2003). Synergism of nitric oxide and maturation signals on human dendritic cells occurs through a cyclic GMP-dependent pathway. *J. Leukoc. Biol.* **73,** 253–262.

Papayannopoulou, T., Priestley, G. V., Bonig, H., and Nakamoto, B. (2003). The role of G-protein signaling in hematopoietic stem/progenitor cell mobilization. *Blood* **101,** 4739–4747.

Park, J. I., Choi, H. S., Jeong, J. S., Han, J. Y., and Kim, I. H. (2001). Involvement of p38 kinase in hydroxyurea-induced differentiation of K562 cells. *Cell Growth Differ.* **12,** 481–486.

Park, S. W., Lee, S. G., Song, S. H., Heo, D. S., Park, B. J., Lee, D. W., Kim, K. H., and Sung, M. W. (2003). The effect of nitric oxide on cyclooxygenase-2 (COX-2) overexpression in head and neck cancer cell lines. *Int. J. Cancer* **107,** 729–738.

Perner, A., Andresen, L., Normark, M., and Rask-Madsen, J. (2002). Constitutive expression of inducible nitric oxide synthase in the normal human colonic epithelium. *Scand. J. Gastroenterol.* **37,** 944–948.

Pluta, R. M., Dejam, A., Grimes, G., Gladwin, M. T., and Oldfield, E. H. (2005). Nitrite infusions to prevent delayed cerebral vasospasm in a primate model of subarachnoid hemorrhage. *JAMA* **293,** 1477–1484.

Poderoso, J. J., Carreras, M. C., Lisdero, C., Riobo, N., Schopfer, F., and Boveris, A. (1996). Nitric oxide inhibits electron transfer and increases superoxide radical production in rat heart mitochondria and submitochondrial particles. *Arch. Biochem. Biophys.* **328,** 85–92.

Queen, L. R., Xu, B., Horinouchi, K., Fisher, I., and Ferro, A. (2000). Beta(2)-adrenoceptors activate nitric oxide synthase in human platelets. *Circ. Res.* **87,** 39–44.

Rafferty, S. P., Domachowske, J. B., and Malech, H. L. (1996). Inhibition of hemoglobin expression by heterologous production of nitric oxide synthase in the K562 erythroleukemic cell line. *Blood* **88,** 1070–1078.

Rees, D. D., Palmer, R. M., and Moncada, S. (1989). Role of endothelium-derived nitric oxide in the regulation of blood pressure. *Proc. Natl. Acad. Sci. USA* **86,** 3375–3378.

Reiser, G. (1992). Nitric oxide formation caused by Ca^{2+} release from internal stores in neuronal cell line is enhanced by cyclic AMP. *Eur. J. Pharmacol.* **227,** 89–93.

Reiter, C. D., Wang, X. D., Tanus-Santos, J. E., Hogg, N., Cannon, R. O., Schechter, A. N., and Gladwin, M. T. (2002). Cell-free hemoglobin limits nitric oxide bioavailability in sickle-cell disease. *Nat. Med.* **8,** 1383–1389.

Robinson, L. J., Weremowicz, S., Morton, C. C., and Michel, T. (1994). Isolation and chromosomal localization of the human endothelial nitric oxide synthase (NOS3) gene. *Genomics* **19,** 350–357.

Rodgers, G. P., Dover, G. J., Uyesaka, N., Noguchi, C. T., Schechter, A. N., and Nienhuis, A. W. (1993). Augmentation by erythropoietin of the fetal-hemoglobin response to hydroxyurea in sickle cell disease. *N. Engl. J. Med.* **328,** 73–80.

Rood, P. M., Gerritsen, W. R., Kramer, D., Ranzijn, C., von dem Borne, A. E., and van der Schoot, C. E. (1999). Adhesion of hematopoietic progenitor cells to human bone marrow or umbilical vein derived endothelial cell lines: A comparison. *Exp. Hematol.* **27,** 1306–1314.

Rother, R. P., Bell, L., Hillmen, P., and Gladwin, M. T. (2005). The clinical sequelae of intravascular hemolysis and extracellular plasma hemoglobin—a novel mechanism of human disease. *JAMA* **293,** 1653–1662.

Russwurm, M., Behrends, S., Harteneck, C., and Koesling, D. (1998). Functional properties of a naturally occurring isoform of soluble guanylyl cyclase. *Biochem. J.* **335,** 125–130.

Rybicki, A. C., Schwartz, R. S., Hustedt, E. J., and Cobb, C. E. (1996). Increased rotational mobility and extractability of band 3 from protein 4.2-deficient erythrocyte membranes: Evidence of a role for protein 4.2 in strengthening the band 3-cytoskeleton linkage. *Blood* **88,** 2745–2753.

Safaya, S., Ibrahim, A., and Rieder, R. F. (1994). Augmentation of gamma-globin gene promoter activity by carboxylic acids and components of the human beta-globin locus control region. *Blood* **84,** 3929–3935.

Sakahira, H., Enari, M., and Nagata, S. (1998). Cleavage of CAD inhibitor in CAD activation and DNA degradation during apoptosis. *Nature* **391,** 96–99.

Salgo, M. G., Squadrito, G. L., and Pryor, W. A. (1995). Peroxynitrite causes apoptosis in rat thymocytes. *Biochem. Biophys. Res. Commun.* **215,** 1111–1118.

Sawada, K., Krantz, S. B., Dai, C. H., Koury, S. T., Horn, S. T., Glick, A. D., and Civin, C. I. (1990). Purification of human blood burst-forming units-erythroid and demonstration of the evolution of erythropoietin receptor. *J. Cell. Physiol.* **142,** 219–230.

Saxena, R. K., and Chandrasekhar, B. (2000). A novel nonphagocytic mechanism of erythrocyte destruction involving direct cell-mediated cytotoxicity. *Int. J. Hematol.* **71,** 227–237.

Sessa, W. C., Garcia-Gardena, G., Liu, J., Keh, A., Pollock, J. S., Bradley, J., Thiru, S., Braverman, I. M., and Desai, K. M. (1995). The Golgi association of endothelial nitric oxide synthase is necessary for the efficient synthesis of nitric oxide. *J. Biol. Chem.* **270,** 17641–17644.

Sethi, S., and Dikshit, M. (2000). Modulation of polymorphonuclear leukocytes function by nitric oxide. *Thromb. Res.* **100,** 223–247.

Shami, P. J., Moore, J. O., Gockerman, J. P., Hathorn, J. W., Misukonis, M. A., and Weinberg, J. B. (1995). Nitric oxide modulation of the growth and differentiation of freshly isolated acute non-lymphocytic leukemia cells. *Leuk. Res.* **19,** 527–533.

Shami, P. J., and Weinberg, J. B. (1996). Differential effects of nitric oxide on erythroid and myeloid colony growth from CD34+ human bone marrow cells. *Blood* **87,** 977–982.

Shaul, P. W., Smart, E. J., Robinson, L. J., German, Z., Yuhanna, I. S., Ying, Y., Anderson, R. G., and Michel, T. (1996). Acetylation targets endothelial nitric-oxide synthase to plasmalemmal caveolae. *J. Biol. Chem.* **271,** 6518–6522.

Stamler, J. S., Jia, L., Eu, J. P., McMahon, T. J., Demchenko, I. T., Bonaventura, J., Gernert, K., and Piantadosi, C. A. (1997). Blood flow regulation by S-nitrosohemoglobin in the physiological oxygen gradient. *Science* **276,** 2034–2037.

Stangl, V., Lorenz, M., Meiners, M., Ludwig, A., Bartsch, C., Moobed, M., Vietzke, A., Kinkel, H. T., Baumann, G., and Stangl, K. (2004). Long-term up-regulation of eNOS and improvement of endothelial function by inhibition of the ubiquitin-proteasome pathway. *FASEB J.* **18,** 272–279.

Stevenson, M. M., and Riley, E. M. (2004). Innate immunity to malaria. *Nat. Rev. Immunol.* **4,** 169–180.

Stone, J. R., and Marletta, M. A. (1994). Soluble guanylate cyclase from bovine lung: Activation with nitric oxide and carbon monoxide and spectral characterization of the ferrous and ferric states. *Biochemistry* **33,** 5636–5640.

Stone, J. R., and Marletta, M. A. (1995). Heme stoichiometry of heterodimeric soluble guanylate cyclase. *Biochemistry* **34,** 14668–14674.

Stuehr, D. (1999). Mammalian nitric oxide synthases. *Biochim. Biophys. Acta* **1411,** 217–230.

Swana, H. S., Smith, S. D., Perrotta, P. L., Saito, N., Wheeler, M. A., and Weiss, R. M. (1999). Inducible nitric oxide synthase with transitional cell carcinoma of the bladder. *J. Urol.* **161,** 630–634.

Tamatani, M., Ogawa, S., Niitsu, Y., and Tohyama, M. (1998). Involvement ofBcl-2 family and caspase-3-like protease inNO-mediated neuronal apoptosis. *J. Neurochem.* **71,** 1588–1596.

Tatoyan, A., and Giulivi, C. (1998). Purification and characterization of a nitric-oxide synthase from rat liver mitochondria. *J. Biol. Chem.* **273,** 11044–11048.

Taylor, B. S., and Geller, D. A. (2000). Molecular regulation of the human inducible nitric oxide synthase (iNOS) gene. *Shock* **13,** 413–424.

Thomsen, L. L., Lawton, F. G., Knowles, R. G., Beesley, J. E., Riveros-Moreno, V., and Moncada, S. (1994). Nitric oxide synthase activity in human gynecological cancer. *Cancer Res.* **54,** 1352–1354.

Thomsen, L. L., Miles, D. W., Happerfield, L., Bobrow, L. G., Knowles, R. G., and Moncada, S. (1995). Nitric oxide synthase activity in human breast cancer. *Br. J. Cancer* **72,** 41–44.

Tournier, C., Hess, P., Yang, D. D., Xu, J., Turner, T. K., Nimnual, A., Bar-Sagi, D., Jones, S. N., Flavell, R. A., and Davis, R. J. (2000). Requirement of JNK for stress-induced activation of the cytochrome c-mediated death pathway. *Science* **288,** 870–874.

Toyoshima, H., Nasa, Y., Hashizume, Y., Koseki, Y., Isayama, Y., Kohsaka, Y., Yamada, T., and Takeo, S. (1998). Modulation of cAMP-mediated vasorelaxation by endothelial nitric oxide and basal cGMP in vascular smooth muscle. *J. Cardiovasc. Pharmacol.* **32,** 543–551.

Tsoukias, N. M., and Popel, A. S. (2002). Erythrocyte consumption of nitric oxide in presence and absence of plasma-based hemoglobin. *Am. J. Physiol. Heart Circ. Physiol.* **282,** H2265–H2277.

Urbich, C., Fleming, I., Förstermann, U., Zeiher, A. M., and Dimmeler, S. (2002). Inhibitors of histone deacetylation downregulate the expression of endothelial nitric oxide synthase and compromise endothelial cell function in vasorelaxation and angiogenesis. *Circ. Res.* **91,** 837–844.

Vaandrager, A. B., and de Jonge, H. R. (1996). Signalling by cGMP-dependent protein kinases. *Mol. Cell. Biochem.* **157,** 23–30.

van der Zee, R., Murohara, T., Luo, Z., Zollmann, F., Passeri, J., Lekutat, C., and Isner, J. M. (1997). Vascular endothelial growth factor/vascular permeability factor augments nitric oxide release from quiescent rabbit and human vascular endothelium. *Circulation* **95,** 1030–1037.

Vaughn, M. W., Kuo, L., and Liao, J. C. (1998). Effective diffusion distance of nitric oxide in the microcirculation. *Am. J. Physiol.* **274,** H1705–H1714.

Vaughn, M. W., Huang, K. T., Kuo, L., and Liao, J. C. (2000). Erythrocytes possess an intrinsic barrier to nitric oxide consumption. *J. Biol. Chem.* **275,** 2342–2348.

Veith, R., Galanello, R., Papayannopoulou, T., and Stamatoyannopoulos, G. (1985). Stimulation of F-cell production in patients with sickle-cell anemia treated with cytarabine or hydroxyurea. *N. Engl. J. Med.* **313,** 1571–1575.

Wang, S., Wang, W., Wesley, R. A., and Danner, R. L. (1999). A Sp1 binding site of the tumor necrosis factor alpha promoter functions as a nitric oxide response element. *J. Biol. Chem.* **274,** 33190–33193.

Wang, Y. Z., Cao, Y. Q., Wu, J. N., Chen, M., and Cha, X. Y. (2005). Expression of nitric oxide synthase in human gastric carcinoma and its relation to p53,PCNA. *World J. Gastroenterol.* **11,** 46–50.

Wedel, B., Humbert, P., Harteneck, C., Foerster, J., Malkewitz, J., Böhme, E., Schultz, G., and Koesling, D. (1994). Mutation of His-105 in the β1 subunit yields a nitric oxide-insensitive form of soluble guanylyl cyclase. *Proc. Natl. Acad. Sci. USA* **91,** 2592–2596.

Weninger, W., Rendl, M., Pammer, J., Mildner, M., Tschugguel, W., Schneeberger, C., Sturzl, M., and Tschachler, E. (1998). Nitric oxide synthases in Kaposi sarcoma are expressed predominantly by vessels and tissue macrophages. *Lab. Invest.* **78,** 949–955.

Wilson, J. G., and Tavassoli, M. (1994). Microenvironmental factors involved in the establishment of erythropoiesis in bone marrow. *Ann. NY Acad. Sci.* **718,** 271–283.

Wilson, K. T., Fu, S., Ramanujam, K. S., and Meltzer, S. J. (1998). Increased expression of inducible nitric oxide synthase andcyclooxygenase-2 in Barrett s esophagus and associated adenocarcinomas. *Cancer Res.* **58,** 2929–2934.

Wu, H., Liu, X., Jaenisch, R., and Lodish, H. F. (1995). Generation of committed erythroid BFU-E and CFU-E progenitors does not require erythropoietin or the erythropoietin receptor. *Cell* **83,** 59–67.

Wu, H. M., Wen, H. C., and Lin, W. W. (2002). Proteasome inhibitors stimulate interleukin-8 expression via Ras and apoptosis signal-regulating kinase-dependent extracellular signal-related kinase and c-Jun N-terminal kinase activation. *Am. J. Respir. Cell Mol. Biol.* **27,** 234–243.

Xu, B., Li, J., Gao, L., and Ferro, A. (2000). Nitric oxide-dependent vasodilatation of rabbit femoral artery by beta(2)-adrenergic stimulation or cyclic AMP elevation *in vivo*. *Br. J. Pharmacol.* **129,** 969–974.

Xu, X., Cho, M., Spencer, N. Y., Patel, N., Huang, Z., Shields, H., King, S. B., Gladwin, M. T., Hogg, N., and Kim-Shapiro, D. B. (2003). Measurements of nitric oxide on the heme iron and beta-93 thiol of human hemoglobin during cycles of oxygenation and deoxygenation. *Proc. Natl. Acad. Sci. USA* **100,** 11303–11308.

Yamaguchi, H., Ishii, E., Saito, S., Tashiro, K., Fujita, I., Yoshidomi, S., Ohtubo, M., Akazawa, K., and Miyazaki, S. (1996). Umbilical vein endothelial cells are an important source of c-kit and stem cell factor which regulate the proliferation of haemopoietic progenitor cells. *Br. J. Haematol.* **94,** 606–611.

Yamaguchi, T., Chattopadhyay, N., Kifor, O., Sanders, J. L., and Brown, E. M. (2000). Activation of p42/44 and p38 mitogen-activated protein kinases by extracellular calcium-sensing receptor agonists induces mitogenic responses in the mouse osteoblastic MC3T3-E1 cell line. *Biochem. Biophys. Res. Commun.* **279,** 363–368.

Yang, Y. M., Pace, B., Kitchens, D., Ballas, S. K., Shah, A., and Baliga, B. S. (1997). BFU-E colony growth in response to hydroxyurea: Correlation between *in vitro* and *in vivo* fetal hemoglobin induction. *Am. J. Hematol.* **56,** 252–258.

Yuen, P. S., Potter, L. R., and Garbers, D. L. (1990). A new form of guanylyl cyclase is preferentially expressed in rat kidney. *Biochemistry* **29,** 10872–10878.

Yui, Y., Hattori, R., Kosuga, K., Eizawa, H., Hiki, K., and Kawai, C. (1991). Purification of nitric oxide synthase from rat macrophages. *J. Biol. Chem.* **266,** 12544–12547.

Zech, B., Kohl, R., von Knethen, A., and Brune, B. (2003). Nitric oxide donors inhibit formation of the Apaf-1/caspase-9 apoptosome and activation of caspases. *Biochem. J.* **371,** 1055–1064.

Zuhrie, S. R., Pearson, J. D., and Wickramasinghe, S. N. (1988). Haemoglobin synthesis in K562 erythroleukaemia cells is affected by intimate contact with monolayers of various human cell types. *Leuk. Res.* **12,** 567–574.

CHAPTER EIGHT

Diamond Blackfan Anemia: A Disorder of Red Blood Cell Development

Steven R. Ellis* *and* Jeffrey M. Lipton[†,‡]

Contents

Abstract

Diamond Blackfan anemia (DBA) is an inherited hypoplastic anemia that typically presents in the first year of life. The genes identified to date that are mutated in DBA encode ribosomal proteins, and in these cases ribosomal protein haploinsufficiency gives rise to the disease. The developmental timing of DBA presentation suggests that the changes in red blood cell production that occur around the time of birth trigger a pathophysiological mechanism, likely linked to defective ribosome synthesis, which precipitates the hematopoietic

* Department of Biochemistry and Molecular Biology, University of Louisville, Louisville, Kentucky
† Division of Hematology/Oncology and Stem Cell Transplantation, Schneider Children's Hospital, Albert Einstein College of Medicine, Long Island Jewish Medical Center, New Hyde Park, New York
‡ The Feinstein Institute for Medical Research, Manhasset, New York

Current Topics in Developmental Biology, Volume 82
ISSN 0070-2153, DOI: 10.1016/S0070-2153(07)00008-7

phenotype. Variable presentation of other clinical phenotypes in DBA patients indicates that other developmental pathways may also be affected by ribosomal protein haploinsufficiency and that the involvement of these pathways is influenced by modifier genes. Understanding the molecular basis for the developmental timing of DBA presentation promises to shed light on a number of baffling features of this disease. This chapter also attempts to demonstrate how the marriage of laboratory and clinical science may enhance each and permit insights into human disease that neither alone can accomplish.

1. Introduction

Diamond Blackfan anemia (DBA; OMIM #205900) is one of a rare group of genetic disorders known as the inherited bone marrow failure syndromes (IBMFS) (Young and Alter, 1994). These disorders have in common proapoptotic hematopoiesis, bone marrow failure, birth defects (Gripp *et al.*, 2001), and a predisposition to cancer (Lipton *et al.*, 2001). Interest in these disorders has grown dramatically as the study of each has clarified, or revealed for the first time, new molecular events in development or cellular function. Significant expectations await the complete elucidation of these events.

DBA was first reported in 1936 (Josephs, 1936) and more completely described by Diamond and Blackfan (1938). The diagnostic criteria for DBA published in 1976 consist of presentation of anemia prior to the first birthday with near normal or slightly decreased neutrophil counts, variable platelet counts, reticulocytopenia, macrocytic anemia, and normal marrow cellularity with a paucity of red blood cell precursors (Diamond *et al.*, 1976). These criteria have, until recently, remained the accepted standard. In addition to macrocytosis, the presence of elevated fetal hemoglobin levels and an elevation in red blood cell adenosine deaminase enzyme activity are important supporting features associated with DBA. The presence of macrocytosis and elevated fetal hemoglobin levels each felt to be a consequence of "stress erythropoiesis" and skipped erythroid cell divisions is not unique to DBA but is observed in most incidences of bone marrow failure. These features are also found in recovery from anemia such as that caused by iron deficiency when erythropoietin levels are elevated. The reason for the elevated eADA activity in DBA has remained obscure nearly 25 years after the original observation (Glader *et al.*, 1983).

The careful analysis of DBA-affected pedigrees for the presence of members with the DBA-associated findings of macrocytosis, elevated fetal hemoglobin (HbF) levels and/or increased erythrocyte adenosine deaminase (eADA) activity (elevated in 85% of patients with DBA) and congenital anomalies strongly suggested a greater number of autosomal dominant cases than previously thought (Vlachos *et al.*, 2001). With the discovery of the first gene mutated in DBA (Draptchinskaia *et al.*, 1999; Gustavsson *et al.*, 1997), it has become evident that the penetrance of autosomal dominant DBA is quite variable with regard to both hematologic and non-hematologic manifestations. Indeed the estimate of 10–15% of DBA cases being familial has increased to around 45% as a consequence of mutation analysis of family members of probands (Orfali *et al.*, 2004). Although described classically in children less than one year of age, the disorder may present in adults, more frequently than first recognized (Balaban *et al.*, 1985).

As mentioned, birth defects have long been known to be a feature of DBA. A distinct facial appearance and triphalangeal thumbs have been classically described in DBA as the Cathie facies (Cathie, 1950) and Aase syndrome (Aase and Smith, 1969), respectively. A cute snub nose and wide-spaced eyes originally described by Cathie, as well as possibly severe cranio-facial anomalies are the most common physical anomalies described in DBA. Abnormal thumbs are classic (Alter, 1978). In all, congenital anomalies were found in 30–47% of the patients in the Italian (Campagnoli *et al.*, 2004), French (Willig *et al.*, 1999), UK (Orfali *et al.*, 2004), and North American registries (Lipton *et al.*, 2006). Additional anomalies of the upper limb and hand, genitourinary system and heart each are described in as many as 30–40% of patients. It is important to note that the prevalence of genitourinary and cardiac anomalies may be underestimated when abdominal/pelvic and cardiac ultrasonography are not routinely performed in asymptomatic patients. More than one anomaly is described in about a quarter of all patients. Short stature is clearly constitutional in many patients. However, an accurate assessment of linear growth retardation is complicated in patients who may be anemic, iron-overloaded or taking corticosteroids all from a very young age. A representative table of malformations has been published (Willig *et al.*, 1999). These data indicate that the mutations that give rise to DBA likely affect other developmental pathways. The heterogeneity with which these other pathways are affected in DBA strongly suggests the influence of modifier genes manifested by genomic background affects.

A combination of laboratory and clinical investigations has begun to reveal the relationships between the moribund erythron, the biology of erythropoiesis, and the non-hematologic manifestations of DBA. Moreover, recent advances in the understanding of DBA, in part as a result of data from international DBA registries (Campagnoli *et al.*, 2004; Ohga *et al.*, 2004; Orfali *et al.*, 2004; Vlachos *et al.*, 2001a; Willig *et al.*, 1999), are resulting in more sophisticated diagnostic criteria and improvements in clinical care (Vlachos *et al.*, submitted for publication).

2. Pathophysiology and Genetics: DBA, a Disorder of Ribosome Biosynthesis

2.1. DBA genes identified to date encode ribosomal proteins

A number of theories regarding the pathophysiology of DBA have been proposed and discarded (Ershler *et al.*, 1980; Hoffman *et al.*, 1976; Ortega *et al.*, 1975; Sawada *et al.*, 1985). Evidence debunking the notion that DBA is the result of immune-mediated red blood cell failure in favor of the concept of DBA as an intrinsic hematopoietic progenitor defect first appeared in 1976 when the group from Toronto (Freedman *et al.*, 1976) suggested that some patients with DBA had decreased numbers of erythroid colony-forming units (CFU-E). Investigators in Boston (Nathan *et al.*, 1978) extended this observation and suggested a block in erythroid maturation prior to the burst-forming unit-erythroid (BFU-E) stage. These findings may not, however, be true in all instances as later studies showed that both BFU-E and CFU-E colonies are present, often in normal numbers, in the marrow of young DBA patients, but their differentiation to mature erythrocytes is defective (Lipton *et al.*, 1986). Interestingly, Chan *et al.* (1982) demonstrated that the growth *in vitro* of DBA progenitor cells in semisolid media could be enhanced by the addition of corticosteroids similar to the *in vivo* response (Nathan *et al.*, 1978). Studies exploring the pathophysiology of DBA have been hampered by the fact that there are currently no animal models for this disease. Sadly, the knockout of the first DBA gene described below is a homozygous lethal and the hemizygote lacks a DBA phenotype (Matsson *et al.*, 2004). Furthermore, no naturally occurring animal models of DBA exist as the anemic and macrocytic W/W^v and Sl/Sl^d mice do not respond to steroids and have mutations in genes for c-kit and kit ligand (now known not to be involved in the molecular pathology of DBA), respectively (Alter *et al.*, 1993). Based on available data, it is now widely accepted that DBA results from an intrinsic cellular defect in which erythroid progenitors and precursors are highly sensitive to death by apoptosis (Lipton *et al.*, 1986; Ohene-Abuakwa *et al.*, 2005; Perdahl *et al.*, 1994; Tsai *et al.*, 1989).

Familial DBA appears to be inherited as an autosomal dominant trait. No instance of autosomal recessive inheritance has been confirmed and apparent recessive inheritance is likely a consequence of nonpenetrance (Gripp *et al.*, 2001) or rarely, gonadal mosaicism (Cmejla *et al.*, 2000). The first DBA gene, *DBA1*, is located at chromosome 19q13.2 and is mutated in approximately 25% of patients. *DBA1* has been cloned and identified as *RPS19*, a gene that codes for a ribosomal protein (Draptchinskaia *et al.*, 1999; Gustavsson *et al.*, 1997). Two classes of *RPS19* mutations have been identified in DBA patients. In the first class, expression of the *RPS*19 protein is disrupted by insertions, deletions, splice site mutations, and nonsense mutations with nonsense-mediated mRNA decay. Many missense mutations, which comprise the second class of *RPS19* mutations, disrupt the assembly of the protein into 40S ribosomal subunits. This often results in the rapid turnover of the free protein (Angelini *et al.*, 2007; Da Costa *et al.*, 2003; Gazda *et al.*, 2004). Thus, current evidence suggests that in those patients for whom *RPS19* is mutated the disease results from Rps19 protein haploinsufficiency.

In general, ribosomal proteins lack specific roles in translation, but instead function primarily in the assembly of ribosomal subunits. The assembly process can be monitored by analyzing intermediates in the pathway by which mature ribosomal RNAs (rRNAs) are liberated from a primary rRNA transcript containing three of the four mature rRNAs (Fig. 8.1). It has been recently demonstrated that the yeast homologue of Rps19 is required for the maturation of the 3′-end of 18S rRNA. In cells depleted of Rps19 immature subunits containing a 3′-extended 18S rRNA precursor accumulate in the nucleus leading to a deficit of functional 40S subunits in the cytoplasm (Leger-Silvestre *et al.*, 2005). Related defects in

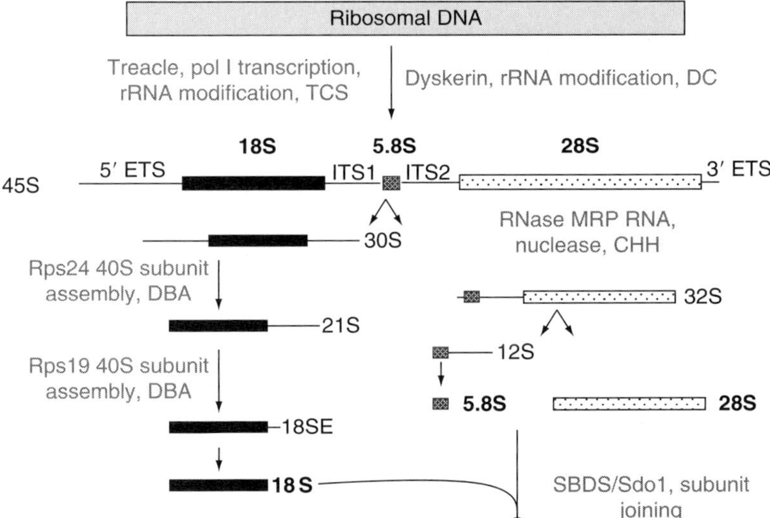

Figure 8.1 Putative biochemical function for Treacle, Dyskerin, Rps19/Rps24, RMRP, and Sbds in ribosomal biogenesis. A simplified version of the human pre-rRNA processing pathway is shown. Functions are deduced from data for either yeast or human orthologues. Mature rRNA species (denoted in bold) are liberated from the 45S pre-rRNA primary transcript by a series of processing steps. 5′ and 3′ ETS represent external transcribed sequences found on the 5′ and 3′ ends of the primary transcript, respectively. ITS represents internal transcribed sequences present in the primary transcript that are not retained in mature rRNAs. The 18S rRNA is a component of the 40S ribosomal subunit. The 5.8S and 28S rRNAs, together with the 5S rRNA (not shown), are components of the 60S ribosomal subunit.

the maturation of the 3′-end of 18S rRNA have been reported in human cell lines where Rps19 expression was reduced by siRNA technology or in bone marrow and lymphoblastoid cells derived from DBA patients with mutations in the *RPS19* gene (Choesmel *et al.*, 2007; Flygare *et al.*, 2007; Idol *et al.*, 2007). These data point to a defect in 40S ribosomal subunit production in this subset of DBA patients.

It is also possible, however, that a defect in an extraribosomal function attributable to Rps19 may ultimately be responsible for the pathophysiology of DBA. In this regard, Rps19 has been implicated in the postapoptotic mediation of monocyte chemotaxis (Yamamoto, 2000). Other possible functions for Rps19 may be a consequence of interactions with a novel nucleolar protein S19-binding protein (Maeda *et al.*, 2006), fibroblast growth factor 2 (Soulet *et al.*, 2001), and the PIM-1 oncoprotein (Chiocchetti *et al.*, 2005). These alternative scenarios now seem less likely given the identification of a second DBA gene as encoding another small subunit ribosomal protein. This gene, *RPS24*, is mutated in approximately 2% of DBA patients (Gazda *et al.*, 2006).

The identification of *RPS24* as a second DBA gene strongly suggests that defective ribosome synthesis and/or function are responsible for the disease pathophysiology. The yeast orthologue of Rps24, like Rps19, is required for 18S rRNA processing and maturation of 40S ribosomal subunits (Ferreira-Cerca *et al.*, 2005). In contrast to Rps19, which is required for efficient processing of the 3′ end of 18S rRNA however, Rps24 is required for an early step in the pathway giving rise to the mature 5′ end of 18S rRNA (Fig. 8.1). Thus, these two proteins have distinct roles in 40S subunit maturation suggesting that the specific nature of the processing defect may not be as important as a net effect on the production of 40S ribosomal subunits. An alternative hypothesis that Rps19 and Rps24 have a common function that sets them apart from most other ribosomal proteins is supported by the observation that antibodies to Rps19, Rps24, and two other small subunit proteins block the binding of initiation factor 2 (eIF2) to 40S ribosomal subunits (Bommer *et al.*, 1988). Enthusiasm for this hypothesis is tempered by the fact that since both proteins are required for subunit biogenesis, subunits lacking these proteins are unlikely to be transported to the cytoplasm where this specific function in translation could be impaired. Most recently, the existence of a third DBA gene, *RPS17*, has been reported (Cmejla *et al.*, 2007). Thus, the current data point toward a defect in ribosome synthesis as the underlying molecular basis for DBA.

2.2. Ribosomal protein haploinsufficiency can have differential effects on cell fates during development

Recent studies by Sulic *et al.* (2005) provide insight into the effect of defective ribosome synthesis on cell growth and development. In this study, they conditionally deleted one of two copies of the gene encoding ribosomal protein S6 in the T-cell lineage of transgenic mice. These mice showed normal T-cell maturation in the thymus. RPS6 mRNA levels were decreased in thymic T cells from hemizygous mice relative to wild-type animals but Rps6 protein and ribosome levels were unaffected. The disparate effects of *RPS6* hemizygocity on RPS6 mRNA and protein levels may be explained by the recent observation in HeLa cells that ribosomal proteins are synthesized in excess of that typically needed for ribosome assembly, with excess protein rapidly degraded by the proteasome (Lam *et al.*, 2007). If this condition were met in thymic T cells, a reduction in Rps6 protein synthesis as a result of decreased mRNA levels may not be reflected in the steady-state level of the Rps6 protein as long as the synthesis rate did not fall to the point where Rps6 became limiting for ribosome synthesis. The observation that ribosome levels in thymic T cells were unchanged between wild-type and hemizygous mice further suggests that Rps6 protein levels in the hemizygous state are not limiting for ribosome synthesis during normal homeostatic development.

In contrast to thymic T cells, peripheral T cells were reduced in *RPS6* hemizygous mice relative to controls. In order to study the effect of *RPS6* hemizygocity on T-cell activation, cells were stimulated with anti-CD3 or CD28 antibodies and growth and proliferation analyzed. T cells from the hemizygous mice showed a normal increase in size with activation despite a decrease in the steady-state level of Rps6 protein and defective ribosome biogenesis. Conversely, the proliferative response in activated T cells with lower Rps6 and ribosome synthesis was severely compromised relative to wild type. This defective proliferative response was linked to a p53-dependent G1/S arrest and increased apoptosis. These data indicate that within a single cell lineage defects in ribosome synthesis and associated physiological changes can be precipitated at specific stages of development in response to changing demands for protein synthetic capacity. Furthermore, the relationship of the nucleolus and defective ribosome synthesis to p53-mediated apoptosis suggests a role for interdicting mutations in p53 and distal pathways in oncogenesis in DBA (see below).

3. Other Diseases Linked to Defects in Ribosome Synthesis

3.1. Hematologic disorders

DBA is one of a growing family of human diseases linked to defects in ribosome biogenesis (Liu and Ellis, 2006). Most of these disorders such as Shwachman Diamond syndrome (SDS), dyskeratosis congenita (DC), and cartilage hair hypoplasia (CHH) are congenital bone marrow failure syndromes like DBA. Interestingly, the genes affected in DBA to date encode structural components of the ribosome, whereas the other bone marrow failure syndromes listed above are caused by mutations in genes encoding factors present in the myriad of extraribosomal factors required for proper subunit assembly, pre-rRNA maturation, and subunit transport from nucleolus to cytoplasm (Fig. 8.1). In some cases, the ribosome biogenesis defect may be secondary to other functions ascribed to affected genes such as telomerase and mRNA turnover defects in DC and CHH, respectively (Thiel *et al.*, 2005; Vulliamy and Dokal, 2006). In contrast, there are intriguing similarities and differences between SDS and DBA, where in both cases the primary defect appears to reside in ribosome synthesis and/or function.

Classical SDS is characterized by neutropenia (although other hematopoietic lineages may be affected), exocrine pancreas dysfunction, leukemia predisposition, and a heterogeneous collection of other congenital anomalies (Shimamura, 2006). A perplexing problem is trying to understand the absence of nonhematologic cancers in SDS, a feature quite distinct from the other IBMFS associated with the predisposition to malignancy. *SBDS*, the gene affected in SDS, encodes a protein associated with 60S subunits that shuttles between the nucleolus and cytoplasm (Austin *et al.*, 2005; Ganapathi *et al.*, 2007). The yeast orthologue of *SBDS*, *SDO1*, is needed for the functional activation of 60S subunits through its involvement in the release of antisubunit association factor Tif6 from pre-60S subunits, newly transported to the cytoplasm (Menne *et al.*, 2007). Secondary effects on recycling of Tif6 back to the nucleolus likely also affect earlier steps in the ribosome synthesis pathway. Thus, in contrast to mutations in *RPS19*, *RPS24*, and *RPS17* that affect the production of 40S ribosomal subunits, mutations in *SBDS* affect the functional activation of 60S subunits. While mutations in all three genes affect the production of functional ribosomes, the differences in clinical features between DBA and SDS suggest that defects in the synthesis of ribosomal subunits can impact cell growth and development in a manner distinct from defects in the functional activation of ribosomal subunits. Alternatively, the putative extraribosomal functions of these proteins are more important than previously surmised. Further complicating the issue, recent evidence supports a mutation in a gene encoding a large subunit ribosomal protein in a subgroup of DBA patients (Robert Arceci, personal communication). Sorting out similarities and differences in the pathophysiology of DBA and SDS should help us understand more about both diseases.

3.2. Nonhematologic disorders: Relationships between DBA and Treacher Collins syndrome

A number of patients with unequivocal DBA have a phenotype indistinguishable from Treacher Collins Syndrome (TCS; OMIM #154500), another developmental disorder linked to diminished ribosome biogenesis

(Gripp *et al.*, 2001; Hasan and Inoue, 1993). TCS is characterized in classical cases by "bilateral down slanting palpebral fissures, frequently accompanied by colobomas of the lower eyelids and a paucity of eyelashes medial to the defect, abnormalities of the external ears, atresia of the external auditory canal and bilateral conductive hearing loss, hypoplasia of the zygomatic complex and mandible and cleft palate" (Teber *et al.*, 2004). As in DBA, the disorder is inherited as an autosomal dominant with approximately 60% of cases occurring spontaneously. The incidence of both disorders is of the same order of magnitude at about 1 in 50,000 live births for TCS (Fazen *et al.*, 1967) and 1 in 100,000 for DBA (Willig *et al.*, 1999). Case ascertainment of mild phenotypes in both disorders is clearly a cause for underreporting.

Additional similarities between TCS and DBA also exist. Orofacial clefts have been reported in 3% of cases of DBA in the literature (Lipton and Alter, 1993) and in 5.7% of cases from the Diamond Blackfan Anemia Registry (DBAR) (Vlachos *et al.*, 2001a). Gripp *et al.* (2001) identified first cousins, male children of sisters, with bilateral microtia and cleft palate consistent with TCS, but with hematologic abnormalities consistent with DBA. The proband had been previously enrolled in the DBAR of North America, a database of over 400 patients (Vlachos *et al.*, 2001a), while his cousin with similar facial anomalies had not developed classical DBA at the time of the report. The cousin's relevant hematologic parameters were significant for macrocytosis, elevated HbF level, and increased eADA activity, and therefore consistent with a nonclassical hematologic DBA phenotype. Physical and hematologic evaluations of the patients' mothers were normal. Therefore, both mothers appear to be obligate heterozygotes for this dominantly inherited disorder, with neither demonstrating any hematologic evidence of DBA nor even minimal congenital anomalies. This multiplex family demonstrates the extremely variable penetrance and expressivity of this DBA gene. In all, the described proband and 2 other of the 21 patients with DBA and a cleft palate reported to the DBAR were originally diagnosed with "classical" TCS. However, none of these three patients had mutations at *TCOF1* (5q32-q33.1), the gene mutated in 90% of patients with TCS (Wise *et al.*, 1997). Moreover, although these patients met the criteria for a diagnosis of DBA, none of the patients with craniofacial anomalies had *RPS19* mutations (Vlachos *et al.*, 2002). Thus, although not yet statistically significant, a possible genotype–phenotype correlation is suggested in which the "TCS-like" phenotype in DBA is not associated with mutations in *RPS19* (Vlachos *et al.*, 2002). Whether genes encoding other ribosomal proteins are affected in this subset of DBA patients is currently unknown.

TCOF1 encodes the nucleolar phosphoprotein treacle (Dixon *et al.*, 1997; So *et al.*, 2004; Wise *et al.*, 1997). Analogous to *RPS19* in DBA, TCS most likely results from treacle haploinsufficiency (Dixon, 1996).

Treacle has been reported to interact with upstream binding factor (UBF), a known transcription regulator of RNA polymerase I (Pol I), which participates in the initiating event in ribosomal DNA (rDNA) transcription, generating the 47S pre-rRNA (Valdez *et al.*, 2004). Treacle is also involved in other steps in ribosome synthesis, linking rDNA transcription to pre-rRNA posttranscriptional methylation (Gonzales *et al.*, 2005). Thus, treacle haploinsufficiency would inhibit multiple events in the ribosome biosynthetic pathway, having a profound effect upon ribosome synthesis. Comparable to proapoptotic erythropoiesis in DBA, the mechanism for the disruption of craniofacial structures in TCS appears to be increased apoptosis in the neural crest cells of the prefusion neural folds just prior to fusion during embryogenesis (Dixon *et al.*, 2000).

Here again we are left with some intriguing similarities and differences between DBA and another disease. This discussion must be qualified to indicate that it presupposes that the "TCS-like" DBA phenotype also results from a mutation resulting in dysfunctional ribosome biogenesis. Mutations in *RPS19* and *RPS24* interfere with the maturation of 40S subunits, whereas mutations in *TCOF1* have effects on Pol I transcription and rRNA methylation, the latter of which likely affects downstream processing events. Mutations in all three genes, again, interfere with the production of functional ribosomes. Yet only some of the clinical features of the diseases overlap and then not with all patients. In this regard, only about 30% of DBA patients have craniofacial abnormalities, many identical to those observed in TCS. Interestingly, the effects of *TCOF1* mutations in mice show that they are highly dependent on genetic background indicating a powerful influence of modifier genes on *TCOF1* phenotype. We need to consider if these background effects are related to genes controlling the expression of structural components of both the ribosome and the extraribosomal factors involved in ribosome synthesis. The tissue-specific influences of these genes on the expression of genes involved in ribosome synthesis could account for tissue specificity, whereas polymorphisms in these controlling genes could potentially explain how genetic background affects the penetrance of disease-related mutations.

4. Ribosome Dysfunction and Red Blood Cell Development

Conventional wisdom describes selective red blood cell hypoplasia as the defining characteristic of DBA, which seems at odds with the fact that ribosomes are a ubiquitous feature of all cell types with the exception of mature erythrocytes. However, this uniquely hematologic perspective disregards the existence of growth retardation and other congenital anomalies in DBA patients. Furthermore, in rare instances other significant hematologic cytopenias are also observed. How then does haploinsufficiency for Rps19, Rps24, and Rps17 manifest clinically by selectively affecting only certain tissues and most evidently red blood cell production? It has been argued that the high demand for ribosome synthesis associated with the proliferation and differentiation of red blood cell precursors may make these precursors unusually sensitive to the effects of a reduction in ribosome synthesis (Liu and Ellis, 2006; Morimoto *et al.*, 2007; Ohene-Abuakwa *et al.*, 2005). It is important to consider this point of view from the perspective of distinct features of red blood cell development between the fetus, the neonatal period, and the transition to adult erythropoiesis. Red blood cell production in rapidly growing fetuses in the third trimester is reported to be approximately three to five times that in the adult steady state (Palis and Segel, 1998). If DBA is solely a consequence of the unusually high demands for ribosome synthesis in red blood cell progenitors, one would predict DBA to manifest initially during fetal development. Although there are instances of early fetal loss (perhaps attributable to erythroid failure) and *hydrops fetalis* in DBA, the median age at presentation of classical DBA is 8 weeks (range, birth to 6 years) with 93% of DBA patients presenting during the first year of life.

A number of developmental and physiological changes conspire to decrease red blood cell production shortly after birth. After birth, erythropoietin production decreases in response to high partial pressures of oxygen, a high hemoglobin level, and the switch to the lower O_2-affinity adult hemoglobin allowing more oxygen delivery to tissues (Palis and Segel, 1998). Because erythropoietin provides proliferative, survival, and differentiation signals to erythroid progenitors, the fall in its production with birth contributes to a diminished erythron. There is also evidence to suggest that erythroid progenitors become less sensitive to erythropoietin in the transition from the fetal/neonatal state to the adult state (Weinberg *et al.*, 1992). Together, these changes lead to a transient physiological anemia at 4–8 weeks after birth until a new steady state of red blood cell production is reached. The failure to reach a new steady state would result in an anemic presentation of DBA at 8 weeks, the time when new red blood cell production is required. An important question relevant to DBA pathophysiology is, after 8 weeks of postnatal life, what effect these changes in developmental and physiological signals have on ribosome synthesis and whether, as conditions change beyond this 8 week time point, a state is triggered where mutational inactivation of an allele of either *RPS19*, *RPS24*, or *RPS17* limits ribosome synthesis in erythroid progenitors. Ribosome synthesis defects, in turn, could result in p53-dependent cell cycle arrest and enhanced apoptosis similar to that reported for activated T cells (Sulic *et al.*, 2005).

The analogy between DBA and the results obtained by Sulic *et al.* (2005) on T-cell development is limited by the fact that in the latter study haploinsufficiency was induced in a single-cell type through genetic engineering whereas in DBA the affected genes are found in all nucleated cells. Yet DBA manifests primarily as a red blood cell aplasia with growth retardation and select birth defects. To account for this tissue specificity, it has been hypothesized that ribosomal proteins are expressed in amounts differing relative to one another in a tissue-specific manner, and that haploinsufficiency for a particular protein may make that protein limiting for ribosome assembly in some tissues and not others (Ellis and Massey, 2006). Whether a ribosomal protein would limit assembly when an allele is mutationally inactivated would be dependent on the level of the protein relative to the factor that limits ribosome assembly under normal conditions. The amount of a ribosomal protein relative to the factor-limiting assembly could in principle differ from one cell type to another and within a particular cell type at different stages of development. Consequently, if a ribosomal protein was expressed at levels comparable to that of the limiting factor in a particular cell type under normal conditions, diminished expression linked to mutational inactivation of one allele could decrease expression to the point where the ribosomal protein now becomes limiting for assembly.

Under normal conditions, synthesis of the primary pre-rRNA transcript is thought to be the major factor defining ribosome levels within cells (Mayer and Grummt, 2006). Limiting ribosome synthesis at the level of pre-rRNA synthesis avoids the accumulation of partially assembled intermediates in the subunit maturation pathway that have been shown to cause nucleolar stress and activate p53 checkpoints (Pestov *et al.*, 2001). Such partially assembled intermediates have been shown to accumulate in cells where ribosomal proteins limit assembly (Choesmel *et al.*, 2007; Flygare *et al.*, 2007; Idol *et al.*, 2007). Related mechanisms could explain the p53-mediated effects on cell cycle progression and apoptosis reported by Sulic *et al.* (2005) in T cells.

To account for the timing of DBA presentation, we would propose that under the conditions of rapid growth found in the third trimester fetus, ribosomal proteins are expressed at high enough levels relative to pre-rRNA that even with the mutational inactivation of one allele they do not become limiting for ribosome assembly. After birth, when the demand for red blood cell production decreases, the demand for ribosome synthesis also likely decreases. Then upon the reestablishment of an erythroid drive and the establishment of a new steady state of ribosome synthesis and red blood cell production, the presence of haploinsufficiency for critical ribosomal proteins becomes manifest. Whether all of the factors involved in ribosome synthesis decrease proportionally from fetal levels to this new steady state is unknown but would seem unlikely given the involvement of somewhere between 200 and 300 gene products in the synthesis of functional ribosomes. Thus, in the new steady state defined after birth, the expression of certain ribosomal proteins relative to pre-rRNA may change to such an extent that mutational inactivation of a ribosomal protein allele could now limit ribosome assembly with ensuing effects on cell proliferation and enhanced apoptosis (Fig. 8.2). The fact that the level of some ribosomal proteins may change relative to pre-rRNA and others may not would indicate that only a subset of ribosomal protein genes when mutated could give rise to DBA. We can therefore surmise that it is the level of expression of individual ribosomal proteins relative to factors that determine the number of ribosomes per cell that determines the consequence of germ line haploinsufficiency in a particular cell type. Factors that contribute to first and second trimester fetal loss and a prenatal presentation of anemia (*hydrops fetalis*) as well as the remission of anemia in approximately 20% of cases (Lipton *et al.*, 2006) will need to be elucidated to validate this hypothesis.

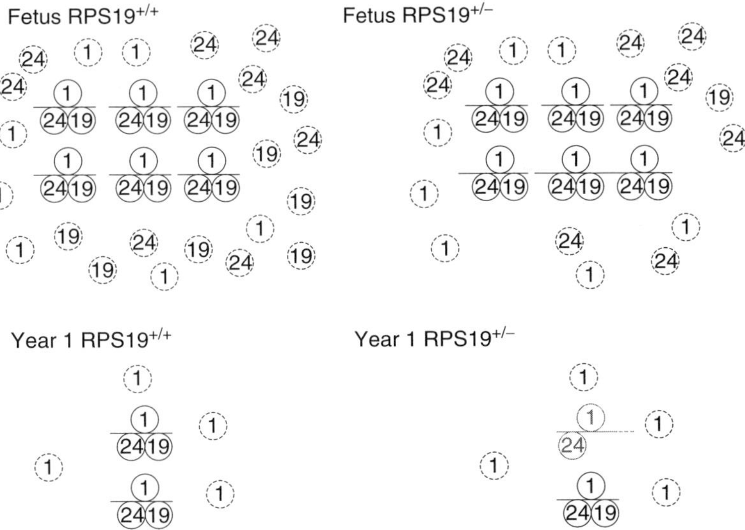

Figure 8.2 Proposed mechanism for developmental timing of *RPS19* haploinsufficiency. Ribosomal proteins are shown as numbered circles with only a small number of ribosomal proteins shown. Dashed circles represent rapidly degraded proteins synthesized in excess of that needed for ribosome assembly. Circles with filled numbers are proteins assembled into ribosomal subunits that have relatively long half-lives. Lines represent rRNA. The pre-40S precursor that accumulates when Rps19 becomes limiting for assembly is shown in red. Developmental state and *RPS19* genotypes are listed for each diagram. Based on its level of expression relative to pre-rRNA at year 1, *RPS1* would presumably not be a DBA gene.

5. DBA Is a Cancer Predisposition Syndrome

At least 26 cases of cancer in patients with DBA have been reported in the literature. Sixteen were hematopoietic malignancies. Of these, 10 were cases of acute myeloid leukemia (AML), 2 myelodysplastic syndrome (MDS), 2 Hodgkin disease, 1 non-Hodgkin lymphoma, and 1 acute lymphoblastic leukemia. Ten solid tumors have been reported; three osteogenic sarcoma, two breast cancer, two hepatocellular carcinoma, and one each of gastric carcinoma, vaginal melanoma, and malignant fibrous histiocytoma (Lipton *et al.*, 2001). A number of cases have been reported from international registries and a large institutional cohort (Janov *et al.*, 1996; Willig *et al.*, 1999). Of the 420 patients registered in the DBAR at the time of the most recent analysis, there were 8 patients who were found to have solid tumors, hematopoietic malignancies, or MDS. Some of these patients also appear as case reports in the literature. Three patients were diagnosed with osteogenic sarcoma (total of five known cases) and one each with AML, MDS, and myelofibrosis with myeloid metaplasia, colon cancer, and soft tissue sarcoma. There is an individual who is a DBA-affected relative of a registered patient with melanoma, who is not yet enrolled in the DBAR. Although cancer is relatively rare in DBA as compared to Fanconi anemia, the incidence appears to be well in excess of what would be expected for the age group represented. In particular, cases of breast and colon cancers have been described in very young adults (Lipton *et al.*, 2001; Willig *et al.*, 1999). The prognosis for patients with DBA and cancer is extremely poor. This is due in part from severe chemotherapy-induced myelosuppression (Lipton *et al.*, 2001). The presence of these malignancies and the young age at diagnosis of many of these cancers appears to define DBA as a cancer predisposition syndrome. However, before a true cancer incidence is determined, data from international DBA registries will need to undergo a careful analysis as has been applied to Fanconi anemia (Rosenberg *et al.*, 2003) and severe congenital neutropenia (Rosenberg *et al.*, 2005).

A mechanism leading to AML and MDS in Fanconi anemia has been postulated (Lensch *et al.*, 1999). In this model, the existence of pro-apoptotic hematopoiesis exerts a selective pressure on the myeloid compartment. This results in the emergence of clones with apoptosis-contravening mutations, resulting in leukemia. It seems reasonable to suggest that a similar mechanism could exist for DBA (Fig. 8.3). Several recent studies, including work already cited, have shown that nucleolar stress can induce p53-dependent cell cycle arrest and/or apoptosis (Lindstrom *et al.*, 2007; Pestov *et al.*, 2001). Mutations in the p53 gene and *MDM2* could potentially subvert this process providing a growth advantage to clones harboring p53 mutations, which at the same time

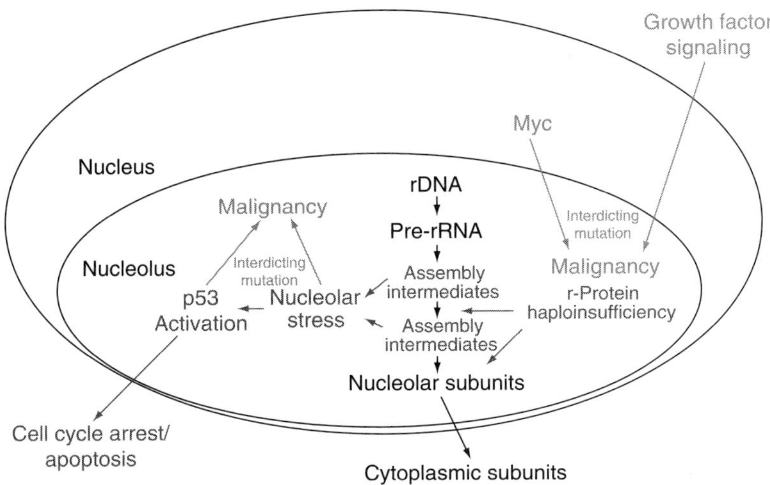

Figure 8.3 Proposed mechanism for cancer predisposition in DBA. Ribosomal protein haploinsufficiency results in the accumulation of partially assembly intermediates that result in nucleolar stress signaling. Nucleolar stress, in turn, leads to p53 activation and either cell cycle arrest or apoptosis. These steps are outlined in red. Interdicting mutations that subvert the p53 pathway and favor the outgrowth of malignant clones are shown in green. Also shown in green are interdicting mutations that could enhance ribosomal protein synthesis and diminish nucleolar stress but because of the oncogenic nature of the genes involved could again favor the outgrowth of malignant clones. (See Color Insert.)

could favor malignant outgrowth. Other mechanisms also exist that could explain the increased cancer incidence in DBA patients. Because DBA patients characterized to date have one remaining active ribosomal gene, mutations that enhance ribosomal protein gene expression could potentially compensate for the inactive allele. The oncogene, c-Myc, is known to upregulate many components of the translational machinery in its role in stimulating cell growth and proliferation (Boon *et al.*, 2001; Oskarsson and Trumpp, 2005). Activating mutations in c-Myc could therefore compensate for a ribosomal protein deficiency allowing for the emergence of clones with a survival advantage predisposed to malignancy. As many signaling pathways promoting cell growth and division regulate ribosome synthesis as a downstream target, numerous other targets also exist to explain the increased cancer incidence in DBA patients (Ruggero and Pandolfi, 2003). Finally, heterozygous mutations in 11 different ribosomal protein genes in zebra fish have recently been found in fish with peripheral nerve sheath tumors (Amsterdam *et al.*, 2004). These dominantly inherited mutations appear to result from loss of function, raising the specter that under certain circumstances ribosomal protein genes may act as tumor suppressor loci.

6. Treatment and Outcomes

Since 1938 when first recognized as a distinct clinical entity, treatment strategies for DBA have changed only as the consequence of the introduction of corticosteroids in 1951 and a developing understanding of steroid toxicity, ongoing improvements in blood product safety, the introduction of and improvements in the administration of iron chelation since the mid-1970s, and finally improvements in hematopoietic stem cell transplantation since the first successful transplants reported in 1968. As well advances in genetics and have made possible prevention in the form of newly available reproductive strategies. Although these developments have lead to major increments in both the quantity and the quality of life for patients with DBA, they are based on non-DBA-specific advances in medical knowledge. Thus, DBA shares with many disorders, perhaps most notably the hemoglobinopathies, the frustration that improvements in care do not derive from the burgeoning understanding of the molecular pathology of the disorder. The following sections briefly describe the current state of the art in the hope that readers will be inspired toward a more enlightened path.

6.1. Corticosteroids and red blood cell transfusions

Corticosteroids and red blood cell transfusions are the mainstays of therapy for DBA. Since 1951 (Gasser, 1951), it has been known that the anemia of DBA can be ameliorated by corticosteroids. It seems that the suggestion to use corticosteroid therapy was based on the theory that DBA was somehow an "allergic" disorder. The response to corticosteroids further perpetuated the erroneous notion of DBA as an autoimmune disease, even when only miniscule doses were required to maintain adequate erythropoiesis. Corticosteroids do not appear to be efficacious through increasing the expression of ribosomal proteins, thereby compensating for haploinsufficiency (Ebert *et al.*, 2005). Thus, despite the widespread use of corticosteroids in treating DBA, their mechanism of action remains obscure.

Clinically the almost unlimited use of corticosteroids, even when toxic doses were required, continued into the early twenty-first century. This appears to be due largely to the persistent fear, in patients and physicians, of transfusion-acquired HIV and hepatitis C as well as the difficulty encountered in the almost daily subcutaneous administration of the iron chelator, deferoxamine (Desferal®, Novartis) (Olivieri and Brittenham, 1997). Data from international registries have clarified the efficacy and toxicity of corticosteroid therapy. Data from the DBAR are representative (Lipton *et al.*, 2006). As has been reported in the literature, 79% of patients are initially responsive to steroids, 17% were nonresponsive, and 4% were never treated with steroids. Somewhat unpredicted was the high incidence of significant untoward effects. Not surprising, nearly half of the patients had developed cushingoid features. However, with the finding that 22% and 12% of patients ever treated with steroids developed pathologic fractures and cataracts, respectively, the use of steroids is being modified. Currently, 37% of the patients are receiving corticosteroids and 31% are receiving red blood cell transfusions. Of the transfusion-dependent patients, 35% were never steroid responsive, 22% became steroid refractory over time, and 33% could not be weaned to an acceptable dose. Five percent never received steroid therapy and 5% are being transfused for unknown reasons. Although too soon to statistically validate, there appears to be a recent increase in the number of patients judged as "unweanable" as well as an increase in patients who never received steroids. The ability to wean a patient is dependent to some extent upon the side effects that the patient and the physician are willing to tolerate. In general, a corticosteroid dose-equivalent of 0.5 mg/kg/day of prednisone is suggested as a maximum "maintenance" dose after an initial dose of 2 mg/kg/day (Vlachos *et al.*, in preparation). Patients who fail to respond within a month are considered steroid refractory. Clearly the recognition that the blood supply is safe combined with the recent availability of the oral iron chelator Deferasirox (ICL670, Exjade®, Novartis) (Nisbet-Brown *et al.*, 2003) has allowed for the more liberal use of packed red blood cell transfusions in patients rather than toxic doses of corticosteroids. The number of patients never on steroids has increased, reflecting the older and safer age at which steroid treatment is being commenced (Lipton *et al.*, 2006).

Although remissions had been reported in DBA, data on the fraction of patients who were able to sustain erythropoiesis for over 6 months without treatment has been sparse. The DBAR has provided some actuarial data (Lipton *et al.*, 2006). About 20% of patients will enter remission, the majority sustained. Of these, about 75% do so prior to their 10th birthday. These data may reflect the bias to a younger age of patients in the DBAR as anecdotal observations suggest a surge in remissions in adolescent males and occasional remissions in older patients. Proportionally an equal number of patients remit from both transfusion and steroid therapy. However, there are very few remitters who had never responded to corticosteroids. Pregnancy and the use of birth control pills contribute to relapse (Alter *et al.*, 1999). It is possible that many of the same mechanisms offered for cancer predisposition in DBA patients could also factor into mechanisms of remission or vice versa. Clearly more studies are needed on this important subject.

6.2. Hematopoietic stem cell transplant

Although curative in DBA, hematopoietic stem cell transplantation remains the most controversial therapy (Alter, 1998; Vlachos *et al.*, 2001b; Vlachos *et al.*, 2005; Willig *et al.*, 1999). A recent series from the IBMTR and a compilation from the literature are consistent with findings from the DBAR

(Roy *et al.*, 2005; Vlachos *et al.*, 2001b). In this database, 36 patients have undergone stem cell transplant (SCT), 21 HLA-matched related, and 15 alternative donor SCT. The major indication for SCT was transfusion dependence. In addition, two patients developed severe aplastic anemia and one had significant thrombocytopenia. Two patients transplanted using a nonmyeloablative conditioning regimen, one receiving a matched related umbilical cord and one receiving unrelated bone marrow, are alive and well. The majority of alternative donor transplants were performed using total body irradiation for conditioning whereas busulfan/cyclophosphamide containing regimens were typical of the matched-related transplants. Sixteen of the 21 HLA-matched sibling donor transplants are alive and red blood cell transfusion independent. Of the 15 alternative donor SCT, 2 patients received mismatched related bone marrow, 4 patients unrelated cord blood, 8 unrelated bone marrow, and 1 unrelated peripheral blood stem cells. Four of these 15 patients are alive. Of the 16 deaths, 15 were related to infection, graft versus host disease, and/or veno-occlusive disease of the liver with only 1 death, in the alternative donor group, occurring as a consequence of graft failure. In contrast to Fanconi anemia, DC, and SDS, these deaths do not appear to be the consequence of intolerance to typical transplant conditioning regimens. The survival for allogeneic sibling versus alternative donor transplant is 72.7% ± 10.7% versus 19.1% ± 11.9% at greater than 5 years from SCT ($p = 0.01$) or 17.1% ± 10.8% (including a patient diagnosed with osteogenic sarcoma posttransplant, $p = 0.012$). The survival rate for patients less than age of 10 years receiving matched related SCT is greater than 90%. This success has led to an increase in families looking to preimplantation genetic diagnosis with *in vitro* fertilization to "create" HLA-matched, non-*RPS19*-mutated sibling donors. A number of patients worldwide have been successfully transplanted using umbilical cord-derived stem cells from donors produced in this way. A discussion of the complicated religious, ethical, and economic questions generated by this approach is beyond the scope of this chapter (Kuliev *et al.*, 2005; Wagner *et al.*, 2004).

6.3. Outcomes

Of the 36 deaths reported to the DBAR, 25 (70%) are treatment-related: 5 from infections (2 *Pneumocystis jiroveci* pneumonia, 1 varicella pneumonia, 1 pseudomonas pneumonia/sepsis, 1 unknown infection), 5 from complications of iron overload, 1 from cardiac tamponade secondary to a vascular access device complication, and 14 from stem cell transplant complications. Only 3 of the deaths related to stem cell transplantation were in patients with life-threatening cytopenias. Just 8 deaths were directly related to DBA: 1 from severe aplastic anemia, 7 from malignancy (2 of whom also received a stem cell transplant), and 3 were from undetermined causes.

Thus, although SCT is potentially curative and reduces the risk of aplastic anemia, AML and MDS, the poor outcomes for unrelated donor SCT suggest restraint in the use of this modality. The probability of remission must also be taken in to account when high-risk SCT is being considered.

The overall actuarial survival at greater than 40 years of age is 75.1% ± 4.8; 86.7% ± 7.0% for corticosteroid-maintainable patients and 57.2% ± 8.9% for transfusion-dependent patients (Lipton *et al.*, 2006). There is a trend to increased survival in patients with sustained remission versus corticosteroid-maintainable patients ($p = 0.08$) and a statistically significant survival advantage for steroid-maintainable patients as compared to transfusion-dependent patients ($p = 0.007$). Seven transfusion-dependent patients died as a consequence of stem cell transplant-related complications.

7. Summary

The data summarized in this chapter define DBA as a developmental disorder. The timing of DBA presentation in most patients in the first year of life strongly suggests that the changes in red blood cell production that accompany birth trigger a pathophysiological mechanism, likely linked to defective ribosome synthesis, which precipitates the clinical phenotype. As with many features of DBA, this timing is not absolute with some patients presenting *in utero* and others later in life. The response to corticosteroids and a fairly high remission rate remain unexplained. DBA is not solely a disease of red blood cell development as witnessed by the heterogeneous array of congenital anomalies found in many patients. The heterogeneous nature of these anomalies together with the variation in timing of hematologic features of DBA indicates that the pathophysiological mechanisms underlying this disease are strongly influenced by modifier genes. Understanding the underlying mechanisms responsible for the developmental features of DBA is not only critical to identifying the underlying mechanisms for the disease itself leading to a more rationale treatment strategy but also for drawing further attention to the role of ribosome synthesis defects in a growing number of developmental disorders.

REFERENCES

Aase, J. M., and Smith, D. (1969). Congenital anemia and triphalangeal thumbs: A new syndrome. *J. Pediatr.* **74,** 471–472.

Alter, B. P. (1978). Thumbs and Anemia. *Pediatrics* **62,** 613–614.

Alter, B. P. (1998). Bone marrow transplant in Diamond-Blackfan anemia. *Bone Marrow Transplant* **21,** 965–966.

Alter, B. P., Gaston, T., and Lipton, J. M. (1993). Lack of effect of corticosteroids in W/Wv and S1/S1d mice: These strains are not a model for steroid-responsive Diamond Blackfan anemia. *Eur. J. Haematol.* **50,** 275–278.

Alter, B. P., Kumar, M., Lockhart, L. L., Sprinz, P. G., and Rowe, T. F. (1999). Pregnancy in bone marrow failure syndromes: Diamond-Blackfan anaemia and Shwachman-Diamond syndrome. *Br. J. Haematol.* **107,** 49–54.

Amsterdam, A., Sadler, K. C., Lai, K., Farrington, S., Bronson, R. T., Lees, J. A., and Hopkins, N. (2004). Many ribosomal protein genes are cancer genes in zebrafish. *PLOS Biol.* **2,** 690–698.

Angelini, M., Cannata, S., Mercaldo, V., Gibello, L., Santoro, C., Dianzani, I., and Loreni, F. (2007). Missense mutations associated with Diamond-Blackfan anemia affect the assembly of ribosomal protein S19 into the ribosome. *Hum. Mol. Genet.* **16,** 1720–1727.

Austin, K. M., Leary, R. J., and Shimamura, A. (2005). The Shwachman-Diamond SBDS protein localizes to the nucleolus. *Blood* **106,** 1253–1258.

Balaban, E. P., Buchanan, G. R., Graham, M., and Frenkel, E. P. (1985). Diamond-Blackfan syndrome in adult patients. *Am. J. Med.* **78,** 533–538.

Bommer, U. A., Stahl, J., Henske, A., Lutsch, G., and Bielka, H. (1988). Identification of proteins of the 40 S ribosomal subunit involved in interaction with initiation factor eIF-2 in the quaternary initiation complex by means of monospecific antibodies. *FEBS Lett.* **233,** 114–118.

Boon, K., Caron, H. N., van Asperen, R., Valentijn, L., Hermus, M. C., van Sluis, P., Roobeek, I., Weis, I., Voute, P. A., Schwab, M., and Versteeg, R. (2001). N-myc enhances the expression of a large set of genes functioning in ribosome biogenesis and protein synthesis. *EMBO J.* **20,** 1383–1393.

Campagnoli, M. F., Garelli, E., Quarello, P., Carando, A., Varotto, S., Nobili, B., Longoni, D., Pecile, V., Zecca, M., Dufour, C., Ramenghi, U., and Dianzani, I. (2004). Molecular basis of Diamond-Blackfan anemia: New findings from the Italian registry and a review of the literature. *Haematologica* **89,** 480–489.

Cathie, I. A. B. (1950). Erythrogenesis imperfecta. *Arch. Dis.* **25,** 313–324.

Chan, H. S., Saunders, E. F., and Freedman, M. H. (1982). Diamond Blackfan anemia: Erythropoiesis in prednisone responsive and resistant disease. *Pediatr. Res.* **16,** 474–476.

Chiocchetti, A., Gibello, L., Carando, A., Aspesi, A., Secco, P., Garelli, E., Loreni, F., Angelini, M., Biava, A., Dahl, N., Dianzani, U., Ramenghi, U., *et al.* (2005). Interactions between RPS19, mutated in Diamond-Blackfan anemia, and the PIM-1 oncoprotein. *Haematol. Hematol. J.* **90,** 1453–1462.

Choesmel, V., Bacqueville, D., Rouquette, J., Noaillac-Depeyre, J., Fribourg, S., Cretien, A., Leblanc, T., Tchernia, G., Da Costa, L., and Gleizes, P. E. (2007). Impaired ribosome biogenesis in Diamond-Blackfan anemia. *Blood* **109,** 1275–1283.

Cmejla, R., Blafkova, J., Stopka, T., Zavadil, J., Pospisilova, D., Mihal, V., Petrtylova, K., and Jelinek, J. (2000). Ribosomal protein S19 gene mutations in patients with Diamond-Blackfan anemia and identification of ribosomal protein S19 pseudogenes. *Blood Cells Mol. Dis.* **26,** 124–132.

Cmejla, R., Cmejlova, J., Handrkova, H., Petrak, J., and Pospisilova, D. (2007). Ribosomal protein S17 is mutated in Diamond-Blackfan anemia. *Hum. Mutat.* **12,** 1178–1182.

Da Costa, L., Tchernia, G., Gascard, P., Lo, A., Meerpohl, J., Niemeyer, C., Chasis, J. A., Fixler, J., and Mohandas, N. (2003). Nucleolar localization of RPS19 protein in normal cells and mislocalization due to mutations in the nucleolar localization signals in 2 Diamond-Blackfan anemia patients: Potential insights into pathophysiology. *Blood* **101,** 5039–5045.

Diamond, L. K., and Blackfan, K. D. (1938). Hypoplastic anemia. *Am. J. Dis. Child.* **56,** 464–467.

Diamond, L. K., Wang, W. C., and Alter, B. P. (1976). Congenital hypoplastic anemia. *Adv. Pediatr.* **22,** 349–378.

Dixon, M. J. (1996). Treacher Collins syndrome. *Hum. Mol. Genet.* **5,** 1391–1396.

Dixon, J., Edwards, S. J., Anderson, I., Brass, A., Scambler, P. J., and Dixon, M. J. (1997). Identification of the complete coding sequence and genomic organization of the Treacher Collins syndrome gene. *Genome Res.* **7,** 223–234.

Dixon, J., Brakebusch, C., Fassler, R., and Dixon, M. J. (2000). Increased levels of apoptosis in the prefusion neural folds underlie the craniofacial disorder, Treacher Collins syndrome. *Hum. Mol. Genet.* **9,** 1473–1480.

Draptchinskaia, N., Gustavsson, P., Andersson, B., Pettersson, M., Willig, T. N., Dianzani, I., Ball, S., Tchernia, G., Klar, J., Matsson, H., Tentler, D., Mohandas, N., *et al.* (1999). The gene encoding ribosomal protein S19 is mutated in Diamond-Blackfan anaemia. *Nat. Genet.* **21,** 169–175.

Ebert, B. L., Lee, M. M., Pretz, J. L., Subramanian, A., Mak, R., Golub, T. R., and Sieff, C. A. (2005). An RNA interference model of RPS19 deficiency in Diamond-Blackfan anemia recapitulates defective hematopoiesis and rescue by dexamethasone: Identification of dexamethasone-responsive genes by microarray. *Blood* **105,** 4620–4626.

Ellis, S. R., and Massey, A. T. (2006). Diamond Blackfan anemia: A paradigm for a ribosome-based disease. *Med. Hypotheses* **66,** 643–648.

Ershler, W. B., Ross, J., Finlay, J. L., and Shahidi, N. T. (1980). Bone-marrow micro-environment defect in congenital hypoplastic-anemia. *N. Engl. J. Med.* **302,** 1321–1327.

Fazen, L. E., Elmore, J., and Nadler, H. L. (1967). Mandibulo-facial dystosis. (Treacher Collins syndrome. *Am. J. Dis. Child.* **113,** 405–410.

Ferreira-Cerca, S., Poll, G., Gleizes, P. E., Tschochner, H., and Milkereit, P. (2005). Roles of eukaryotic ribosomal proteins in maturation and transport of pre-18S rRNA and ribosome function. *Mol. Cell* **20,** 263–275.

Flygare, J., Aspesi, A., Bailey, J. C., Miyake, K., Caffrey, J. M., Karlsson, S., and Ellis, S. R. (2007). Human RPS19, the gene mutated in Diamond-Blackfan anemia, encodes a ribosomal protein required for the maturation of 40S ribosomal subunits. *Blood* **109,** 980–986.

Freedman, M. H., Ainata, D., and Saunders, E. F. (1976). Erythroid colony formation in congenital hypoplastic anemia. *J. Clin. Invest.* **57,** 673–677.

Ganapathi, K. A., Austin, K. M., Lee, C. S., Dias, A., Malsch, M. M., Reed, R., and Shimamura, A. (2007). The human Shwachman-Diamond syndrome protein, SBDS, associates with ribosomal RNA. *Blood* **110**(5), 1458–1465.

Gasser, C. (1951). Aplastische anamie (chronische erythroblastophthise) und cortison. *Schweiz. Med. Wochenschr.* **81,** 1241.

Gazda, H. T., Zhong, R., Long, L., Niewiadomska, E., Lipton, J. M., Ploszynska, A., Zaucha, J. M., Vlachos, A., Atsidaftos, E., Viskochil, D. H., Niemeyer, C. M., Meerpohl, J. J., *et al.* (2004). RNA and protein evidence for haplo-insufficiency in Diamond-Blackfan anaemia patients with RPS19 mutations. *Br. J. Haematol.* **127,** 105–113.

Gazda, H. T., Grabowska, A., Merida-Long, L. B., Latawiec, E., Schneider, H. E., Lipton, J. M., Vlachos, A., Atsidaftos, E., Ball, S. E., Orfali, K. A., Niewiadomska, E., Da Costa, L., *et al.* (2006). Ribosomal protein S24 gene is mutated in Diamond-Blackfan anemia. *Am. J. Hum. Genet.* **79,** 1110–1118.

Glader, B. E., Backer, K., and Diamond, L. K. (1983). Elevated erythrocyte adenosine deaminase activity in congenital hypoplastic anemia. *New Engl. J. Med.* **309,** 1486–1490.

Gonzales, B., Henning, D., So, R. B., Dixon, J., Dixon, M. J., and Valdez, B. C. (2005). The Treacher Collins syndrome (TCOF1) gene product is involved in pre-rRNA methylation. *Hum. Mol. Genet.* **14,** 2035–2043.

Gripp, K. W., Donald-McGinn, D. M., La Rossa, D., McGain, D., Federman, N., Vlachos, A., Glader, B. E., McKenzie, S. E., Lipton, J. M., and Zackai, E. H. (2001). Bilateral microtia and cleft palate in cousins with Diamond-Blackfan anemia. *Am. J. Med. Genet.* **101,** 268–274.

Gustavsson, P., Willig, T. N., vanHaeringen, A., Tchernia, G., Dianzani, I., Donner, M., Elinder, G., Henter, J. I., Nilsson, P. G., Gordon, L., Skeppner, G., vantVeerKorthof, L., *et al.* (1997). Diamond-Blackfan anaemia: Genetic homogeneity for a gene on chromosome 19q13 restricted to 1.8 Mb. *Nat. Genet.* **16,** 368–371.

Hasan, R., and Inoue, S. (1993). Diamond-Blackfan anemia associated with Treacher-Collins syndrome. *Pediatr. Hematol. Oncol.* **10,** 261–265.

Hoffman, R., Zanjani, E. D., Vila, J., Zalusky, R., Lutton, J. D., and Wasserman, L. R. (1976). Diamond-Blackfan syndrome—lymphocyte-mediated suppression of erythropoiesis. *Science* **193,** 899–900.

Idol, R. A., Robledo, S., Du, H. Y., Crimmins, D. L., Wilson, D. B., Ladenson, J. H., Bessler, M., and Mason, P. J. (2007). Cells depleted for RPS19, a protein associated with Diamond Blackfan anemia, show defects in 18S ribosomal RNA synthesis and small ribosomal subunit production. *Blood Cells Mol. Dis.* **39,** 35–43.

Janov, A. J., Leong, T., Nathan, D. G., and Guinan, E. C. (1996). Diamond-Blackfan anemia—natural history and sequelae of treatment. *Medicine* **75,** 77–87.

Josephs, H. W. (1936). Anemia of infancy and early childhood. *Medicine* **15,** 307–451.

Kuliev, A., Rechitsky, S., Tur-Kaspa, I., and Verlinsky, Y. (2005). Preimplantation genetics—improving access to stem cell therapy. *Cooley'S Anemia Eighth Symposium* **1054,** 223–227.

Lam, Y. W., Lamond, A. I., Mann, M., and Andersen, J. S. (2007). Analysis of nucleolar protein dynamics reveals the nuclear degradation of ribosomal proteins. *Curr. Biol.* **17,** 749–760.

Leger-Silvestre, I., Caffrey, J. M., Dawaliby, R., varez-Arias, D. A., Gas, N., Bertolone, S. J., Gleizes, P. E., and Ellis, S. R. (2005). Specific role for yeast homologs of the Diamond Blackfan anemia-associated Rps19 protein in ribosome synthesis. *J. Biol. Chem.* **280,** 38177–38185.

Lensch, M. W., Rathbun, R. K., Olson, S. B., Jones, G. R., and Bagby, G. C. (1999). Selective pressure as an essential force in molecular evolution of myeloid leukemic clones: A view from the window of Fanconi anemia. *Leukemia* **13,** 1784–1789.

Lindstrom, M. S., Jin, A. W., Deisenroth, C., Wolf, G. W., and Zhang, Y. P. (2007). Cancer-associated mutations in the MDM2 zinc finger domain disrupt ribosomal protein interaction and attenuate MDM2-induced p53 degradation. *Mol. Cell. Biol.* **27,** 1056–1068.

Lipton, J. M., and Alter, B. P. (1993). Diamond Blackfan anemia. *In* "Clinical Disorders and Experimental Models of Erythropoietic Failure." (S. A. Feig, and M. H. Freedman, Eds.), pp. 39–67. CRC Press, Boca Raton, FL.

Lipton, J. M., Kudisch, M., Gross, R., and Nathan, D. G. (1986). Defective erythroid progenitor differentiation system in congenital hypoplastic (Diamond-Blackfan) anemia. *Blood* **67,** 962–968.

Lipton, J. M., Federman, N., Khabbaze, Y., Schwartz, C. L., Hilliard, L. M., Clark, J. I., and Vlachos, A. (2001). Osteogenic sarcoma associated with Diamond-Blackfan anemia: A report from the Diamond-Blackfan Anemia Registry. *J. Pediatr. Hematol. Oncol.* **23,** 39–44.

Lipton, J. M., Atsidaftos, E., Zyskind, I., and Vlachos, A. (2006). Improving clinical care and elucidating the pathophysiology of Diamond Blackfan anemia: An update from the Diamond Blackfan Anemia Registry. *Pedriatr. Blood Cancer* **46,** 558–564.

Liu, J. M., and Ellis, S. R. (2006). Ribosomes and marrow failure: Coincidental association or molecular paradigm?. *Blood* **107,** 4583–4588.

Maeda, N., Toku, S., Kemmochi, N., and Tanaka, T. (2006). A novel nucleolar protein interacts with ribosomal protein S19. *Biochem. Biophys. Res. Commun.* **339,** 41–46.

Matsson, H., Davey, E. J., Draptchinskaia, N., Hamaguchi, I., Ooka, A., Leveen, P., Forsberg, E., Karlsson, S., and Dahl, N. (2004). Targeted disruption of the ribosomal protein S19 gene is lethal prior to implantation. *Mol. Cell. Biol.* **24,** 4032–4037.

Mayer, C., and Grummt, I. (2006). Ribosome biogenesis and cell growth: mTOR coordinates transcription by all three classes of nuclear RNA polymerases. *Oncogene* **25,** 6384–6391.

Menne, T. F., Goyenechea, B., Sanchez-Puig, N., Wong, C. C., Tonkin, L. M., Ancliff, P. J., Brost, R. L., Costanzo, M., Boone, C., and Warren, A. J. (2007). The Shwachman-Bodian-Diamond syndrome protein mediates translational activation of ribosomes in yeast. *Nat. Genet.* **39,** 486–495.

Morimoto, K., Lin, S., and Sakamoto, K. (2007). The functions of RPS19 and their relationship to Diamond-Blackfan anemia: A review. *Mol. Genet. Metab.* **90,** 358–362.

Nathan, D. G., Clarke, B. J., Hillman, D. G., Alter, B. P., and Houseman, D. E. (1978). Erythroid precursors in congenital hypoplastic (Diamond Blackfan) anemia. *J. Clin. Invest.* **61,** 489–498.

Nisbet-Brown, E., Olivieri, N. F., Giardina, P. J., Grady, R. W., Neufeld, E. J., Sechaud, R., Krebs-Brown, A. J., Anderson, J. R., Alberti, D., Sizer, K. C., and Nathan, D. G. (2003). Effectiveness and safety of ICL670 in iron-loaded patients with thalassaemia: A randomised, double-blind, placebo-controlled, dose-escalation trial. *Lancet* **361,** 1597–1602.

Ohene-Abuakwa, Y., Orfali, K. A., Marius, C., and Ball, S. E. (2005). Two-phase culture in Diamond Blackfan anemia: Localization of erythroid defect. *Blood* **105,** 838–846.

Ohga, S., Mugishima, H., Ohara, A., Kojima, S., Fujisawa, K., Yagi, K., Higashigawa, M., and Tsukimoto, I. (2004). Diamond-Blackfan anemia in Japan: Clinical outcomes of prednisolone therapy and hematopoietic stem cell transplantation. *Int. J. Hematol.* **79,** 22–30.

Olivieri, N. F., and Brittenham, G. M. (1997). Iron-chelating therapy and the treatment of thalassemia. *Blood* **89,** 739–761.

Orfali, K. A., Ohene-Abuakwa, Y., and Ball, S. E. (2004). Diamond Blackfan anaemia in the UK: Clinical and genetic heterogeneity. *Br. J. Haematol.* **125,** 243–252.

Ortega, J. A., Shore, N. A., Dukes, P. P., and Hammond, D. (1975). Congenital hypoplastic anemia inhibition of erythropoiesis by sera from patients with congenital hypoplastic anemia. *Blood* **45,** 83–89.

Oskarsson, T., and Trumpp, A. (2005). The Myc trilogy: Lord of RNA polymerases. *Nat. Cell Biol.* **7,** 215–217.

Palis, J., and Segel, G. B. (1998). Developmental biology of erythropoiesis. *Blood Rev.* **12,** 106–114.

Perdahl, E. B., Naprstek, B. L., Wallace, W. C., and Lipton, J. M. (1994). Erythroid failure in Diamond-Blackfan anemia is characterized by apoptosis. *Blood* **83,** 645–650.

Pestov, D. G., Strezoska, Z., and Lau, L. F. (2001). Evidence of p53-dependent cross-talk between ribosome biogenesis and the cell cycle: Effects of nucleolar protein Bop1 on G1/S transition. *Mol. Cell. Biol.* **21,** 4246–4255.

Rosenberg, P. S., Greene, M. H., and Alter, B. P. (2003). Cancer incidence in persons with Fanconi anemia. *Blood* **101,** 822–826.

Rosenberg, P. S., Alter, B. P., Bolyard, A. A., Bonilla, M. A., Boxer, L. A., Cham, B., Fier, C., Freedman, M., Kannourakis, G., Kinsey, S., Schwinzer, B., Zeidler, C., *et al.* (2006). The incidence of leukemia and mortality from sepsis in patients with severe congenital neutropenia receiving long-term G-CSF therapy. *Blood* **107,** 4628–4635.

Roy, V., Perez, W. S., Eapen, M., Marsh, J. C. W., Pasquini, M., Pasquini, R., Mustafa, M. M., and Bredeson, C. N. (2005). Bone marrow transplantation for Diamond-Blackfan anemia. *Biol. Blood Marrow Transplant.* **11,** 600–608.

Ruggero, D., and Pandolfi, P. P. (2003). Does the ribosome translate cancer? *Nat. Rev. Cancer* **3,** 179–192.

Sawada, K., Koyanagawa, Y., Sakurama, S., Nakagawa, S., and Konno, T. (1985). Diamond-Blackfan syndrome—a possible role of cellular factors for erythropoietic suppression. *Scand. J. Haematol.* **35,** 158–165.

Shimamura, A. (2006). Shwachman-diamond syndrome. *Semin. Hematol* **43,** 178–188.

So, R. B., Gonzales, B., Henning, D., Dixon, J., Dixon, M. J., and Valdez, B. C. (2004). Another face of the Treacher Collins syndrome (TCOF1) gene: Identification of additional exons. *Gene* **328,** 49–57.

Soulet, F., Al Saati, T., Roga, S., Amalric, F., and Bouche, G. (2001). Fibroblast growth factor-2 interacts with free ribosomal protein S19. *Biochem. Biophys. Res. Commun.* **289,** 591–596.

Sulic, S., Panic, L., Barkic, M., Mercep, M., Uzelac, M., and Volarevic, S. (2005). Inactivation of S6 ribosomal protein gene in T lymphocytes activates a p53-dependent checkpoint response. *Genes Dev.* **19,** 3070–3082.

Teber, O. A., Gillessen-Kaesbach, G., Fischer, S., Bohringer, S., Albrecht, B., Albert, A., Rslan-Kirchner, M., Haan, E., Hagedorn-Greiwe, M., Hammans, C., Henn, W., Hinkel, G. K., *et al.* (2004). Genotyping in 46 patients with tentative diagnosis of Treacher Collins syndrome revealed unexpected phenotypic variation. *Eur. J. Hum. Genet.* **12,** 879–890.

Thiel, C. T., Horn, D., Zabel, B., Ekici, A. B., Salinas, K., Gebhart, E., Ruschendorf, F., Sticht, H., Spranger, J., Muller, D., Zweier, C., Schmitt, M. E., *et al.* (2005). Severely incapacitating mutations in patients with extreme short stature identify RNA-processing endoribonuclease RMRP as an essential cell growth regulator. *Am. J. Hum. Genet.* **77,** 795–806.

Tsai, P. H., Arkin, S., and Lipton, J. M. (1989). An intrinsic progenitor defect in Diamond-Blackfan anemia. *Br. J. Haematol.* **73,** 112–120.

Valdez, B. C., Henning, D., So, R. B., Dixon, J., and Dixon, M. J. (2004). The Treacher Collins syndrome (TCOF1) gene product is involved in ribosomal DNA gene transcription by interacting with upstream binding factor. *Proc. Natl. Acad. Sci. USA* **101,** 10709–10714.

Vlachos, A., Klein, G. W., and Lipton, J. M. (2001a). The Diamond Blackfan anemia registry: Tool for investigating the epidemiology and biology of Diamond-Blackfan anemia. *J. Pediatr. Hematol. Oncol.* **23,** 377–382.

Vlachos, A., Federman, N., Reyes-Haley, C., Abramson, J., and Lipton, J. M. (2001b). Hematopoietic stem cell transplantation for Diamond Blackfan anemia: A report from the Diamond Blackfan anemia registry. *Bone Marrow Transplant.* **27,** 381–386.

Vlachos, A., Klein, G. W., Avarello, J. T., Rivera, R. D., Atsidaftos, E., Gripp, K. W., McKenzie, S., and Lipton, J. M. (2002). Diamond Blackfan Anemia (DBA) and associated orofacial clefts (OFC): A Diamond Blackfan Anemia Registry (DBAR) analysis. *Pediatr. Res.* **51,** 245A.

Vlachos, A., Atsidaftos, E., and Lipton, J. M. (2005). Matched related vs alternative donor stem cell tranplantation (SCT) for Diamond Blackfan anemia: A report from the Diamond Blackfan anemia registry. *Blood* **106,** 583A–584A.

Vulliamy, T., and Dokal, I. (2006). Dyskeratosis congenita. *Semin. Hematol.* **43,** 157–166.

Wagner, J. E., Kahn, J. P., Wolf, S. M., and Lipton, J. M. (2004). Preimplantation testing to produce an HLA-matched donor infant. *JAMA* **292,** 803–804.

Weinberg, R. S., He, L., and Alter, B. P. (1992). Erythropoiesis is distinct at each stage of ontogeny. *Pediatr. Res.* **31,** 170–175.

Willig, T. N., Niemeyer, C. M., Leblanc, T., Tiemann, C., Robert, A., Budde, J., Lambiliotte, A., Kohne, E., Souillet, G., Eber, S., Stephan, J. L., Girot, R., *et al.* (1999). Identification of new prognosis factors from the clinical and epidemiologic analysis of a registry of 229 Diamond-Blackfan anemia patients. *Pediatr. Res.* **46,** 553–561.

Wise, C. A., Chiang, L. C., Paznekas, W. A., Sharma, M., Musy, M. M., Ashley, J. A., Lovett, M., and Jabs, E. W. (1997). TCOF1 gene encodes a putative nucleolar phosphoprotein that exhibits mutations in Treacher Collins syndrome throughout its coding region. *Proc. Natl. Acad. Sci. USA* **94,** 3110–3115.

Yamamoto, T. (2000). Molecular mechanism of monocyte predominant infiltration in chronic inflammation: Mediation by a novel monocyte chemotactic factor, S19 ribosomal protein dimer. *Pathol. Int.* **50,** 863–871.

Young, N. S., and Alter, B. P. (1994). Inherited bone marrow failure syndromes: Introduction. *In* "Aplastic Anemia Acquired and Inherited." pp. 271–274. WB Saunders Company, Philadelphia, PA.

Index

T

U

V

X

Y

Contents of Previous Volumes

Volume 47

Volume 48

Volume 49

Volume 50

Volume 51

Volume 52

Volume 53

Volume 54

Volume 55

Volume 56

Volume 57

Volume 60

1. **Therapeutic Cloning and Tissue Engineering**
   Chester J. Koh and Anthony Atala
2. **α-Synuclein: Normal Function and Role in Neurodegenerative Diseases**
   Erin H. Norris, Benoit I. Giasson, and Virginia M.-Y. Lee
3. **Structure and Function of Eukaryotic DNA Methyltransferases**
   Taiping Chen and En Li
4. **Mechanical Signals as Regulators of Stem Cell Fate**
   Bradley T. Estes, Jeffrey M. Gimble, and Farshid Guilak
5. **Origins of Mammalian Hematopoiesis: *In Vivo* Paradigms and *In Vitro* Models**
   M. William Lensch and George Q. Daley
6. **Regulation of Gene Activity and Repression: A Consideration of Unifying Themes**
   Anne C. Ferguson-Smith, Shau-Ping Lin, and Neil Youngson
7. **Molecular Basis for the Chloride Channel Activity of Cystic Fibrosis Transmembrane Conductance Regulator and the Consequences of Disease-Causing Mutations**
   Jackie F. Kidd, Ilana Kogan, and Christine E. Bear

Volume 61

1. **Hepatic Oval Cells: Helping Redefine a Paradigm in Stem Cell Biology**
   P. N. Newsome, M. A. Hussain, and N. D. Theise
2. **Meiotic DNA Replication**
   Randy Strich
3. **Pollen Tube Guidance: The Role of Adhesion and Chemotropic Molecules**
   Sunran Kim, Juan Dong, and Elizabeth M. Lord
4. **The Biology and Diagnostic Applications of Fetal DNA and RNA in Maternal Plasma**
   Rossa W. K. Chiu and Y. M. Dennis Lo
5. **Advances in Tissue Engineering**
   Shulamit Levenberg and Robert Langer

Volume 62

Volume 63

Volume 64

Volume 65

Volume 66

Volume 67

Volume 68

Volume 69

Volume 70

Volume 71

Volume 72

Volume 73

Volume 74

Volume 75

Volume 76

Volume 77

Volume 78

Volume 79

Volume 80

Volume 81

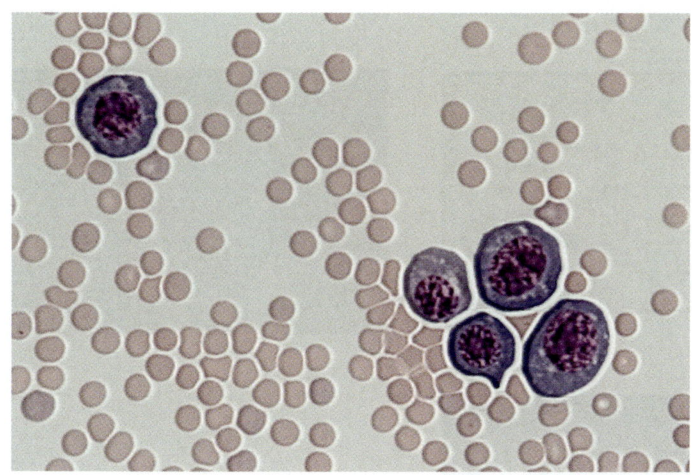

Kathleen McGrath and James Palis, Figure 1.1 Visual comparison of primitive and definitive erythroid cells in the mouse. Circulating blood cells were isolated both from E9.5 embryos and from adult mice, mixed together, cytospun, and stained with Wright-Giemsa. The nucleated primitive erythroid cells are at the basophilic stage of maturation and markedly larger than the definitive erythrocytes.

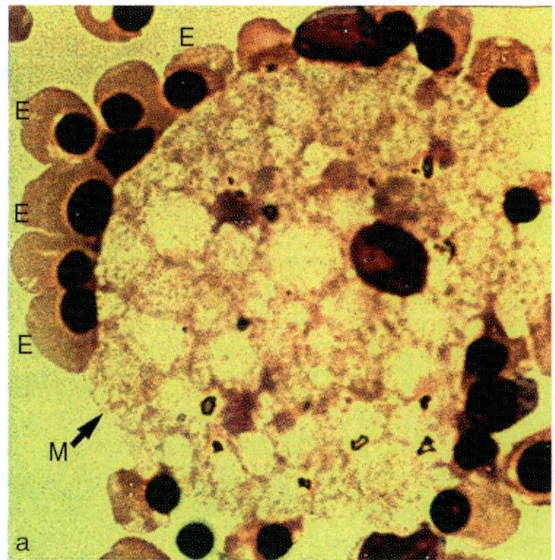

Deepa Manwani and James J. Bieker, Figure 2.2 Characteristic appearance of human erythroblastic island. A Wright-Giemsa stained, cytocentrifuged preparation of cells obtained from erythroblasts generated by *in vitro* culture of peripheral blood-derived mononuclear cells using two-phase liquid culture system was analyzed. An erythroblastic island consisting of a central macrophage (M) surrounded by a ring of late erythroblasts (E) on day 12 of the second phase of a macrophage-containing culture is shown. This research was originally published in *Blood* (Hanspal *et al.*, 1998). © The American Society of Hematology.

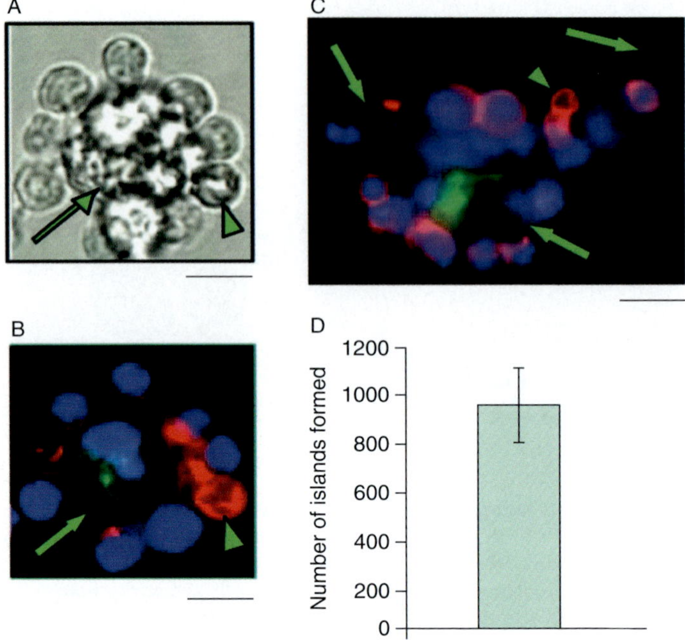

Deepa Manwani and James J. Bieker, Figure 2.3 Reconstituted erythroblastic islands. Bright field (A), immunofluorescent standard (B), and confocal (C) micrographs of typical erythroblastic islands formed from single-cell suspensions of MacGreen mouse bone marrow. Macrophages from MacGreen mice express the macrophage colony-stimulating factor (M-CSF) receptor-green fluorescent protein transgene, thereby providing a useful macrophage identifier. Results from an assay for reforming islands from single-cell suspensions of freshly harvested mouse bone marrow are depicted. Adults 3–5 months of age were used. A single-cell suspension was prepared, then cells were incubated for carefully controlled times in media containing manganese. Islands and their cellular components were identified by three-color immunofluorescence microscopy. Immunofluorescent micrographs of islands show cells stained for erythroid-specific marker GPA (Ter119; red), macrophage marker M-CSF receptor GFP transgene expression (green), and DNA (Hoechst 33342; blue). Because surface expression of glycophorin A increases during terminal differentiation, the intensity of Ter119 staining served as an effective indicator of erythroblast stage. A faint blush of Ter119 fluorescence was present in early erythroblasts and increasing degrees of staining were observed in progressively more differentiated cells. The fluorescence intensity of Ter119 label varied among erythroblasts in an individual island, indicating that islands were composed of erythroblasts at various stages of differentiation. Young, multilobulated reticulocytes were present in many islands, again consistent with prior descriptions of erythroblastic islands formed *in vivo*. In the confocal image, some of the cells appear blurred because they are not in the plane of focus. However, macrophage staining is apparent in various regions of the island. Reticulocytes, arrowheads; macrophage, arrows; bars represent 10 μm. (D) Histogram shows number of erythroblastic islands formed from 1×105 single cells; $n=10$. Results are shown as mean $\pm$ SD. This research was originally published in *Blood* (Lee *et al.*, 2006).

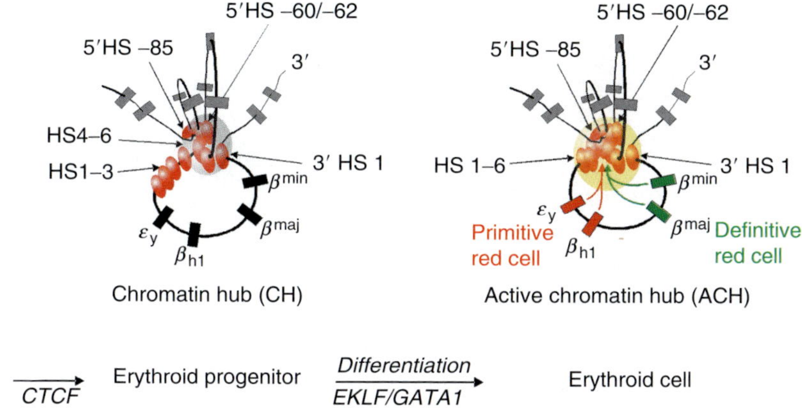

Wouter de Laat *et al.*, Figure 5.3 Formation of the β-globin active chromatin hub during erythroid differentiation. Indicated are the transcription factors that are required for each conformational transition.

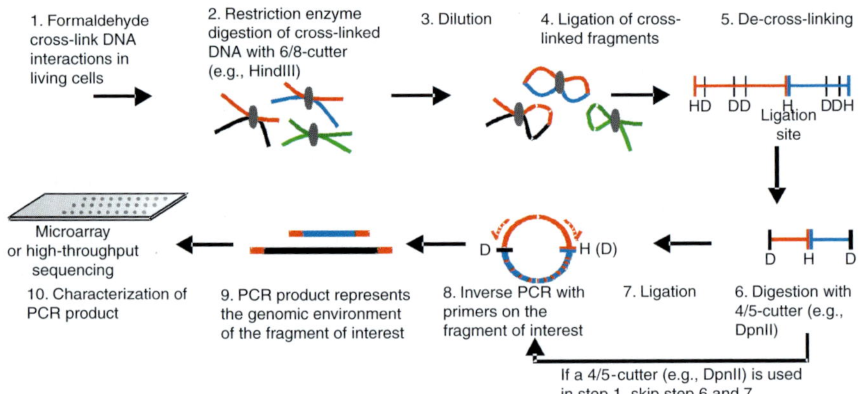

Wouter de Laat *et al.*, Figure 5.4 4C technology. Outline of the 4C procedure; see text for details.

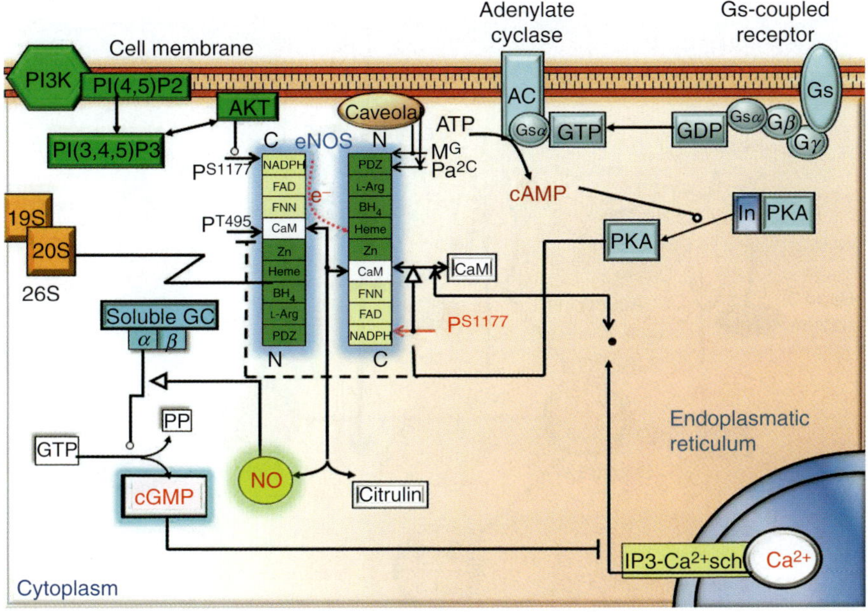

Vladan P. Čokić and Alan N. Schechter, Figure 7.3 Activation of eNOS enzyme. eNOS enzyme is a homodimer composed of two identical subunits that contain the reductase (yellow), oxygenase (green), and calmodulin (CaM)-binding domains. cAMP-dependent protein kinase A (PKA) signaling acts by increasing phosphorylation of residue Ser-1177 and dephosphorylation of Thr-495 to activate eNOS. cAMP-mediated eNOS activation depends on Ca^{2+} release from internal stores. Ca^{2+}/CaM binding aligns the domains and stimulates electron flow, accounting for enzyme activation. eNOS oxidizes L-arginine in a process that consumes NADPH and produces L-citrulline and NO. NO activates soluble GC to produce cGMP. An elevated cGMP level attenuates the store-operated Ca^{2+} entry. The phosphatidylinositol 3′-kinase (PI3K)/protein kinase B (PKB/Akt) pathway can also activate eNOS. Caveolin-1 targets eNOS to the caveolae, vesicular organelles located at the plasma membrane. Myristylation at the N-terminal glycine (glycine-2 before removal of the initiation methionine by a specific aminopeptidase) is mainly responsible for the membrane association of the eNOS enzyme; palmitoylation may also contribute. 26S proteasome are large multi-subunit complexes that selectively degrade eNOS protein. The proteolytic core of this complex is the 20S proteasome.

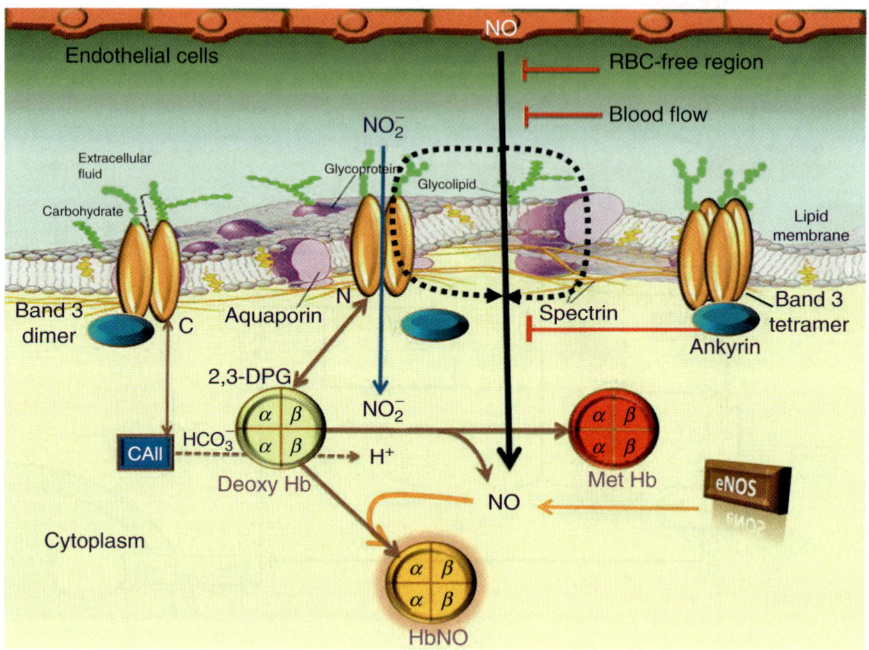

Vladan P. Čokić and Alan N. Schechter, Figure 7.4 Nitric oxide uptake through the membrane skeleton of RBC. On the cytosolic side of the RBC membrane, there is a cytoskeletal network of proteins, including spectrin and band 3 tetramers association with ankyrin, which significantly slow the diffusion of NO. In addition to diffusing through the lipid membrane, it is thought that NO and nitrite ions may use band 3 and/or aquaporin to pass through the RBC membrane. The band 3 protein binds both deoxyhemoglobin and carbonic anhydrase II (CAII, forms the H^+ necessary for nitrite reduction by deoxyhemoglobin), and thus may channel nitrite. The products of deoxyhemoglobin (deoxyHb, $HbFe^{II}$) and nitrite (NO_2^-) reaction are methemoglobin (metHb, $HbFe^{III}$) and NO, which rapidly reacts with deoxyhemoglobin to form iron-nitrosyl-hemoglobin (HbNO, $HbFe^{II}NO$). The RBC-free region, near the vessel wall, and laminar intravascular flow limit NO delivery from the endothelium to RBC-enclosed Hb, in contrast to extracellular Hb, which is more rapidly exposed to NO and appears to destroy its activity to a much greater extent than intracellular Hb. 2,3-disphosphoglycerate (2,3-DPG).

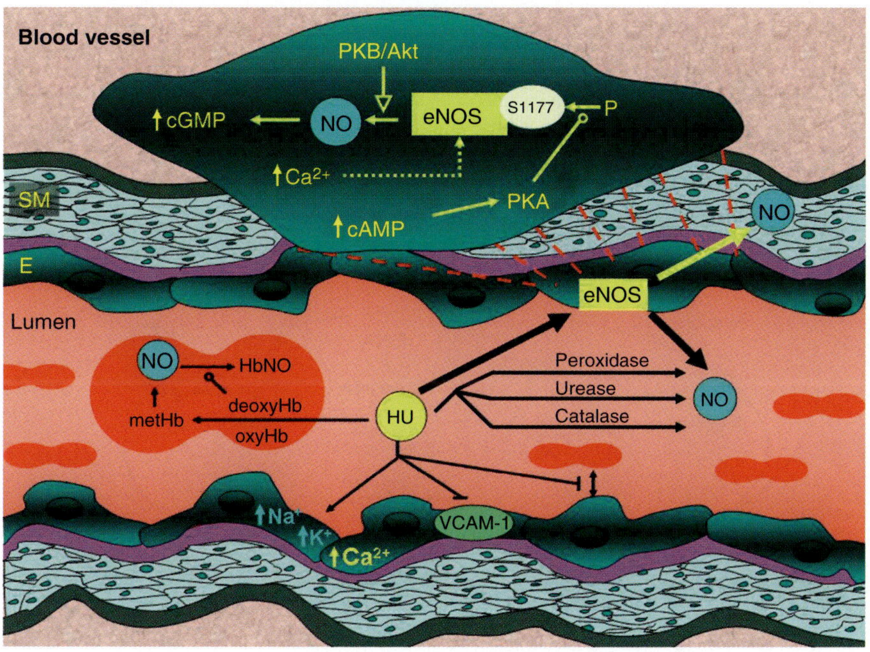

Vladan P. Čokić and Alan N. Schechter, Figure 7.6 Hydroxyurea effects on blood vessels. Hydroxyurea oxidizes oxy- and deoxyhemoglobin to methemoglobin (metHb), which may react directly with other molecules of hydroxurea (or via hydroxylamine) to form nitrosylhemoglobin (HbNO) which can slowly release NO. Hydroxyurea may also chemically or enzymatically (peroxidase, urease, catalase) decompose to NO, as well as directly effect intracellular levels of cations, production of VCAM-1, RBC adhesion, etc. Our data show that hydroxyurea also stimulates the phosphorylation and thus activation of endothelial NOS (eNOS) with resultant production of NO. This effect is modulated by increased intracellular calcium levels and increases in cAMP that stimulates phosphorylation at Ser-1177 by PKA. NO production is regulated by both PKA and PKB/Akt. The increase in eNOS activity presumably leads to increases in cGMP in endothelial and surrounding cells, which result in many of the pleiotropic effects of endothelial produced NO. SMm smooth muscle cells; E, endothelial cells. This figure was originally published in *Blood* (Cokic *et al.*, 2006) and is reproduced with permission.

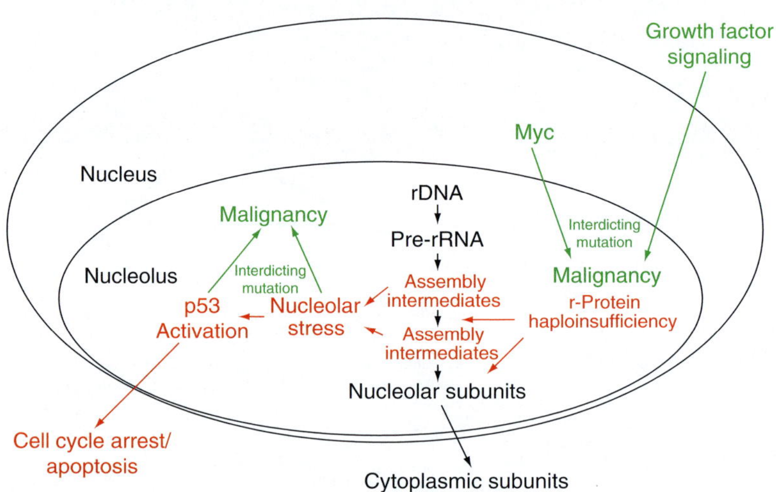

Steven R. Ellis and Jeffrey M. Lipton, Figure 8.3 Proposed mechanism for cancer predisposition in DBA. Ribosomal protein haploinsufficiency results in the accumulation of partially assembly intermediates that result in nucleolar stress signaling. Nucleolar stress, in turn, leads to p53 activation and either cell cycle arrest or apoptosis. These steps are outlined in red. Interdicting mutations that subvert the p53 pathway and favor the outgrowth of malignant clones are shown in green. Also shown in green are interdicting mutations that could enhance ribosomal protein synthesis and diminish nucleolar stress but because of the oncogenic nature of the genes involved could again favor the outgrowth of malignant clones.